World Changes

World Changes

Thomas Kuhn and the Nature of Science

Edited by Paul Horwich

A Bradford Book
The MIT Press, Cambridge, Massachusetts, and London, England

This book was set in Baskerville by DEKR Corporation and printed and bound in the United States of America.

First printing.

Library of Congress Cataloging-in-Publication Data

World changes : Thomas Kuhn and the nature of science / edited by Paul Horwich.
     p.   cm.
"A Bradford book."
Includes bibliographical references and index.
ISBN 0-262-08216-0
1. Science—Philosophy. 2. Science—History. 3. Kuhn, Thomas S. I. Kuhn, Thomas S. II. Horwich, Paul.
Q175.3.T48   1993
501—dc20   92-518
                                              CIP

Contents

World Changes

Introduction

Paul Horwich

A good scientist will experiment, gather data, explain them with simple theoretical hypotheses, and thereby progress, rationally and inexorably, toward the truth. This compelling picture of how things are supposed to work in science was once generally taken for granted, but for the last thirty years, thanks to Thomas Kuhn and *The Structure of Scientific Revolutions*, that has no longer been so. Kuhn's critique called into question many of the central elements of the traditional picture—the concept of absolute truth, the observation/theory distinction, the determinacy of rational choice, and the normative function of philosophy of science—and it provided an alternative model of scientific change that dispensed with these notions altogether.

Kuhn's radical views have been the focus of much debate not only by philosophers, historians, and sociologists of science but also by large numbers of practicing scientists. Nevertheless, many questions remain unsettled regarding their precise nature and validity, so it was thought desirable to bring together a group of experts to consider these questions and to see where Kuhn himself stands with respect to them. To that end, on 18 and 19 May 1990 a conference took place on the campus of the Massachusetts Institute of Technology (sponsored by the Sloan Foundation). The philosophical and historical papers presented at these meetings have now been revised and are collected in the present volume. Each contribution is inspired, in one way or another, by the ideas set out in *Structure* and in Kuhn's subsequent writings; each one illuminates central aspects of his work; and together they testify to its continuing interest and influence. They are not intended, however, to form a comprehensive or balanced treatment of Kuhn's legacy. The history of science takes up only one third of the book, and there is little here to suggest Kuhn's considerable impact on the social studies of science. The emphasis, rather, is on philosophy, and this reflects the locus of Kuhn's thinking during the last ten years and the nature of his current project.

An underlying theme of the essays is the difference between the old philosophy of science, associated with logical empiricism and constituting the mainstream of thought until the 1960s, and the new philosophy of science, articulated in Kuhn's book (and promoted also in the work of Hanson, Feyerabend, Toulmin, and Lakatos). According to the old picture, each scientist proceeds by increasing his stock of observed facts, employing a set of nondeductive logical principles to determine which set of theoretical sentences best explains these facts, and thereby accumulating theoretical knowledge. This rather tempting conception contains the various elements, mentioned above, on which Kuhn focused critical scrutiny. To elaborate a little, most variants of logical empiricism include (1) the idea that philosophy can and should specify how science *ought* to be done, (2) the assumption that there is such a thing as the absolute truth independent of language or theory, (3) the view that progress in science consists in finding theories that more and more closely approximate to the truth, (4) the idea that there are canons of rationality that *determine*, from the available data, the appropriate precise degree of confidence in any given theory, (5) the view that scientific development may be modeled within the thought processes of an individual scientist, (6) the assumption that there exists a theory-neutral body of observable facts, and (7) the conception of theory as a set of symbolic generalizations with individual empirical contents.

Let me try, in the sketchiest of terms, to get some sense of how Kuhn has challenged each of these ideas, in order to provide a context for the following essays:

1 Logical empiricism laid down a priori norms of scientific propriety. Its aim was to specify how science *ought* to be done so that it would be clearly distinguishable from metaphysics. Yet Kuhn's focus, as Hempel notes, is on the *actual* structure of scientific change, and therefore he emphasizes the need for a closer relation between the philosophy and history of science. It is striking and significant, as Noel Swerdlow shows, that astrology, our typical pseudoscience, was regarded by the most sophisticated thinkers of the mid fifteenth century as the very best of the sciences. An adequate philosophy of science must accommodate such historical phenomena. Moreover, Michael Friedman indicates how the development of philosophy as a whole has been influenced to a significant degree by the evolution of science and argues for an extension of Kuhn's methodology into philosophy more generally.

2   The meaning of a theoretical term depends on its use and therefore on the theoretical framework in which it is deployed. Consequently, different theories are, to greater or lesser degrees, *incommensurable*: many of the terms in which a theory is formulated cannot be translated into expressions of other theories. For example, the modern word "star" does not have the same sense as *any* term of ancient astronomy. Indeed, John Heilbron explains how even the names of the scientific disciplines—"physics," "chemistry," etc.—are highly susceptible to such changes in meaning. Thus different theories involve different partially nonintertranslatable languages: the set of meaningful propositions (i.e., what can be said) varies as a function of one's theoretical perspective. Consequently, the set of *true* propositions (i.e., the set of *facts*) also varies. Thus there is no absolute truth: as science evolves "the world changes." Ian Hacking shows that, understood properly, this notorious Kuhnian doctrine is by no means as paradoxical as it may at first appear to be.

3   In that case, truth cannot be the aim of science, and scientific progress cannot consist in the construction of theories that approximate more and more closely to the truth. And as John Earman points out in his comparison of Kuhn with Carnap, we cannot even assess these theories for their *probable truth*. Rather, we must recognize an *instrumental* conception of progress whereby later theories are able to solve more problems than earlier ones, that is, explain a broader array of phenomena. This somewhat sceptical form of antirealism is the concern of Ernan McMullin's article.

4   The familiar criteria of theory choice (scope, accuracy, simplicity, consistency, etc.) are immediately entailed and explained by the in-

strumental objective of science. As governed by these criteria, theory choice is evidently rational. We have no reason to suppose, however, that there is a uniquely correct way to weigh the relative importance of these various desiderata. Therefore, different rational individuals might well disagree on particular questions of theory choice. So it would be a mistake to think there could be a strict algorithm (such as a Carnapian confirmation function) that would determine, for any theory in any evidential situation, its precise degree of plausibility in comparison with other theories.

5   Since the theory choices of rational individuals will diverge, a scientific community will split into advocates of competing theories pursuing different research projects. As one group achieves more success in solving problems than the others, it will grow at the expense of the others by attracting students and converts. Thus the development of science is somewhat analogous to the evolution of species. Inferior theories disappear from the scene in the same way as species unfitted to their environment. Consequently, a model of scientific development cannot be wholly individualistic; some attention to the dynamics of scientific communities is indispensable. Norton Wise implements this approach by providing an analysis of how local groups combine into larger scientific cultures.

6   The developmental conception of science helps one to appreciate that a theory confronts questions of justification only in comparison to the available competition; there is no demand for absolute legitimation. Thus the traditional requirement that a theory be certifiable as probable in relation to the known absolute data was misconceived, and efforts to articulate this sort of requirement were doomed to failure. Not only has it indeed proved impossible to make sense of the notion of "the absolute data," but, as Jed Buchwald demonstrates, the history of science has no room for it. For the real questions of theory choice are always contextual. We are wondering not whether theory T_1 is good simpliciter but whether T_1 would be better than T_2. And relative to this context one can locate a common body of claims, $C(T_1, T_2)$, that are not in dispute. So the question is which of T_1 or T_2 best accommodates $C(T_1, T_2)$. No notion of absolute data is involved.

7   The predominant concern of logical empiricism was to separate science from metaphysical nonsense, and this was to be done by means of the supposition that each scientifically respectable hypothesis (including each symbolic generalization) possesses a certain empirical meaning or content, a body of observable predictions whose observed truth or falsity would settle whether or not the hypothesis should be accepted. With the demise of the observation/theory distinction, a new conception of content for the symbolic generalization of a theory was called for, and Kuhn's idea (elaborated in Nancy Cartwright's paper) was that it is through the ways in which such formulas are *applied* in paradigmatic problem solutions that their meanings are constituted. Such applications often vary as a science evolves, so even though a given equation may be preserved in the context of a theory change, its content could well be altered. Hence the above-mentioned incommensurability of theories.

Needless to say, Kuhn's philosophy of science has not remained fixed since the first edition of *Structure*. Ideas have been clarified, misreadings corrected, emphases shifted. In the final essay we have a preview of his current way of seeing things. The philosophy program at MIT has been fortunate to be Tom's intellectual home while these new ideas were brewing. We have seen his struggle to get things right and admired the humility, passion, and good humor that went into it. I am glad that the contributions to this volume, in their variety, depth, and quality, reflect something of the virtues to which Tom's students and colleagues have been treated over the years.

Thomas Kuhn, Colleague and Friend

Carl G. Hempel

When I first met Tom Kuhn in 1963 at the Center for Advanced Study in the Behavioral Sciences, I approached his ideas with diffident curiosity. My views at the time were strongly influenced by the antinaturalism of Carnap, Popper, and similarly minded thinkers within or close to the Vienna Circle, who held that the proper task of the methodology and philosophy of science was to provide "explications" or "rational reconstructions" of the form and function of scientific reasoning. Such explications were to furnish norms or standards of rationality for the pursuit of scientific inquiry and were to be formulated with rigorous precision by means of the conceptual apparatus of logic. Those norms were thus definitely not intended to provide a descriptive, "naturalistic" account of actual scientific-research practice in its diverse psychological, historical, and sociocultural aspects. Rather, they were propounded as standards for rationally sound procedure in science, standards considered to be sometimes woefully violated in actual scientific practice.

Kuhn's approach to the methodology of science was of a radically different kind: it aimed at examining the modes of thinking that informed and directed research, theory formation, and theory change in the practice of scientific inquiry, past and present. As for the standards of rationality propounded by logical empiricism, Kuhn took the view that if those standards should be infringed here and there in instances of research that were viewed as sound and productive by the pertinent community of scientific specialists, then we had better change our conception of proper scientific procedure rather than reject the research in question as irrational. Kuhn's perspective came increasingly to appeal to me.

When, in 1964, Tom moved to Princeton, the exchange of ideas between Tom and me was intensified and enriched to the point of active academic collaboration. We offered a joint course in the philosophy of science in which we took turns lecturing in a spirit not at all antagonistic but cooperative, aimed at enriching the students' total perspective on the subject. Tom and I were in constant communication about the issues to be discussed in class and about the questions we planned to raise about each other's approaches.

Tom's ideas have influenced my thinking in various ways and have certainly contributed to my shift from an antinaturalistic stance to a naturalistic one. In the light of Tom's recent taxonomic turn, that may not be an important, or even quite the right, lesson to have learned, but the present occasion surely is not one for speculation about that.

Whatever position your colleagues may take, Tom, I am sure that they all feel a large debt of gratitude to you for your provocative and illuminating ideas, and all of us in this audience await with keen interest your thoughts on the proceedings of this "Kuhnfest."

The Philosophers Look Back

Carnap, Kuhn, and the Philosophy of Scientific Methodology

John Earman

1 Introduction

For the past two decades logical positivism has served as a whipping boy. By emphasizing the shortcomings of this failed philosophical program, the virtues of the new postpositivist philosophy of science are made to seem more lustrous. It is, of course, not surprising to find such polemical devices employed, since they are common to the rhetoric of revolutions, whether political, scientific, or philosophical. Or so the standard assessment would go. What I find askew in this assessment is the notion that a philosophical revolution as opposed to an evolution has taken place. For although I am no apologist for logical positivism, it does seem to me that many of the themes of the so-called postpositivist philosophy of science are extensions of ideas found in the writings of Carnap and other leading logical positivists and logical empiricists.[1] But my purpose here is not to contribute to a revisionist history of philosophy. Rather, I aim to pay homage to both Carnap and Kuhn by noting some striking similarities and also some striking differences. These similarities and differences are useful in helping to focus some of the still unresolved issues about the nature of scientific methodology.

2 Logical Positivism, Logical Empiricism, and Kuhn's *Structure of Scientific Revolutions*

The members of the Vienna Circle often took votes on the issues they debated. While I have no documentary evidence to offer of an actual vote, I am morally certain of what the result would have been for the question, Verification is a relation between what and what? In "x is verified by y," x is presumably a sentence. And it is tempting to take y to be a fact, state of affairs, or something in the world that makes x true and can be directly ascertained. But an attempt to compare language and the world would have struck the Circle members as of a piece with the metaphysics that the verifiability principle of meaning was supposed to banish. Their alternative was to take y to be another sentence, for then the relation between x and y is an unproblematic logical relation.[2] This move, however, seems to leave us in the same metaphysical thicket since verification would seem to require that y is a true sentence, and is not a true sentence one that corresponds to the facts?

The escape that some of the circle members sought was a resort to a coherence theory of truth. Eventually, however, Carnap abandoned this resort, presumably because of a combination of the drawbacks of the coherence account of truth and the attractiveness of Tarski's theory of truth. But what I wish to emphasize here are the qualifications that Carnap put on any talk about comparing statements with facts. In "Truth and Confirmation" (1949) he emphasized that he preferred to speak of confronting statements with facts:

There has been a good deal of dispute as to whether in the procedure of scientific testing *statements must be compared with facts* or as to whether such comparisons be unnecessary, if not impossible. If "comparison of statement with fact" means the procedure which we called the first operation[3] then it must be admitted that this procedure is not only possible, but even indispensable for scientific testing. Yet it must be remarked that the formulation "comparison of statement and fact" is not unobjectionable. First, the concept "comparison" is not quite appropriate here. Two objects can be compared in regard to a property which may characterize them in various ways. . . . We therefore prefer to speak of "confrontation" rather than "comparison." Confrontation is understood to consist in finding out as to whether . . . the fact is such as is described in the statement, or, to express it differently, as to whether the statement is true to fact. (1949, 125)

Carnap then continued with a passage that might have served as an advertisement for Kuhn's *Structure of Scientific Revolutions*.

Furthermore, the formulation in terms of "comparison," in speaking of "facts" or "realities" easily tempts one into the absolutistic view according to which we are said to search for an absolute reality whose nature is assumed as fixed independently of the language chosen for its description. The answer to a question concerning reality however depends not only upon that "reality" or upon the facts, but also upon the structure (and the set concepts) of the language used for the description. In translating one language into another the factual content of an empirical statement cannot always be preserved unchanged. Such changes are inevitable if the structures of the two languages differ in essential points. (1949, 125–126)

Here we have two of the key theses of the "postpositivist" philosophy of science: the nonexistence of neutral facts and incommensurabilty in the form of failure of intertranslatability. Note that these theses were propounded in the mid 1930s, for although I have, for the sake of convenience, quoted from the 1949 version of Carnap's paper, the relevant passage is also in "Wahrheit und Bewährung" (1936).

Given these Kuhnian themes—or should we rather say Carnapian themes?—one might predict that Carnap would have found *Structure* philosophically congenial. That this was indeed the case has been documented by Reisch (1991). *Structure* was published as part of the International Encyclopedia of Unified Science, of which Carnap was an associate editor. After reading the completed manuscript for *Structure*, Carnap wrote to Kuhn in April of 1962. The text of the letter is reproduced in Reisch 1991. I will quote from notes written in Carnap's archaic shorthand. Carnap begins with a piece of Darwinian evolution and then adds, concerning Kuhn's thesis,

In analogy to this one has to understand the development of scientific theories: not directed to an ideal [true] theory, the *one* true theory of the world, but evolution as a step to a better form, by selection of one out of several competing forms. The selection is made on the basis of preference in the community of scientists. Many factors, sociological, cultural, . . . , are involved. Not: we are approaching truth, but: we are improving an instrument.[1]

Another important though largely tacit theme of *Structure*, a holistic view of meaning, can also be seen as emerging from the writings of the logical empiricists in the 1950s. Hempel, for example, took to heart Quine's attack on the analytic/synthetic distinction, which may be taken to embody the thesis that there is a sharp distinction to be drawn between two functions of language: one to specify meaning, the other to make empirical assertions. Applying the moral to scientific theories, one arrives at the conclusion that there is no principled way to distinguish those postulates of the theory that may properly be dubbed meaning postulates. It is then but a seemingly short and tempting step to the further conclusion that all the postulates of the theory function to specify the meaning of the constituent terms and thus that any significant change in the theory implies a change in meaning.

This route to semantic holism cannot be traced to any of Carnap's writings. Indeed, in his response to Hempel's contribution to the Schilpp volume (1963), Carnap attempted to use the notion of the Ramsey sentence of the theory to identify the postulates of the theory that "merely represent meaning relations" (1963a, 965). And in his last book, *Philosophical Foundations of Physics* (1966), Carnap maintained that "a sharp analytic-synthetic distinction is of supreme importance for the philosophy of science" (1966, 257).[5] There is, however, another Carnapian route to semantic holism, but that route must be traced all the way back to Carnap's attempt in the *Aufbau* (1928) to explain how scientific objectivity can emerge from a reconstruction that starts from a phenomenalistic basis. An exploration of this matter would take me too far afield; I will simply refer the reader to Michael Friedman's (1987) illuminating discussion.

3 Carnap's Relativism

In comparison with Carnap of the 1930s, many contemporary relativists seem anemic. *The Logical Syntax of Language* (1934) and "Philosophy and Logical Syntax" (1935) proclaimed the relativity of all philosophical theses to language. This relativity was supposed to hold the key to solving, or rather dissolving, traditional philosophical disputes. Suppose, for example, one philosopher asserts, "Numbers are primitive entities," while another proclaims, "Numbers are classes of classes." "They may," Carnap writes, "philosophize without end about the question of what numbers really are, but in this way they will never come to an agreement" (1935, 450). If, however, they are acute enough to recognize Carnap's relativity principle, they will quickly realize that one is asserting, "In Language L_1 (Peano), numerical expressions are elementary expressions," while the other is maintaining, "In Language L_2 (Russell) numerical expressions are class expressions of the second order." "Now these assertions are compatible with each other and both are true; the controversy has ceased to exist" (1935, 451). This model for resolving philosophical disputes in the philosophy of mathematics was supposed by Carnap to be extendible quite broadly to philosophical disputes, such as phenomenalism versus materialism and the question of whether space-time points have an existence independent of physical events.

Some, like Donald Davidson (1973), have found an air of paradox in the fact that saying in one breath (as Carnap thought he could) that S is true in L_1, L_5, L_{28}, . . . but false in L_2, L_4, . . . seems to presuppose a neutral metaframe within which all the language frames can be treated.[6] Others, like Michael Friedman (1992), have argued that Carnap's relativism is undercut by Gödel's incompleteness theorems, which show that no such neutral metaframe is available. My objections are more local and tactical.

My first complaint is that Carnap assumes what needs to be proved. According to Carnap's "principle of tolerance," we are free to choose whatever language system we like. The decision is largely a pragmatic affair, turning on such matters as efficiency and fruitfulness for the purposes at hand. But to apply the slogan of "free to choose" to dissolve, say, the phenomenalism versus materialism debate assumes that a phenomenalistic language has been produced that shows how physical object talk can be reduced to talk about sensa data or, as Carnap preferred, momentary total experiences ("Elementarerlebnisse"). This, of course, is exactly what Carnap tried to do in the *Aufbau*. But by his own admission, the attempt has to counted as a failure if, as he originally assumed, the reduction has to proceed via explicit definitions and explicit translations. Similarly, to apply the "free to choose" slogan to dissolve disputes about the ontological status of space-time points assumes that it can be shown how space-time points can be constructed out of events, such as coincidences of particles. Advocates of relational theories of space and time repeatedly claim that this can be done and even that it has been done. But none of the claims stands up to scrutiny.[7] I am emphatically not claiming that materialism is correct or that space-time points construed as irreducible entities are essential to physics. Rather, I am claiming that the dissolution of the traditional disputes on these matters is not as easy as Carnap made it seem.

Another complaint arises from the breathtaking scope of the intended application of Carnap's dissolving strategy. Among the "philosophical" disputes that Carnap proposed to treat in this way were such matters as whether time is finite or infinite and whether the world is deterministic. There is obviously a very slippery slope here. If the question about the finitude of time is a philosophical question in the relevant sense, then why not the question of whether the world began from a big-bang singularity? And if this latter question is a philosophical question in the relevant sense, then why not other deep questions in cosmology? But more important, one does not have to go down the slope to recognize the implausibility of Carnap's procedure. Even if one agrees to talk about truth in L rather than truth period, there is no plausibility to the idea that whether time is finite and whether determinism holds are matters to be settled in L by adopting linguistic rules for L rather than by consulting the facts. Carnap, not surprisingly, acknowledged the point. Speaking of the determinism issue, he said,

The objection may perhaps be raised at this point that the form of physical laws depends upon experimental results of physical investigation, and that it is not determined by a merely theoretical syntactical consideration.[8] This assertion is quite right, but we must bear in mind the fact that the empirical results at which physicists arrive by way of their laboratory experiments by no means *dictate their choice* between the deterministic and the statistical form of laws. The form in which a law is to be stated has to be decided by an act of volition. This decision, it is true, depends upon the empirical results, but not logically, only practically. The results of the experiments show merely that one mode of formulation would be more suitable than another. (1935, 455)

These sentiments resonate with those Carnap expressed three decades later in commenting on Kuhn's *Structure*. But here the sentiments are not to the point. The issue is not whether, for Duhemian or other reasons, the results of experiments, say the recent Einstein, Rosen, Podolsky, and Bell type of experiments, fail to dictate the acceptance of indeterministic laws. Rather the issue is whether determinism is a scientific claim to be argued over the way one argues over other deep scientific claims, none of which ever gets definitively settled by the dictates of experimental evidence; *or* whether determinism is a claim that can be made true by linguistic fiat in L_1, L_6,

$L_{35}, \ldots$ and false by linguistic fiat in $L_2, L_7, L_{37}, \ldots$, and then we just pays our money and takes our choice of language. Again I refuse to give a global answer to this query and favor instead a tactical response. All indications are that the debate over the implications for determinism of the Bell inequalities and the Aspect experiments belongs to the former rather than to the latter. Indeed, indications are that in any language system adequate for the formulation of theories that save the experimentally verifiable quantum statistics, the laws must be indeterministic.[9]

I suspect that Carnap's relativism began by his being impressed by the achievements of Frege, Russell, and others in the philosophy of mathematics and was furthered by a misplaced zeal for extending his model for resolving philosophical disputes in this area to a broad area of philosophical and scientific questions. Of course, whatever the origins of Carnap's relativism, it or something like it could perhaps be promoted on the basis of his doctrine that language-neutral facts do not exist. I find it difficult to assess this matter, since I do not find in Carnap's writing a helpful explanation of this doctrine. In the following section I will comment on the related doctrine of Feyerabend, Hanson, and Kuhn that observation is theory laden.

In closing this section, I note that Carnap displayed a consistency on the matter at hand—not the consistency that is the hobgoblin of little minds but the magnificent consistency of a grand visionary. In his contribution to the Schilpp volume for Carnap, Herbert Feigl (1963) sketched a mind-body identity theory that he was later to elaborate in his famous essay "The 'Mental' and the 'Physical'"(1958).[10] Clearly, the politically correct thing for Carnap was to endorse Feigl's approach. Instead, he wrote, "it seems preferable to me to formulate the question [of mind-body identity] in the meta language, not as a factual question about the world, but as a question concerning the choice of language. Although we prefer a different language, we must admit that a dualistic language can be constructed and used without coming into conflict with either the laws of logic or with empirically known facts" (1963b, 885–886).

4 Kuhn's Relativism

Kuhn resists being labeled a relativist. I use the label here to refer to three doctrines of *Structure:* the theory ladenness of observation, the incommensurability of theories, and the denial that there is a theory-independent notion of truth.

Part of what was meant by the theory ladenness of observation is embodied in the thesis that what we see depends upon what we believe, a thesis open to challenge (see Fodor 1984). I am concerned rather with the related thesis of the nonexistence of a neutral observation language in which different theories can be compared. My response is once again tactical. That is, without trying to adjudicate the general merits of the thesis, I claim that things aren't so bad for actual historical examples. Even for cases of major scientific revolutions, we can find, without having to go too far downward toward something like foundations for knowledge, an observation base that is *neutral enough for purposes at hand*. A nice example is provided by Allan Franklin (1986, 110–113), who shows how to construct an experiment that is theory-neutral enough between Newtonian and special-relativistic mechanics to unambiguously decide between the predictions of these theories for elastic collisions. The two theories agree on the procedure for measuring the angle between the velocity vectors of the scattered particles, and the two theories predict different angles.

More generally, I claim that in the physical sciences there is in principle always available a neutral observation base in spatial coincidences, such as dots on photographic plates, pointer positions on dials, and the like. If intersubjective agreement on such matters were not routine, then physical science as we know it would not be possible. I reject, of course, the positivistic attempt to reduce physics to such coincidences. And I readily acknowledge that such coincidences by themselves are mute witnesses in the tribunal for judging theories. But what is required to make these mute witnesses articulate is not a Gestalt experience but a constellation of techniques, hypotheses, and theories: techniques of data analysis, hypotheses about the operation of measuring instruments, and auxiliary theories that support bootstrap calculations of values for the relevant theoretical parameters that test the competing theories. But I again assert that to the extent that this process cannot be explicitly articulated but relies on some *sui generis* form of perception, the practice is not science. This is not to say, however, that the vulgar image of science as a blindly impartial enterprise is correct, for the articulation uncovers assumptions to which different scientists may assign very different degrees of confidence. But this sense in which different scientists can (misleadingly) be said to "see" different things when looking at the same phenomenon is one with which a probabilistic or Bayesian epistemology—the kind of epistemology which the later Carnap came to advocate—must cope on a routine basis, even in cases far away from the boundaries of scientific revolutions. How these differences are resolved is part of the Bayesian analogue of Kuhn's problem of community decision on theory choice. Kuhn's problem will be encountered in the following section, and the Bayesian analogue will be discussed in sections 8 and 9.

The matter of incommensurability is much more difficult to discuss for two reasons. First, it is tied to difficult issues about meaning and reference that I cannot broach here. Second, issues about incommensurability present amorphous and shifting targets. In *Structure*, for example, incommensurability was a label for the entire constellation of factors that lead proponents of different paradigms to talk past one another. In recent years Kuhn has come around to a more Carnapian or linguistic formulation in which incommensurability is equated with untranslatability. More specifically, the focus has shifted from paradigms to theories, and two theories are said to be incommensurable just in case "there is no common language into which both can be fully translated" (Kuhn 1989, 10). I have no doubts about Kuhn's claims that theories on different sides of a scientific revolution often use different "lexicons," that differences in lexicons can make for a kind of untranslatability, and that in turn this explains why scientists reading out-of-date texts often encounter passages that "make no sense" (1989, 9). But I deny that there is incommensurability/untranslatability that makes for insuperable difficulties for confirmation or theory choice (a phrase I don't like for reasons to be given below) in the standardly cited cases of scientific revolutions such as the transition from Newtonian to special-relativistic mechanics and the subsequent transition to general relativity. Newtonian, special-relativistic, general-relativistic, and many other theories can all be formulated in a common language, the language of differential geometry on a four-dimensional manifold, and the crucial differences in the theories lie in the differences in the geometric object fields postulated and the manner in which these fields relate to such things as particle orbits. This language is anachronistic and so may not be the best device to use when trying to decide various historical disputes.[11] But it does seem to me to be an appropriate vehicle for framing and answering the sorts of questions of most concern to working physicists and philosophers of science. For example, on the basis of the available evidence, what is it reasonable to believe about the structure of space and time and the nature of gravitation? This is not to say that the common language makes for an easy answer. It is indeed a difficult business, but it is a business that involves the same sorts of difficulties already present when testing theories that lie on the same side of a scientific revolution. Finally, so that there can be no misunderstanding, let me repeat: I am not claiming that what I call a common language provides what Kuhn wants. It does not show, for example, that the Newtonian and the Einsteinian can be brought into agreement about what is and is not a "meaningful" question about simultaneity. But what I do claim is that these residual elements of incommensurability do not undermine standard accounts of theory testing and confirmation.[12]

My response to worries about the applicability of the notion of truth to whole theories is similarly local and tactical. In the Postscript to the second edition of *Structure*, Kuhn writes, "There is, I think, no theory-independent way to reconstruct phrases like 'really there'; the notion of a match between the ontology of a theory and its 'real' counterpart in nature now seems to me illusive in principle" (1970, 206). I need not demur if "theory" is understood in a *very* broad sense to mean something like a conceptual framework so minimal that without it "the world" would be undifferentiated Kantian ooze. But I do demur if "theory" is taken in the ordinary sense, i.e., as Newton's theory or special-relativity theory or general-relativity theory.[13] For scientists are currently working in a frame in which they can say, correctly I think, that the match between the ontology of the theory and its real counterpart in nature is better for the special theory of relativity and even better for the general theory. Of course, to get to this position required two major coneptual revolutions. How such revolutions affect theory choice, or as I would prefer to say, theory testing and confirmation, remains to be discussed.

5 Kuhn's Account of Scientific Revolutions

Carnap, as we have seen, found Kuhn's *Structure* congenial. But many philosophers of the younger generation, including those who prided themselves on having gone beyond the crudities of logical positivism, professed shock and dismay at Kuhn's account of the displacement of an old paradigm by a new one. For those readers who do not have a copy of *Structure* to hand, here are some of the purple passages:

Like the choice between competing political institutions, that between competing paradigms proves to be a choice between incompatible modes of community life. . . . When paradigms enter, as they must, into a debate about paradigm choice, their role is necessarily circular. Each group uses its own paradigm to argue in that paradigm's defense. (P. 94)

As in political revolutions, so in paradigm choice—there is no standard higher than the assent of the relevant community. To discover how scientific revolutions are effected, we shall therefore have to examine not only the impact of nature and logic, but also the techniques of persuasive argumentation within the quite special groups that constitute the community of scientists. (P. 94)

The proponents of competing paradigms practice their trades in different worlds. . . . Practicing in different worlds, the two groups of scientists see different things when they look from the same point in the same direction. (P. 150)

In these matters neither proof nor error is at issue. The transfer of allegiance from paradigm to paradigm is a conversion experience that cannot be forced. (P. 151)

Before they can hope to communicate fully, one group or the other must experience the conversion that we have been calling a paradigm shift. Just because it is a shift between incommensurables, the transition between competing paradigms cannot be made a step at a time, forced by logic and neutral experience. Like a gestalt switch, it must occur all at once (though not necessarily at an instant) or not at all. (P. 150)

Many readers saw in these passages an open invitation to arationality if not outright irrationality. Thus Imre Lakatos took Kuhn to be saying that theory choice is a matter of "mob psychology" (1970, 178), while Dudley Shapere read Kuhn as saying that the decision to adopt a new paradigm "cannot be based on good reasons" (1966, 67).

Kuhn in turn was equally shocked by such criticisms. In the Postscript to the second edition of *Structure* (1970), he professed surprise that readers could have imposed such unintended interpretations on the above quoted passages. I will leave aside the unfruitful question of whether or not Kuhn ought to have anticipated such interpretations and will concentrate instead on what, upon reflection, he intended to say.

Kuhn's own explanation in the Postscript begins with the common-place that "debate over theory-choice cannot be cast in a form that resembles logical or mathematical choice" (1970, 195). But he hastens to add that this commonplace does not imply that "there are no good reasons for being persuaded or that these reasons are not ultimately decisive for the group" (1970, 195). The reasons listed in the Post-script are accuracy, simplicity, and fruitfulness. The later paper "Objectivity, Value Judgments, and Theory Choice" (1977) added two further reasons: consistency and scope. And as Kuhn himself notes, the final list does not differ (with one notable exception to be discussed later) from similar lists drawn from standard philosophy-of-science texts (see also Kuhn 1983).

These soothing sentiments serve to deflate charges of arationality and irrationality, but at the same time they also serve to raise the question of how Kuhn's views are to be distinguished from the orthodoxy that *Structure* was supposed to upset. The answer given in the Postscript contains two themes, which are elaborated in "Objectivity." First, the items on the above list are said to "function as values" that can "be differently applied, individually and collectively by men who concur in honoring them" (1970, 199). Thus, "there is no neutral algorithm for theory choice, no systematic decision procedure which, properly applied, must lead each individual in the group to the same decision." Second, it (supposedly) follows that "it is the community of specialists rather than the individual members that makes the effective decision" (1970, 200).

I think that Kuhn is correct in locating objectivity in the community of specialists, at least in the uncontroversial sense that intersubjective agreement among the relevant experts is a necessary condition for objectivity. But how the community of experts reaches a decision when the individual members differ on the application of shared values is a mystery that to my mind is not adequately resolved by

Structure or by subsequent writings. My strategy will be to explore these and related issues from the perspective of Carnap's epistemology.

6 Carnap and Kuhn: Incommensurability?

The passage Reisch (1991) quotes from Carnap's letter to Kuhn and the passage from Carnap's shorthand notes I quoted in section 2 would seem to indicate that Carnap and Kuhn were in substantial agreement as regards paradigm choice. This is surely the case when "paradigm" is interpreted to mean something like a linguistic framework. At this level Carnap would agree, indeed, would insist, on the need to choose, and he would hold that the choice is a pragmatic one whose dynamics may well involve the sorts of factors emphasized in Kuhn's account. But when the focus shifts to theories, as it does in Kuhn's later writings, the disagreement begins. In the first place, Kuhn's list of criteria for theory choice is conspicuous for its omission of any reference to the degrees of confirmation or probabilities of the theories. This is not an oversight, of course, but derives both from explicit doctrines, such as the nonexistence of a theory-neutral observation language, and the largely tacit but pervasive anti-inductivism of *Structure*. Needless to say, this shunning of confirmation theory is most un-Carnapian. But even more anomalous from Carnap's perspective is Kuhn's emphasis on theory choice or acceptance, for in Carnap's version of epistemology, theories are not chosen or accepted but only probabilified.[14]

Carnap's writings in the 1940s and 1950s portray him as espousing a "logical" conception of probability. But by the late 1950s and early 1960s, he clearly favored a view that can be termed tempered personalism: probability is rational degree of belief.[15] I will have more to say on this matter in section 8, but in the meantime I will present Carnap as a tempered Bayesian personalist.

A shotgun marriage of Kuhn and Carnap could be arranged by taking Carnap to supply the probabilities, Kuhn to supply the values or utilities, and then applying the rule of maximizing the expected utility to render a decision on theory choice.[16] But like most shotgun marriages, this one would be a mistake. For Carnap it would be a mistake because it would involve the pretense that the accepted theory is certain even though one's degree of belief in the theory may be less than one, perhaps substantially so. For Kuhn it would also be a mistake, since the efficacy of his values does not depend on the truth of the theories, so estimates of the probable truth of the theories is irrelevant to Kuhnian theory choice.

Part of the wrangle here derives from the unfortunate phrase "theory choice." Scientists do choose theories, but on behalf of Carnap, I would claim that they choose them only in the innocuous sense that they choose to devote their time and energy to them: to articulating them, to improving them, to drawing out their consequences, to confronting them with the results of observation and experiment. Choice in this sense allows for a reconciliation of Bayes and Kuhn, since this choice is informed by both Bayesian and Kuhnian factors: probability and the values of accuracy, consistency, scope, simplicity, and fruitfulness.

Alas, this reconciliation is rather shallow. Once we are clear that the sort of choice involved in "theory choice" is a practical one, there is nothing sacred about the list of items on Kuhn's list of values. Other values, such as getting an NSF grant or winning the Nobel Prize, can and do enter. Further, the kind of choice in question allows a scientist to be bigamous, since he can choose to work on two or more theories at once, and it allows him to be fickle, since he can oscillate back and forth. The kind of choice *Structure* envisioned was much more permanent; indeed, the impression given there is that normal science is not possible without tying Catholic bonds to a theory, bonds that can only be broken by leaving the Church, i.e., by creating a revolution.

Is there no way to bridge the gap between Carnap and Kuhn on this issue? To see how baffling the Bayesian finds the notion of theory acceptance, consider the case of Einstein's general theory of relativity (GTR), arguably the leading theory of gravitation and thus the top candidate for acceptance. Marie, a research worker in the field familiar with all of the relevant experimental findings, does some introspection and finds that her degree of belief in GTR is p.

Case 1: p is 1, or so near 1 as makes no odds. Here there is a natural sense in which the Bayesian can say that Marie accepts GTR. Such cases, however, are so rare as to constitute anomalies. Of course, one can cite any number of cases from the history of science where scientists seem to be saying for their pet theories that they set $p = 1$. Here I would urge the need to distinguish carefully between scientists as advocates of theories versus scientists as judges of theories. The latter role concerns us here, and in that role scientists know, or should know, that only in very exceptional cases does the evidence rationally support a full belief in a theory. Let us move on to case 2.

Case 2: p is, say, .75. Subsequently Marie decides to "accept" GTR on the basis of her probability assignments and the values she attaches to GTR and its competitors. What could this mean?

Subcase 2a. When she accepts GTR, Marie changes her degree of belief from .75 to 1. This is nothing short of folly, since she has already made a considered judgment about evidential support and no new relevant evidence occasioning a rejudgment has come in.

Subcase 2b. When she accepts GTR, Marie does not change her degree of belief from .75 to 1, but she acts *as if* all doubt were swept away in that she devotes every waking hour to showing that various puzzling astronomical observations can be explained by the theory, she assigns her graduate students research projects that presuppose the correctness of the theory, she writes a textbook on gravitational research that is devoted almost exclusively to GTR, etc. But at this point we have come full circle back to a sense of theory acceptance that is really a misnomer, for what is involved is a practical decision about the allocation of personal and institutional resources and not a decision about the epistemic status of the theory.

This rather pedantic diatribe on theory acceptance would be best forgotten were it not for its implications for our picture of normal science. As we have seen, theory "choice" or "acceptance" can refer either to adopting an epistemic attitude or to making a practical choice. In the former case there is no natural Bayesian explication of theory acceptance save in the case where the probability of the theory is one. Since scientists as judges of theories are almost never in a position to justify such an acceptance, the Bayesian prediction is that rarely is a theory accepted in the epistemic sense. Similarly, when theory choice is a matter of deciding what theory to devote one's time and energy to, the Bayesian prediction is that in typical situations where members of the community assign different utilities to such devotions, they will make different choices. Thus from either the epistemic or practical-decision perspective, the Bayesian prediction is diversity. This prediction is, I think, borne out by actual scientific practice. In section 9 below I will argue that insofar as normal science implies a shared paradigm, the paradigm need not and in fact often is not so specific as to include a particular ("accepted") theory. I will also hazard a proposal for a minimal sense of "shared paradigm" that yields a less strait-jacketed image of normal science and that also diminishes without obliterating the difference between normal and revolutionary science.

By way of closing this section and introducing the next, I will consider a final way of reconciling Kuhn and Carnap on theory choice. Radically new theories, so the story goes, carry with them different linguistic/conceptual frameworks. Thus, even seriously to entertain the new theory involves the decision to adopt, if only tentatively, the new framework, and this decision is for Carnap a pragmatic one that involves the sorts of factors emphasized in Kuhn's account of paradigm replacement. In response I would repeat what I have already said in section 4: major scientific revolutions such as the transition from Newtonian to special-relativistic physics and thence to general relativity needn't be seen as forcing a choice between incommensurable linguistic/conceptual systems, since it is often possible to fit the possibilities into a larger scheme that makes the theories commensurable to the extent that confirmation questions can be posed in terms of an observation base that is neutral enough for assessing the relative confirmation of the theories. However, the recognition of the larger possibility set can produce a drastic change in probability values, a change best described in Kuhnian terms.

7 Revolutions and Belief Shifts

A mild form of scientific revolution occurs with the introduction of a new theory that articulates possibilities that lie within the boundaries of the space of theories to be taken seriously but that, because of the failure of actual scientists to be logically omniscient, had previously been unrecognized as explicit possibilities. The more radical form of revolution occurs when the space of possibilities itself needs to be significantly altered to encompass the new theory. In practice the distinction between the two forms of revolution may be blurred, perhaps even hopelessly so, but I will begin discussion by pretending that we can perform a separation of cases.

Even the mild form of revolution may induce a non-Bayesian shift in belief functions. By non-Bayesian I mean that no form of conditionalization, whether strict or Jeffrey or some natural extension of these, will suffice to explain the change. For conditionalizing (in any recognizable sense of the term) on the information that just now a heretofore unarticulated theory T has been introduced is literally nonsensical, because such a conditionalization presupposes that prior to this time there was a well-defined probability for this information and thus for T, which is exactly what the failure of logical omniscience rules out.

We can try to acknowledge the failure of logical omniscience by means of Abner Shimony's (1970) device of a catch-all hypothesis H_c, which asserts in effect that something, we know not what, beyond the previously formulated theories $T_1, T_2, \ldots, T_q$ is true. Now suppose that a new theory T is introduced and that as a result the old degree-of-belief function Pr is changed to Pr'. The most conservative way the shift from Pr to Pr' could take place is by the process I will call *shaving off*, namely, $\Pr(T_i) = \Pr'(T_i)$ for $i = 1, 2, \ldots, q$ and $\Pr'(T) = r > 0$ and $\Pr'(H_c) = \Pr(H_c) - r$. That is, under shaving off, H_c serves as a well for initial probabilities for as yet unborn theories, and the actual introduction of new theories results only in drawing upon this well without disturbing the probabilities of previously formulated theories. Unfortunately, such conservatism eventually leads to the assignment of ever smaller initial probabilities to successive waves of new theories until a point is reached where the new theory has such a low initial probability as to stand not much of a fighting chance.

Certainly shaving off is a factually inadequate description of what happens in many scientific revolutions, especially of the more radical type. Think of what happened following the introduction of Einstein's special theory of relativity (STR) in 1905. Between 1905 and 1915 little new empirical evidence in favor of STR was recorded, and yet the probability of competing theories, such as those of Lorentz and Abraham, set in classical space and time, fell in the estimates of most of the members of the European physics community, and the probability subtracted from these electron theories was transferred to Einstein's STR. The probabilities of auxiliary hypotheses may also be affected, as illustrated by the introduction of the general theory of relativity (GTR). When Einstein showed that GTR accounted for the exact amount of the anomalous advance of Mercury's perihelion, the hypothesis of an amount of zodiacal matter sufficient to affect Mercury's perihelion dropped dramatically in the estimates of most of the physics community (see Earman and Glymour 1991).

In using the term "non-Bayesian" to describe such nonconditionalization belief changes, whether of the conservative shaving-off type or some more radical form, I do not mean to imply that the changes are not informed by Bayesian considerations. Indeed, the problem of the transition from Pr to Pr' can be thought of as no more and no less than the familiar Bayesian problem of assigning initial probabilities, only now with a new initial situation involving a new set of possibilities and a new information basis. But the problem we are now facing is quite unlike those allegedly solved by classical principles of indifference or modern variants thereof, such as E. T. Jaynes's maximum-entropy principle, where it is assumed we know nothing or very little about the possibilities in question. In typical cases the scientific community will possess a vast store of relevant experimental and theoretical information. Using that information to inform the redistribution of probabilities over the competing theories on the occasion of the introduction of the new theory or theories is a process that, in the strict sense of the term, is *a*rational: it cannot be accomplished by some neat formal rules or, to use Kuhn's term, by an algorithm. On the other hand, the process is far from being *ir*rational, since it is informed by reasons. But the reasons, as Kuhn has emphasized, come in the form of persuasions rather than proof. In Bayesian terms, the reasons are marshaled in the guise of plausibility arguments. The deployment of plausibility arguments is an art form for which there currently exists no taxonomy. And considering the limitless variety of such arguments, it is unlikely that anything more than a superficial taxonomy can be developed. Einstein, the consummate master of this art form, appealed to analogies, symmetry considerations, thought experiments, heuristic principles such as the principle of equivalence, etc. All of these considerations, I am suggesting on behalf of the Bayesians, were deployed to nudge assignments of initial probabilities in favor of the theories Einstein was introducing in the early decades of this century. Einstein's success in this regard is no less important than experimental evidence in explaining the reception of his theories.

To summarize, Kuhn's purple passages do not seem overblown when applied to revolutions in the strong sense distinguished above. The persuasions that lead to the adoption of the new shape for the possibility space cannot amount to proofs. Certainly for the Bayesian they cannot consist of inductive proofs, since the very assignment of degrees of belief presupposes the adoption of such a space. After a revolution has taken place, the new and old theories can often be fitted into a common frame that belies any vicious form of incommensurability (as I tried to illustrate in section 4 for Newtonian and relativistic theories). But this retrospective view tends to disguise the shake-up in our system of beliefs occasioned by the adoption of the new shape for the possibility space. Bayesianism brings the shake-up to light, albeit in a way that undercuts the standard form of the doctrine.

I have no way of knowing whether Carnap would have approved of my Bayesian reading of Kuhn. But I do claim that it is a reading that fits naturally with Carnap's mature views on probability and induction.

8 Objectivity and the Problem of Consensus

I have endorsed a Bayesianized version of Kuhn's claim that in scientific revolutions persuasion rather than proof is the order of the day: revolutions involve the introduction of new possibilities; this introduction causes the redistribution of probabilities; the redistribution is guided by plausibility arguments; and such arguments belong to the art of persuasion.

This endorsement is confined to the first stage of the revolution, when the initial probabilities are established for the expanded possibility set. The Bayesian folklore would have it that after this first stage, something more akin to proof than persuasion operates. The idea is that an evidence-driven consensus emerges as a result of the Bayesian learning model: degrees of belief change by conditionalization on the accumulating evidence of observation and experiment, and the long-run result is to force a merger of posterior opinion for those Bayesian agents who initially assign zeros to the same hypotheses. This folklore can draw on some mathematically impressive merger-of-opinion theorems. But these theorems are of dubious applicability not only to the sorts of cases discussed in *Structure* but also to examples from normal, as opposed to revolutionary, science. For one thing, the mathematical results are in the form of long-run or limit results that give no information about how long the long run is.[17]

If honest theorem proving won't suffice to explain the merger of opinion that, for the Bayesian epistemologist, constitutes the heart of scientific objectivity, then perhaps we can define our way to a solution. That is, why not define "scientific community" in terms of de facto convergence of opinion over a relevant range of hypotheses? The answer is the same as that given by Kuhn in the Postscript to the threatened circularity of taking a paradigm to be what members of the community share while also taking a scientific community to consist of those scientists who share the paradigm. Just as scientific communities "can and should be isolated without prior recourse to paradigms" (Kuhn 1970, 176), so they can and should be isolated without recourse to convergence-of-opinion behavior. The European physics community in the opening decades of this century can be identified by well-established historical and sociological techniques. One wants to know how and why, for example, this community so identified reached a consensus about Einstein's STR. Nevertheless, there does seem to be at least this much truth to the definitional move: repeated failures to achieve merger of opinion on key hypotheses will most likely lead to a split in or a disintegration of the community.

This is the appropriate place to ask whether Carnap's views on probability are of any help. In a letter dated July 30, 1963, Carnap wrote to Bruno De Finetti, the arch Bayesian personalist, that he believed that the constraints of rationality extend beyond the requirement of coherence, which entails that degrees of belief must conform to the standard axioms of probability.[18] According to Carnap the requirements of rationality do not suffice to single out a unique probability function, but they do significantly constrain the choice of

Carnap, Kuhn, and the Philosophy of Scientific Methodology

a probability function. If this were correct, the Bayesian version of the problem of scientific objectivity would be made correspondingly easier. However, I do not think that Carnap managed to stake out a defensible position, as can be brought out by the question of how one recognizes which probability functions are "rational." Carnap's answer was to appeal to what he variously called "inductive intuition" and "inductive common sense." The trouble, of course, is that one person's inductive common sense is another's inductive non-sense. So the appeal to intuition reveals very different opinions as to whether it is rational to learn from experience at all and, if so, at what rate.

At this juncture it will be helpful to review a mechanism proposed by Lehrer and Wagner in *Rational Consensus in Science and Society* (1981) for achieving a group consensus. Their mechanism requires that the members of the community change their degrees of belief in accordance with a weighted-aggregation rule. Suppose that at the initial moment, person i has a degree of belief p_i in the theory in question. Each person i is assumed to assign a weight $w_{ij} \geqslant 0$ to every person j, which can be taken as an index of i's opinion as to the reliability of j's opinions. According to Lehrer and Wagner's rule, i then "improves" her initial opinion p_i^0 by changing it to $p_i^1 = \Sigma_j w_{ij} p_j^0$. If there are still differences of opinion, the aggregation process is repeated with the p_i^1 to obtain further "improved" probabilities p_i^2, etc., until eventually the probabilities for all members fall into line.

Lehrer and Wagner offer a consistency argument for their aggregation rule: "If a person refuses to aggregate, though he does assign a positive weight to other members, he is acting as though he assigned a weight of one to himself and a weight of zero to every other member of the group. If, in fact, he assigns positive weight to other members of the group, then he should not behave as if he assigned zero weight to them" (1981, 22). This argument has the flavor of "When are you going to stop beating your wife." I do assign a positive weight to the opinions of others, but as a Bayesian I do this not by means of weighted aggregation but by conditionalization: I conditionalize on information about the opinions of my peers, and I notice that the result is a shift in my degrees of belief toward the degrees of belief of those I respect. When I was a young student, these shifts brought my opinions closely in line with those belonging to people I regarded as the experts, but as a mature member of the community, I find that such shifts, while still nonnegligible, do not conform my opinions to those of others, at least not on matters where I now regard myself as an expert. And I resist any further attempt to bend my carefully considered opinions.

There are two reasons, independent of Bayesianism, to be unhappy with Lehrer and Wagner's proposal and ones like it. The first is that it is descriptively false, as shown by the very example they use to motivate their proposal. In the 1970s Robert Dicke claimed that optical measurements of the solar disk revealed an oblateness large enough to account for 3″ to 5″ of arc in Mercury's centenary perihelion advance and thus to throw into doubt Einstein's explanation of the advance. When other astrophysicists disagreed with Dicke's conclusions the differences were not smoothed over by producing a consensual probability by means of a weighted aggregation process. The disagreement remains unresolved to this day. The weight of opinion does seem to be going against Dicke's interpretation, but this partial agreement is in fact due not to aggregation but to the acquisition of additional evidence.

Of course, Lehrer and Wagner are perfectly aware of these facts, and the descriptive inadequacies of their proposal do not concern them, since they take themselves to be offering a normative proposal. But even in these terms the proposal should be faulted. It is fundamental to science that opinions be evidence-driven. Differences of opinion need not constitute an embarrassment that needs to be quashed, for these differences can serve as a spur to further theoretical and experimental research, and the new information produced may drive a genuine scientific consensus. The alternative, an attempt to manufacture a consensus by a weighted-aggregation procedure, smacks of the "mob psychology" of which Kuhn was criticized.

This last point generalizes. Bayesianism and other approaches to scientific inference as well suggest that unless there is some evidence-driven process that operates on the level of individual scientists to produce a group consensus, the consensus will amount to something that, if not mob psychology, is nevertheless a social artifact not deserving either of the labels "rational" or "scientific." Thus, contrary to Kuhn's idea, the group cannot decide—it cannot rationally decide to agree if the individuals disagree. I do not see how this conclusion can be escaped unless some yet to be articulated collectivist methodology is shown to be viable.

9 A Partial Resolution to the Problem of Consensus

Part of the answer to the Bayesian version of the problem of consensus is that quite often it does not exist and does not need to exist for normal scientific research to take place. *Structure* warned of the danger of taking textbook science as our image of how real science actually operates, and in particular, it showed how textbook science tends to make scientific revolutions invisible by painting an overly rosy picture of a smoothly accumulating stockpile of scientific knowledge. But I think that *Structure* failed to emphasize how textbook science also disguises the diversity of opinions and approaches that flourish in nonrevolutionary science.

If I had the space, I would offer as a case study the development of relativistic gravitational research over the last seventy-five-years.[19] Textbooks in this area have tended to be books on Einstein's GTR, thus fostering the illusion that GTR has achieved the status of paradigm hegemony. In addition, early textbooks not only downplayed the existence of rival theories but disguised serious difficulties with two of the principal experimental tests of GTR, the red shift and the bending of light. Normal scientific research in this field continued in the face of both a challenge to the third experimental leg of GTR deriving from Dicke's solar-oblateness measurements and also an ever growing number of rival theories of gravitation. This and similar examples suggest that normal science is possible when the community of experts share a paradigm in the weak sense of agreement on the explanatory domain of the field, on the circumscription of the space of possible theories to be considered as serious candidates for covering the explanatory domain, on exemplars of explanatory success, and on key auxiliary hypotheses. (I am tempted to say that this is the minimal sense of paradigm needed to underwrite normal science, but historians of science probably have counterexamples waiting in the wings.)

One could argue that not having a paradigm in the stricter sense of a shared theory of gravitation has lowered the puzzle-solving efficiency of normal science. One can recall Thorne and Will's (1971) lament that, faced with a zoo of alternative theories of gravitation, astrophysicists where hamstrung in their model-building activity. While I think that this is a fair observation, I also think that there is more to progress in normal science than puzzle solving. In particular, I would emphasize the conceptual advances derived from the exploration of the space of possible theories, a point that brings me to the second part of my partial answer to the problem of consensus.

Again, if I had the space, I would argue that insofar as a consensus is established, it is often due to a process akin to the much maligned idea of eliminative induction. This process is often accompanied by a proliferation of theories, not as an exercise in Feyerabendian anarchy or Dadaism, but as a means of probing the possibilities and as a preliminary to developing a classification scheme that makes systematic elimination a tractable exercise.[20] The elimination is not of the simpleminded Sherlock Holmes variety, for it involves Bayesian elements, especially in the assessment of the auxiliary assumptions needed to bring about a confrontation of theory and experiment. Thus the Bayesianized version of the problem of consensus remains. And at the present time I do not see any resolution that does not fall back on something like the definitional solution, which I casually dismissed in the preceding section. Such a fallback undermines scientific objectivity in a way that would not have pleased Carnap and, I presume, does not please Kuhn either.

10 Conclusion

I was a distant student of Carnap and a close student of Kuhn. But the two seemed to me so different in style and concerns that I placed them in different parts of the philosophical firmament. Only now have I begun to appreciate how misguided my placement was and how much philosophy of science can be enriched by considering how the ideas of these two giants interact. I have presented one way to stage the interaction. There are surely better ones. I urge more able hands to take up the task.[21]

Carnap, Kuhn, and the Philosophy of Scientific Methodology

Notes

Sections 6–9 of this paper are based on chapter 8 of Earman 1992. I am grateful to Richard Jeffrey and Wes Salmon for helpful comments on an earlier draft.

1. I will not attempt to characterize the differences between logical positivism and logical empiricism. In terms of adherents, the former includes the members of the Vienna Circle, while the latter includes the members of the Berlin Society for Empirical Philosophy.

2. Or so it was thought until Ayer and others tried to spell out the conditions for verifiability. For a review of the problems encountered, see Hempel 1950, 1951, 1965.

3. That is, confrontation of a statement with observation.

4. *"Analog[isch] ist die Entwicklung der wissenschaftlichen Theorien zu verstehen*: nicht als gerichtet auf [   ] die ideale, wahre [?] Theorie, die *eine* wahre Theorie uber die Welt [   ], sondern Entwicklung al Schritt zu einer besseren Form, durch Auswahl einer aus mehreren kompetierenden die Auswahl geschieht durch Bevorzugung in der community der Wissenschaftler, wobei allerhand soziologi, kulturelle usw. Faktoren mitspielen. Nicht: »Wirkommen der Wahrheit naher,« sondern »Wir verbessern ein Instrument.« Archive for Scientific Philosophy, University of Pittsburgh, document no. RC 082-03-01:1r/1. Quoted by permission of the University of Pittsburgh; all rights reserved. I am grateful to Pirmin Steckler-Wiethofer for providing the English translation.

5. Keep in mind that *Philosophical Foundations of Physics* was compiled by Martin Gardner

6. Davidson writes, "The dominant metaphor of conceptual relativism, that of different points of view, seems to betray an underlying paradox. Different points of view make sense, but only if there is a common coordinate system on which to plot them; yet the existence of a common system belies the claim of a dramatic incomparability" (1974, 6).

7. See Earman 1989 for an evaluation of these claims.

8. In this period Carnap was under the illusion that the logic of science could be discussed purely in terms of logical syntax, but the point I am making here holds with respect to syntax and semantics. Carnap's 1930s strategy for dissolving philosophical problems appears in slightly new garb in his distinction between "internal" and "external questions" (see Carnap 1950).

9. See Earman 1986 for some caveats.

10. Feigl's paper for the Schilpp volume was written in 1954. But due to delays, the volume did not appear until 1963.

11. While the use of this language may not be appropriate for understanding all the historical disputes, it does help to illuminate the long running disputes over absolute versus relational conceptions of space and time (see Earman 1989).

12. Here is a place where a resort to Carnapian subscripting may be healthy. I Use "incommensurability$_1$" to indicate Kuhn's sense of incommensurability that derives

John Earman

from changes in the lexicon, and I use "incommensurability$_2$" to stand for the kind of incommensurability that makes theory choice impossible or difficult by means of relatively neutral observations. My claims are that incommensurability$_1$ does not imply incommensurability$_2$ and that as a matter of actual historical fact incommensurability$_2$ is not so bad in typical cases of scientific revolutions.

13. Since I have never been able to understand what is at issue here, I don't know whether I should demur.

14. See Carnap 1962, 1963c, and 1968. Carnap was not a dogmatist on this matter. In his final published pronouncement on this matter he wrote, "When I say that the end result of inductive reasoning is not the acceptance of a hypothesis, but rather the assignment of a probability to the hypothesis, this is not meant as a description of what is actually done, but rather as a proposal for a rational reconstruction. Therefore, although in the present controversy I agree essentially with Professor Bar-Hillel [who argues against rules of acceptance] against Professor Kyburg [who argues for rules of acceptance], I am quite doubtful about one view in which they seem to agree, that we have to choose between two irreconcilable positions. I do not think, as Kyburg does, that our using or not using rules of acceptance (or detachment) makes a vast difference in our philosophy of science. Nor would I, like Bar-Hillel, totally condemn such rules" (1968, 146), For the sake of a sharp contrast with Kuhn, I am presenting a Carnap who would condemn rules of acceptance. Contrary to Carnap, I think that one's attitude on this matter does make a significant difference for one's image of science.

15. I have borrowed the phrase "tempered personalism" from Shimony (1970).

16. There is a bit of awkwardness here, since in Carnap's systems of inductive logic the probability of theories for infinite domains will be flatly zero. Carnap was thus forced to talk about "instance confirmation" of theories. I will pass over this difficulty, since it is one that is peculiar to Carnap's language-based systems and does not apply to Bayesianism in general.

17. For a detailed discussion of these matters, see chapter 6 of Earman 1992.

18. Archive for Scientific Philosophy, University of Pittsburgh, document no. 084-16-01.

19. See chapter 7 of Earman 1992 for details.

20. Again, see Earman 1992 for a discussion of how this exercise works for relativistic gravitational theory.

21. The able hands of Wesley Salmon have taken up the work (see Salmon 1990).

References

Carnap, R. 1928. *Der logische Aufbau der Welt*. English translation: *The Logical Structure of the World*. Berkeley: University of California Press, 1967.

Carnap, R. 1934. *Logische Syntax der Sprache*. English translation: *The Logical Syntax of Language*. London: Routledge and Kegan Paul, 1937.

Carnap, Kuhn, and the Philosophy of Scientific Methodology

Carnap, R. 1935. "Philosophy and Logical Syntax." Reprinted in W. P. Alston and G. Nakhnikian, eds., *Readings in Twentieth Century Philosophy*. New York: Free Press, 1963.

Carnap, R. 1936. "Warheit und Bewährung." *Actes du Congress International de Philosophie Scientifiques*. Paris: Hermann et Cie.

Carnap, R. 1949. "Truth and Confirmation," in H. Feigl and W. Sellars, eds., *Readings in Philosophical Analysis*. New York: Appleton-Century-Crofts.

Carnap, R. 1950. "Empiricism, Semantics, and Ontology." *Revue Internationale de Philosophie* 4:20 40.

Carnap, R. 1962. "The Aim of Inductive Logic." In E. Nagel, P. Suppes, A. Tarski, eds., *Logic, Methodology, and the Philosophy of Science*. Stanford: Stanford University Press.

Carnap, R. 1963a. "Carl G. Hempel on Scientific Theories." In Schilpp 1963.

Carnap, R. 1963b. "Herbert Feigl on Physicalism." In Schilpp 1963.

Carnap, R. 1963c. "My Basic Conception of Probability and Induction." In Schilpp 1963.

Carnap, R. 1966. *Philosophical Foundations of Physics*. New York: Basic Books.

Carnap, R. 1968. "On Rules of Acceptance." In I. Lakatos, ed., *The Problem of Inductive Logic*. Amsterdam: North Holland.

Davidson, D. 1973. "On the Very Idea of a Conceptual Framework." *Proceedings and Address of the American Philosophical Society* 47:5–12.

Earman, J. 1986. *A Primer on Determinism*. Dordrecht: D. Reidel.

Earman, J. 1989. *World Enough and Space-Time: Absolute versus Relational Theories of Space and Time*. Cambridge: MIT Press.

Earman, J. 1992. *Bayes or Bust: A Critical Examination of Bayesian Confirmation Theory*. Cambridge: MIT Press.

Earman, J. and Glymour, C. 1991. "Einstein's Explanation of the Anomalous Motion of Mercury's Perihelion." *Einstein Studies*, forthcoming.

Feigl, H. 1958. "The 'Mental' and the 'Physical'." In H. Feigl, M. Scriven, and G. Maxwell, eds., *Concepts, Theories, and the Mind-Body Problem*, Minnesota Studies in the Philosophy of Science, vol. 2. Minneapolis: University of Minnesota Press.

Feigl, H. 1963. "Physicalism, Unity of Science and the Foundations of Psychology." In Schilpp 1963.

Fodor, J. 1984. "Observation Reconsidered." *Philosophy of Science* 51:23–41.

Franklin, A. 1986. *The Neglect of Experiment*. Cambridge: Cambridge University Press.

Friedman, M. 1987. "Carnap's *Aufbau* Reconsidered." *Noûs* 21:521–545.

Friedman, M. 1992. "Philosophy and the Exact Sciences: Logical Positivism as a Case Study." In J. Earman, ed., *Inference, Explanation, and Other Frustrations: Essays in the Philosophy of Science.* Los Angeles: University of California Press.

Hempel, C. G. 1950. "Problems and Changes in the Empiricist Criterion of Meaning." *Revue Internationale de Philosophie* 4:41–63.

Hempel, C. G. 1951. "The Concept of Cognitive Significance: A Reconsideration." *Proceedings of the American Academy Arts and Sciences* 80:61–77.

Hempel, C. G. 1965. "Empiricist Criteria of Cognitive Significance." In *Aspects of Scientific Explanation.* New York: Free Press.

Kuhn, T. S. 1962. *The Structure of Scientific Revolutions.* Chicago: University of Chicago Press.

Kuhn, T. S. 1970. *The Structure of Scientific Revolutions.* 2nd ed. Chicago: University of Chicago Press.

Kuhn, T. S. 1977. "Objectivity, Value Judgments, and Theory Choice." In *The Essential Tension.* Chicago: University of Chicago Press.

Kuhn, T. S. 1983. "Rationality and Theory Choice." *Journal of Philosophy* 80:563–570.

Kuhn, T. S. 1989. "Possible Worlds in History of Science." In S. Allen, ed., *Possible Worlds in Humanities, Arts, and Sciences.* Berlin: W. de Gruyter.

Lakatos, I. 1970. "Falsification and the Methodology of Scientific Research Programs." In I. Lakatos and A. Musgrave, eds., *Criticisms and the Growth of Knowledge.* Cambridge: Cambridge University Press.

Lehrer, K. and Wagner, C. 1981. *Rational Consensus in Science and Society.* Dordrecht: D. Reidel.

Reisch, G. A. 1991. "Did Kuhn Kill Logical Empiricism?" *Philosophy of Science* 58:264–277.

Salmon, W. 1990. "Rationality and Objectivity in Science; or, Tom Kuhn Meets Tom Bayes." In C. W. Savage, ed., *Scientific Theories.* Minneapolis: University of Minnesota Press.

Schilpp, P. A. 1963. *The Philosophy of Rudolf Carnap.* La Salle, Ill.: Open Court.

Shapere, D. 1966. "Meaning and Scientific Change." In R. G. Colodny, ed., *Mind and Cosmos.* Pittsburgh: University of Pittsburgh Press.

Shimony, A. 1970. "Scientific Inference." In R. G. Colodny, ed., *The Nature and Function of Scientific Theories.* Pittsburgh: University of Pittsburgh Press.

Thorne, K. S. and Will, C. M. 1971. "Theoretical Frameworks for Testing Relativistic Gravity. I: Foundations." *Astrophysical Journal* 163:595–610.

Remarks on the History of Science and the History of Philosophy

Michael Friedman

Thomas Kuhn's *The Structure of Scientific Revolutions* (1962) forever changed our appreciation of the philosophical importance of the history of science. Reacting against what he perceived as the naively empiricist, formalist, and ahistorical conception of science articulated by the logical positivists, Kuhn presented an alternative conception of science in flux, of science driven not so much by the continuous accumulation of uncontroversial observable facts as by profoundly discontinuous conceptual revolutions in which the very foundations of old frameworks of scientific thought are replaced by radically new ones. When such a revolution occurs, we do not simply replace old "false" beliefs with new "true" beliefs; rather, we fundamentally change the system of concepts within which beliefs, whether true or false, can be meaningfully formulated in the first place. This picture of science thus driven by revolutionary forces of conceptual change has itself sparked a revolution of sorts in the philosophy of science: logical positivism has been decisively discredited, and a "new" philosophy of science concerned more with questions of historical development than with questions of logic, justification, and truth has taken its place. Whatever the fate of this new philosophy of science may be, it is clear beyond the shadow of a doubt, I think, that careful and sensitive attention to the history of science must remain absolutely central in any serious philosophical consideration of science.

I would here like to discuss an extension of this lesson of Kuhn's in a somewhat different direction. The history of science is not only important for the philosophy of science, it is also of the highest importance for a full appreciation of the history of philosophy more generally. For the fact is that many of the main developments in the historical evolution of philosophy as a subject are themselves generated by just the kind of conceptual revolutions in scientific thinking to which Kuhn has called our attention. Moreover, since our present philosophical concepts are, of course, the product of the historical evolution of the subject, it follows that careful and sensitive attention to the history of science is important not only to philosophy of science as such but also, for example, to epistemology and metaphysics as well. Here, however, I will confine myself to the relevance of the history of science to the history of philosophy.

Now I do not suppose that this claim about the relevance of conceptual revolutions in science to the history of philosophy is a controversial one. Indeed, it is now becoming more and more common for historians of philosophy, especially those concerned with the modern period, to emphasize precisely this relevance. Thus it is now a commonplace that the articulation of characteristically modern philosophy by Descartes and his successors must be viewed against the background of the scientific revolution of the sixteenth and seventeenth centuries. By emphasizing Descartes's concern to replace the Aristotelian-Scholastic natural philosophy with the "mechanical natural philosophy" of the new science, we can achieve a fuller and deeper understanding of such characteristically modern preoccupations as, for example, the distinction between primary and secondary qualities, the "veil of perception," the mind-body problem, and so on. Viewed against the background of the scientific revolution that created modern "mechanistic" natural science itself, these modern philosophical preoccupations no longer appear arbitrary and capricious, as stemming, perhaps, from otherwise unaccountable obsessions with certainty or with "mirroring reality." Instead, they can be understood as natural attempts to come to grips with a profound reorganization of the very terms in which we conceptualize ourselves and our world.

A concern with coming to terms with modern mechanistic natural science is, of course, shared by the philosophers of the modern period quite generally. What I would like to emphasize here is the philosophical importance of a more fine-grained study of the interaction between the history of science and the history of philosophy. In particular, I think it has only seldom been appreciated that some of the most important developments within the modern tradition have been at least partly driven by fundamental conceptual problems faced by the new science. The problems I have in mind concern the foundations of the new physical dynamics and of the spatiotemporal framework within which this new dynamics is supposed to be articulated.

To get a preliminary sense of the nature of these problems, it is illuminating to contrast the basis of the new physical dynamics with that of Aristotelian-Scholastic natural philosophy. The Aristotelian universe is a finite sphere whose outermost spherical surface contains the fixed stars, and this outermost celestial sphere is uniformly rotating in a westward direction. The physical dynamics of the Aristotelian universe is governed by two principles. The principle of *natural place* governs change within the sublunary region: all elements are naturally in a state of rest in their natural places (earth, for example, is naturally at rest at the center of the finite sphere defining the universe) unless removed therefrom by some *violent* or *forced motion*; if they are so removed, however, they have a tendency to move toward their natural places in straight lines. The principle of *natural motion* governs the superlunary region: celestial spheres on which appear the various heavenly bodies (that of the moon being innermost) form a nested sequence terminating in the outermost sphere described above, and all persist in states of uniform rotational motion.

The physical dynamics of the Aristotelian universe uniquely determines its spatiotemporal framework. Temporal duration is precisely determined by the uniform rotation of the outermost celestial sphere relative to the earth, which is necessarily in a state of perfect rest at the very center of the spherical universe. Moreover, the state of rest of the earth and the state of uniform rotation of the outermost celestial sphere uniquely determine place and motion as well. There is a privileged spatial position at the center of the universe, a privileged line through the center defining the axis of rotation of the outermost celestial sphere, a privileged equatorial plane orthogonal to this axis, and a privileged state of diurnal rotation. Thus any state of translational or rotational motion is uniquely defined and absolute. (Indeed, even the notions of up-down and right-left are uniquely defined and absolute!) Another way to put the point from a modern point of view is that the physical dynamics of the Aristotelian universe admit no nontrivial spatiotemporal symmetries: spatial translations move the privileged central point; spatial rotations about any other axis than that defined by the rotation of the outermost celestial sphere change the orientation of this latter axis; spatiotemporal rotations change the state of rotation of the outermost sphere relative to the earth.

The physical dynamics of the new mechanistic science, by contrast, is based on the law of inertia. The dichotomy between sublunary and superlunary realms is abolished, together with the correlative distinction between natural spatial position in the sublunary region and natural uniform rotational motion in the superlunary region. Instead, the natural state of *all* matter in the universe is one of uniform rectilinear translation: every body in the universe persists in its state of uniform rectilinear motion unless some *force* compels it to deviate from this natural state. This means that the space of the new universe must be infinite Euclidean space, for an unimpeded body will naturally progress along a Euclidean straight line to infinity. It also means that temporal duration in the new universe is measured by this same inertial motion: equal temporal intervals are those during which an inertially moving body traverses equal distances. The basis of the new physical dynamics is thus apparently given by the very simplest elements of Euclidean geometry. Hidden behind this simplicity, however, lies a veritable Pandora's box of fundamental problems.

For how do we apply this framework of infinite Euclidean space and inertially measured time to the nature with which we are empirically confronted? Unlike the rotational motion of the outermost celestial sphere in the Aristotelian universe, we are not in fact given any examples of actual inertial motion at all: all the motions we observe are cases of noninertial motion. So we need to supplement our physical dynamics with a theory of *force* or deviation from the natural inertial state. We need to do this, that is, before we can empirically define the time of the new dynamics in the first place. Moreover, even if we were given actual cases of inertial motion, we would still not be able empirically to determine the notion of place. Nor would it be clear how to determine the notion of change of place over time. Again unlike the finite spherical universe of Aristotelian natural philosophy, infinite Euclidean space has no distinguished positions; in modern terms, any spatial translation is a symmetry of the spatial structure (as are spatial rotations and spatiotemporal rotations of course). It is therefore unclear how change of place or motion is itself to be empirically defined. It follows that the empirical meaning of the law of inertia is also radically unclear: relative to *what* do naturally moving bodies move rectilinearly and uniformly?

I suggest that some of the principal stages in the evolution of the modern philosophical tradition can be understood as successive attempts to come to terms with this fundamental problem lying at the basis of the new physical dynamics. And doing this will, I believe, result in a conception of the evolution of modern philosophy that is more illuminating than the conventional picture of a succession of largely futile attempts to solve the problem of skepticism about the external world.

Thus, for example, Descartes, who is clearly followed in this regard by Spinoza, conceives of the new physical dynamics as resting on a purely geometrical basis. Inertial motion along a Euclidean straight line is the dominant idea, and deviations from this inertial state are initially conceived in the simplest possible geometrical terms, that is, as perfect reflections or mere changes of direction on impact (this is especially clear in axiom 2 following lemma 3 in Part 2 of Spinoza's *Ethics*). Bodies persist along Euclidean straight lines until they come into contact with one another, whereupon they change their direction of motion but preserve their quantity of motion. We thereby obtain the purely geometrical picture of nature characteristic of "mechanical philosophy": all natural change is to be understood in terms of the sizes, figures, and motions of the parts of matter. This geometrical picture underlies some of the most basic ideas of Descartes's and Spinoza's philosophy, for example, the idea that the substance or essence of material things consists of extension alone (that is, of three-dimensional Euclidean space), the correlative doctrine that the Scholastic notion of substance or "substantial form" must be entirely abandoned, the idea that nature unfolds with purely geometrical necessity.

Yet as is well known, this Cartesian version of the mechanical philosophy fails to articulate an adequate physical dynamics. First, the above conception of deviation from the inertial state based on pure geometrical reflection works only for perfectly elastic impact viewed from the center-of-mass frame of the interaction. Notoriously,

Descartes completely failed in attempting to extend the laws of impact to the general case. Indeed, without any adequate conception of mass, it was even impossible to specify the conditions under which geometrical reflection gives a correct description of impact. (Descartes understands quantity of motion as "size" multiplied by speed.) More generally, however, the Cartesian mechanical philosophy completely fails to articulate coherently the spatiotemporal framework within which the motions it describes are to take place. Descartes defines motion as change of position with respect to the "surrounding bodies," and this leaves us with no consistent conception of motion at all: the same body may be moving with respect to some surrounding bodies and at rest with respect to others; the same body may even be changing its state of motion (moving noninertially) with respect to some surrounding bodies and moving inertially with respect to others; and so on. Finally, even if we ignore both of the above problems, Cartesian mechanical philosophy fails to indicate how time is to be empirically defined in this new universe, for we are so far entirely unable to determine whether the motions we actually observe (such as the motions of the heavenly bodies, for example) are uniform.

Some of the key philosophical moves made by Leibniz can then be understood in terms of his quite deep and penetrating insight into these fundamental problems faced by Cartesian mechanical philosophy. Leibniz was convinced of the inadequacy of a purely geometrical physical dynamics from at least 1671 ("The Theory of Abstract Motion"), but the decisive step was made in his "Brief Demonstration of a Notable Error of Descartes and Others Concerning a Natural Law" of 1686. Following Huygens's general theory of perfectly elastic impact, Leibniz sees that the basic dynamical quantity governing such impact is not the Cartesian quantity of motion but rather "living force" or *vis viva* (mv^2, twice our kinetic energy), that is, living force, or *vis viva* is the operative conserved quantity. Moreover, Leibniz sees that this living force is not a purely geometrical/kinematic quantity: it cannot be constructed purely from the sizes, figures, and motions of the interacting bodies in question. Hence mechanics itself cannot be based on purely geometrical/kinematic concepts and requires a more basic conception of *active force*.

This insight in turn leads Leibniz to a thoroughgoing reevaluation of Cartesian and Spinozistic philosophy. First, Leibniz argues that the substance or essence of material beings does not consist of pure geometrical extension but rather of precisely active force. And this means that Leibniz no longer defines the notion of substance purely in terms of independent existence (as do Descartes and especially Spinoza) but rather reverts to the more Aristotelian idea of substance as that which persists through time and controls the temporal evolution of a thing (as the substantial form of an acorn determines its evolution into an oak). Leibniz, in other words, sees that temporal evolution has been entirely left out of Cartesian mechanical philosophy and seeks to restore its primary place. And in seeking thus to restore the primacy of temporal evolution, Leibniz argues that Descartes's and Spinoza's criticisms of Aristotelian-Scholastic natural philosophy has been pressed too far: substantial forms and therefore an appeal to final causation enter into the very foundations of the new physical dynamics itself. As a result, Leibniz also argues that purely geometrical necessity cannot underly the actual laws of nature; a kind of contingency closely tied to final causes and teleological considerations is in fact essential to the actual world.

I am suggesting, then, that this attempt by Leibniz to reconcile the new mechanistic science with elements of the older Aristotelian-Scholastic philosophy is itself fueled by his discovery of the fundamental dynamic importance of *vis viva* or active force. This comes out very clearly in the "Discourse on Metaphysics" of 1786, where, immediately after a summary of his "Brief Demonstration" of the errors of Descartes in section 17, he appends section 18 on the metaphysical importance of active force. This section is worth quoting in full:

This consideration of the force, distinguished from the quantity of motion is of importance, not only in physics and mechanics for finding the real laws of nature and the principles of motion, and even for correcting many practical errors which have crept into the writings of certain able mathematicians, but also in metaphysics it is of importance for the better understanding of principles. Because motion, if we regard only its exact and formal meaning, that is, change of place, is not something entirely real, and when several bodies change their places reciprocally, it is not possible to determine by considering the bodies alone to which among them movement or repose is to be attributed, as I could demonstrate geometrically, if I wished to stop for it now. But the force, or the proximate cause of these changes is something more real, and there are sufficient grounds for attributing it to one body rather than another, and it is only through this latter investigation that we can determine to which one the movement must appertain. Now this force is something different from size, from form or from motion, and it can be seen from this consideration that the whole meaning of a body is not exhausted in its extension together with its modifications as our moderns persuade themselves. We are therefore obliged to restore certain beings or forms which they have banished. It appears more and more clear that although all the particular phenomena of nature can be explained mathematically or mechanically by those who understand them, yet nevertheless, the general principles of corporeal nature and even of mechanics are metaphysical rather than geometrical, and belong rather to certain indivisible forms or natures as the causes of the appearances, than to the corporeal bulk or to extension. This reflection is able to reconcile the mechanical philosophy of the moderns with the circumspection of those intelligent and well-meaning persons who, with a certain justice, fear that we are becoming too far removed from immaterial beings and that we are thus prejudicing piety. (G. R. Montgomery translation)

Leibniz's insight into the foundations of the new physical dynamics thus provides him with just the room he needs to introduce some of his most characteristic philosophical doctrines.

The above passage from section 18 of the "Discourse on Metaphysics" also illustrates another characteristic concern of Leibniz's, a concern for the metaphysical reality of space and motion. As the correspondence with Clarke makes especially clear, Leibniz sees that the existence of nontrivial symmetries makes the spatiotemporal structure of the new science inherently problematic; for it now becomes radically unclear how the notions of spatial position and change of spatial position are to be applied to our experience of nature. Leibniz also sees clearly that the Cartesian definition of motion as change of position with respect to the surrounding bodies is inconsistent with the law of inertia. Unfortunately, however, although Leibniz clearly understands the problem, it does not appear that he ever arrives at a satisfactory solution, for he never makes it clear exactly *how* his notion of active force enables us to determine, in a system of relatively moving bodies, "to which among them movement or repose is to be attributed." He never explains, for example, how the notion of *vis viva* could possibly help us in deciding the issue of heliocentrism. It is no wonder, then, that Leibniz ultimately consigns space, time, and motion to the realm of mere phenomenal appearances, manifestations of a more basic and essentially nonspatiotemporal reality characterized solely in terms of nongeometrical or metaphysical notions of substance and active force.

Just as some of the key philosophical moves of Leibniz can be understood as attempts to resolve the fundamental problems of physical dynamics bequeathed to him by Descartes, some of the central elements of the Kantian philosophy can be understood as attempts to resolve the above problems left open by Leibniz. From the time of his earliest published work of 1747, *Thoughts on the True Estimation of Living Forces,* Kant expresses the Leibnizean conviction that the new physical dynamics must rest on a metaphysical foundation involving the central notions of substance and active force, to which Kant adds the expressly anti-Leibnizean category of interaction or dynamical community. Yet as this last point suggests, Kant is also convinced from the time of his earliest work that the new physical dynamics has been finally correctly formulated by Newton: Kant consistently takes universal gravitation (rather than the conservation of *vis viva*) as his paradigm of physical law. I believe that much of Kant's philosophy can then be understood as an attempt to synthesize Leibniz's metaphysics with Newton's physics.

Kant's first move (in the 1747 essay) is to replace Leibniz's conception of living force with Newton's conception of *impressed force*: force is not an internal property of a single body by which that body determines the temporal evolution of its own future state; rather, force is an action of one body on another essentially distinct body by which the first body changes the state of the second body. Far from expressing the state of motion of a single body, force has nothing at all to do with the state of motion of the body that exerts it. Force expresses a relationship of *real interaction* between two bodies by which one body *changes* the state of motion of the other. Kant sees, therefore, that Newton's conception of force completely overturns Leibniz's monadology and system of preestablished harmony. This is especially obvious if one accepts, as Kant consistently does, Newtonian gravitation, understood as a genuine action-at-a-distance force. (By contrast, this full Newtonian conception of force is necessarily precluded for Leibniz, for a system in which *vis viva* is conserved is a system in which there is zero potential energy and thus no action-at-a-distance forces.)

Kant's next move (in the "New System of Motion and Rest" of 1758) is to grasp the fundamental importance of Newton's third law of motion: the equality of action and reaction. Since every change of the quantity of motion of one body is counterbalanced by a corresponding change in the quantity of motion of a second body, where the first body is the cause of the change of motion of the second body and vice versa, the third law of motion expresses a dynamical community or real interaction of material substances. (Here quantity of motion is, of course, understood in the Newtonian sense of *momentum* or mass multiplied by velocity.) Further and even more significant, Kant sees that this same third law of motion allows us to give genuine empirical meaning to the notion of motion itself: the true motions in a system of interacting bodies are just those described in the *center-of-mass frame* of the system, that is, exactly those motions that make the third law of motion true. In this way Kant grasps the dynamical significance of the Newtonian conception of mass and sees in particular that the problem of empirically applying the notion of motion in our new spatiotemporal framework can now be solved by regarding Newton's laws of motion as, in effect, amounting to a *definition* of the concept of true (or absolute) motion. (In modern terms, true accelerations are those described in an inertial frame of reference, and an inertial frame of reference is one in which the laws of motion are satisfied.)

Kant's final move is to put the above ideas together in a fundamental reinterpretation of the argument of Newton's *Philosphiae Naturalis Principia Mathematica*. This is depicted most clearly in the *Metaphysical Foundations of Natural Science* of 1786. In particular, Kant understands the argument of the *Principia* not, as Newton does, as an argument by which the true (or absolute) motions in the solar system are *found* or *discovered* but rather as a procedure by which the notion of true (or absolute) motion is given objective empirical meaning in the first place. For what Newton actually does in the *Principia* is to apply his laws of motion to the observed, so far merely relative or apparent motions in the solar system in order to derive therefrom the law of universal gravitation *and at the same time rigorously to determine the center-of-mass frame of the solar system wherein the true motions in the solar system can be first rigorously described.* This center-of-mass frame turns out to be centered sometimes within and sometimes without the surface of the sun but never very far from the sun's center, and this is the precise sense in which the heliocentric system is closer to the truth. By contrast, prior to the argument of *Principia* not only are we unable to decide the question of heliocentrism, but the very question itself lacks all objective meaning. In this way, then, Kant understands Newton's conception of force, expressed by Newton's laws of motion and instantiated by universal gravitation, as finally providing a satisfactory basis for the new physical dynamics. Thus, after our determination of a privileged center-of-mass frame via Newton's theory of universal gravitation, we are also in a position rigorously to determine the precise degree of uniformity in the orbital motions of the heavenly bodies.

Kant continues to insist, however, that the new physical dynamics requires a metaphysical foundation, for the Newtonian laws of motion must themselves be philosophically grounded: neither an empirical/inductive account nor a purely postulational/hypothetical account is ultimately satisfactory. The basic idea of this philophical grounding has already been indicated above: the laws of motion do not express facts, as it were, about a notion of true motion antecedently well defined; rather, they constitute the sole conditions under which the notion of true (or absolute) motion has objective empirical meaning in the first place. Hence the laws of motion are true a priori, for they represent the necessary conditions for the new physical dynamics to be applicable to our actual experience of nature. Moreover, since the new physical dynamics itself constitutes a necessary condition for a genuinely objective experience of nature, it follows for Kant that the laws of motion are ultimately grounded in the a priori conditions of the possibility of experience. In particular, Newton's concepts of mass, force, and interaction are ultimately grounded in the a priori categories of substance, causality, and community.

Kant thus ends up by turning Leibniz's conception of the relationship between physics and metaphysics on its head. Metaphysical concepts—such as substance, causality, and community—no longer describe a nonspatiotemporal realm somehow existing "behind" the spatiotemporal phenomena of nature. Rather, the entire point and function of such concepts is precisely to ground the sole spatiotemporal framework within which the phenomena of nature can be empirically and objectively described. In Kantian terms, the categories acquire objective meaning only by being spatiotemporally *schematized*. By the same token, however, our conception of the special a priori status of metaphysical concepts is also fundamentally reinterpreted. Metaphysics is a priori, not because it shares in the geometrical necessity of mathematical reasoning, nor because it describes a realm of nonspatiotemporal entities accessible to pure reason alone, but rather because it describes the a priori conditions that make objective *empirical* thinking possible in the first place. That is, metaphysics articulates the a priori conditions of the possibility of experience.

Now the above account of a particular evolution of thought from Descartes and Spinoza through Leibniz and finally to Kant is, of course, only the briefest sketch, and a full understanding of this evolution would certainly require a much more detailed discussion of both scientific and philosophical developments. I hope that I have said enough, however, to indicate the philosophical interest and importance of this kind of approach. One thing already clear, for example, is that, contrary to the opinion of some contemporary historical writers, the philosophers of the modern tradition are not best understood as attempting to stand outside the new science so as to show, from some mysterious point outside of science itself, that our scientific knowledge somehow "mirrors" an independently existing reality. Rather, the philosophers I have been considering start from the *fact* of modern scientific knowledge as a given fixed point, as it were. Their problem is not so much to justify this knowledge from some "higher" standpoint as to articulate the new *philosophical* conceptions that are forced upon us by the new science. Kant puts the point admirably in section 40 of the *Prolegomena:*

Pure mathematics and pure science of nature had, for their own safety and certainty, no need for such a deduction as we have made of both. For the former rests upon its own evidence, and the latter (though sprung from pure sources of the understanding) upon experience and its thorough confirmation. The pure science of nature cannot altogether refuse and dispense with the testimony of experience; because with all its certainty it can never, as philosophy, imitate mathematics. Both sciences, therefore, stood in need of this inquiry, not for themselves, but for the sake of another science: metaphysics. (L. W. Beck translation)

Kant's inquiry into the a priori conditions of the possibility of mathematics and natural science is not intended to provide an otherwise missing justification for these sciences but rather to establish a reorganization and reinterpretation of metaphysics itself.

I believe that this same kind of approach—with the same philosophical moral—can be fruitfully applied to the evolution of contemporary philosophical concepts and problems as well. In particular, it can be fruitfully applied in coming to a more adequate understanding of the development of twentieth-century logical positivism. These thinkers too are best understood as attempting philosophically to come to terms with the profound conceptual revolutions that initiated twentieth-century science. These thinkers too should be seen not as attempting to justify twentieth-century science from some sterile and futile external vantage point but rather as once again refashioning the basic concepts and principles of philosophy so as to accommodate and comprehend the new scientific developments. That is, their aim is not to justify twentieth-century science from some supposed "higher" standpoint but rather to provide a *rational reconstruction* of that science and to find thereby a new, nonmetaphysical task for philosophy.

More specifically, the logical positivists were attempting to respond philosophically to fundamental conceptual revolutions in both mathematics and physics. The relevant mathematical developments are epitomized by the profound shifts in nineteenth-century thinking resulting in Hilbert's beautiful axiomatic treatment of Euclidean geometry on the one hand and Frege's complete delineation of modern quantificational logic on the other. The relevant physical developments are epitomized by Einstein's theory of relativity, which in turn, of course, rests on nineteenth-century work on non-Euclidean geometries. These two sets of developments imply that Kant's conception of synthetic a priori knowledge, with its corresponding picture of an a priori foundation for scientific thought based on a spatiotemporal schematization of the categories, is no longer tenable. For the above mathematical developments destroy Kant's basic opposition between intuitions and concepts, sensibility and understanding: the mathematical structures in question can now be represented purely logically or conceptually with no reliance whatsoever on spatiotemporal intuition. And the above physical developments destroy the idea that there is a *fixed* a priori structure—consisting of the spatiotemporal framework of Euclidean geometry and Newtonian dynamics—lying at the basis of our empirical science of nature.

It is important to see, however, that the logical positivists did not react to these revolutionary developments by adopting a naively empiricist conception of science. They did not, for example, conceive the spatiotemporal framework of physical dynamics as determined purely empirically. On the contrary, they continue to advocate a *modified* Kantian position according to which there is a fundamental distinction between the spatiotemporal framework of physical dynamics and the empirical laws formulated within this framework. Moreover, they continue to agree with Kant that a prior articulation of a spatiotemporal dynamical framework is a necessary precondition for the empirical meaningfulness of our subsequent description of nature. In this sense, the spatiotemporal framework of physical dynamics can still be conceived of as a priori. Yet the a priori has indeed lost its fixed or absolute status: it can change and develop with the progress of science, and it varies from the context of one scientific theory to another (Euclidean geometry is a priori in the context of Newtonian physics, but Riemannian geometry is a priori in the context of the general theory of relativity). As Reichenbach put it in his *Theory of Relativity and A Priori Knowledge* of 1920, we must distinguish two meanings of the Kantian a priori: the first involves unrevisability and the idea of absolute fixity for all time; the second means "constitutive of the concept of the object of knowledge." In acknowledging the profound philosophical importance of Einstein's theory of relativity, we do not abandon the Kantian a priori altogether. We simply discard the first meaning while continuing to emphasize the crucial scientific role of the second. What relativity shows us, in fact, is that the notion of apriority must itself be relativized.

At the same time, however, the notion of apriority must also be *logicized* or incorporated into the new logical framework due to Frege and Russell. For Frege and Russell have shown us how to dispense with the Kantian synthetic a priori in *pure* mathematics by demonstrating how to embed pure mathematics within the new logic. In this way mathematics retains its a priori status but now becomes purely analytic. (Again, however, I emphasize that a naively empiricist conception of mathematics is simply out of the question for the logical positivists.) Moreover, this achievement of Frege's and Russell's makes it clear that the new mathematical logic must provide the ultimate framework for rigorous philosophical thinking. But it is not yet clear how Frege's and Russell's new logic is related to the relativized notion of apriority lying at the basis of *physical* science. We cannot simply maintain, for example, that the spatiotemporal framework of physical dynamics is itself a part of logic! Once again, therefore, the attempt philosophically to comprehend the totality of the new scientific developments generates fundamental philosophical problems.

The logical positivists were initially tempted to explain the relativized a priori lying at the basis of physical dynamics on the model of Hilbert's axiomatization of geometry. This tendency is especially clear in the work of Schlick and Reichenbach. The idea is that the axioms of geometry *implicitly define* the basic concepts of that science, and this is why geometry is a priori—*analytic,* a priori of course. Points, lines, and planes are any objects that make the axioms of geometry come out true. Thus these axioms do not assert facts, as it were, about some independently given objects. Moreover, if we switch to a different system of geometry, we implicitly define a different system of objects, and so there is no real question of which system of geometry is "true." On the contrary, the choice among such different systems is a mere pragmatic matter of convenience and simplicity. In this sense, it is purely conventional.

Yet this Hilbertian account of the relativized a priori is incompatible with the logicist conception of analyticity due to Frege and Russell. The logicist conception demands *explicit definitions* of basic concepts in purely logical terms, and from this point of view, mere implicit definitions, strictly speaking, cannot count as specifications of meaning at all. Moreover, the method of implicit definition really amounts to nothing more than providing a perspicuous axiomatization of a science. In principle, then, it can be carried out for any science whatsoever, whether the science in question is formal and a priori or empirical and a posteriori. The method of implicit definition is therefore quite incapable of delineating the fundamental distinction between the a priori and the empirical, the analytic and the synthetic, which, I suggest, is of absolutely central concern to the logical positivists.

In any case, however, Carnap, who is clearly the deepest as well as the most rigorous of the positivists, consistently rejects the method of implicit definition. For example, in the *Aufbau* (1928) he attempts to delineate the formal or a priori elements of empirical knowledge by means of a hierarchy of explicit definitions framed within Russellian type theory. The idea is to characterize the content of all empirical concepts by means of what Carnap calls "purely structural definite descriptions." The visual sense modality, for example, is the one and only one sense modality having exactly five dimensions (two of spatial position and three of color quality). Thus each empirical relation is defined or characterized by its purely formal or logical properties, and the empirical or a posteriori content of scientific knowledge is represented by the assertion that there *exist* such relations. In this way Carnap hopes to articulate a distinction between the a priori and the empirical within a conception of logic congenial to Frege's and Russell's logicism. Unfortunately, this attempt ultimately fails for technical reasons—roughly, because even the statement that there *exist* such and such formally characterized relations will turn out to be a truth (or falsehood) of set theory.

Carnap, in his next great work, the *Logical Syntax of Language* of 1934, a work that I take to represent both the high point and the denouement of logical positivism, attempts a radically new approach to the problem. Here Carnap reacts to the important advances in logic and the foundations of mathematics made since the collapse of the original logicist program, specifically, the development of axiomatic set theory, the development of constructivist and intuitionist approaches, and most important, the development of Hilbert's conception of metamathematics. Carnap reponds to these developments by relativizing the logicist conception of logic itself. There is no longer a privileged framework of logic, given by Frege's *Begriffsschrift* or Russellian type theory, for example. Rather, *any* formally or syntactically specified system of formation and transformation rules defines a perfectly legitimate logical system or *linguistic framework* (this, of course, is Carnap's famous principle of tolerance). Yet although there is thus no single privileged logical framework at the level of the *object language*, there is still a privileged framework at the level of the *metalanguage*, namely, the system of logical syntax itself wherein we precisely and purely syntactically describe the infinite multiplicity of possible object-language logical frameworks. This system of logical syntax is, as Gödel's fundamental researches have shown, essentially equivalent to elementary arithmetic, and it is within this relatively neutral and uncontroversial framework that we characterize and specify the logical structure of all richer and more controversial frameworks, for example, the framework of classical mathematical physics.

The ultimate goal of Carnap's program in *Logical Syntax* is then to delineate, for any possible object-language linguistic framework, a purely formal or syntactic distinction between *analytic* and *synthetic* sentences of that framework—a distinction that he hopes will capture the informal and intuitive distinction between a priori and a posteriori or logical and empirical knowledge. In particular, he hopes to show that in a suitable axiomatization of classical mathematical physics, both the purely mathematical part (consisting essentially of real analysis) and the geometric/kinematic part (consisting of the basis of spatiotemporal physical dynamics) will turn out to be analytic, whereas Maxwell's equations, for example, will turn out to be synthetic. In this way Carnap hopes finally to integrate the logicist conception of analyticity due to Frege and Russell (essentially modified under the direct influence of Hilbert) with the relativized a priori lying at the basis of physical science, which, as we have seen derives ultimately from Kantian and neo-Kantian ideas.

Once again, however, Carnap's ingenious construction fails for technical reasons. Roughly, Carnap's conception of analyticity is such that all the truths of classical arithmetic should count as analytic. Yet, as we know, Gödel's incompleteness theorem shows that the notion of arithmetical truth cannot be specified within elementary arithmetic: an essentially richer and more controversial framework is therefore required at the level of the metalanguage. But this means that there is no privileged neutral and uncontroversial meta framework within which the logical syntax of all possible object-language frameworks can be described. Carnap's attempt to relativize the logicist conception of analyticity ultimately collapses in on itself. It is this situation that forms the immediate background of Tarski's articulation of the semantic concept of truth and also, I believe, of Quine's thoroughgoing rejection of any notion of apriority. It is here, I believe, that the ultimate failure of logical positivism must be located.

If this is correct, however, then my attempt briefly to sketch a certain line of historical interaction between philosophy and the sciences has resulted in a rather surprising dialectical twist. For taking the interaction between the history of science and the history of philosophy seriously has led us to a point where it now appears that the currently popular diagnosis of the failure of logical positivism (a diagnosis due largely to the work of Kuhn and his followers) is fundamentally misleading. Indeed, it now appears that the underlying philosophical motivations of the logical positivists cannot happily be described as either naively empiricist, naively formalist, or naively ahistorical. Their empiricism was qualified by, and, I believe, entirely subordinated to, an essentially Kantian preoccupation with the a priori framework within which alone empirical claims have a definite meaning in the first place. Their formalism rested on the idea, which itself evolved naturally from the important developments taking place in the formal sciences themselves, that this a priori framework for empirical knowledge must be specified within the radically new conception of formal logic due to Frege and Russell. Finally, although the logical positivists' preoccupation with the a priori did indeed thereby preclude them from using the history of science as a philosophical tool, this did not prevent them from recognizing the profound philosophical significance of conceptual revolutions in science. On the contrary, their effort to articulate a coherent conception of the *relativized* a priori must, I think, count as the most rigorous attempt we have yet seen philosophically to come to terms with precisely such conceptual revolutions. Of course, as we have also seen, this heroic attempt of the logical positivists was in the end a failure. Yet I do not myself think that we will ever progress beyond this point until we possess a fuller appreciation of the historical evolution of our own philosophical predicament. And this means, as I have tried to emphasize throughout, that we must attend more closely to the history of science, the history of philosophy, *and* to the essential interaction between them.

Rationality and Paradigm Change in Science

Ernan McMullin

As we look back at the first responses of philosophers of science to Thomas Kuhn's classic *The Structure of Scientific Revolutions*, we are struck by their near unanimity toward the challenge that the book posed to the rationality of science. Kuhn's account of the paradigm changes that for him constituted scientific revolutions was taken by many to undermine the rationality of the scientific process itself. The metaphors of conversion and gestalt switch, the insistence that defenders of rival paradigms must inevitably fail to make contact with each other's viewpoints, struck those philosophical readers whose expectations were formed by later logical empiricism as a deliberate rejection of the basic requirements of effective reason giving in the natural sciences.

Kuhn responded to this reading of *SSR* in a lengthy Postscript to the second edition of his book in 1970 and in the reflective essay "Objectivity, Value Judgement, and Theory Choice" in 1977.[1] He labored to show that the implications of his new account of scientific change for the *rationality* of that change were far less radical than his critics were taking them to be. But his disavowals were not, in the main, taken as seriously as he had hoped they would be; the echoes of the rhetoric of *SSR* still lingered in people's minds. It seems worth returning to this ground, familiar though it may seem, in order to assess just what Kuhn *did* have to say about how paradigm change comes about in science. We will see that the radical thrust of his account of science was indeed not directed so much against the rationality of theory choice as against the epistemic, or truthlike, character of the theories so chosen.

1 Good Reasons for Paradigm Change

The theme that recurs in Kuhn's discussions of paradigm change is a two-sided one. On one hand, he wanted to emphasize the fundamental role played by "good reasons" in motivating theory change in science. Notable among these is the perception of anomaly, the growing awareness that something is wrong, which makes it possible for alternatives to be seriously viewed *as* alternatives. On the other hand, these reasons are never coercive in their own right in forcing change; the reasons in favor of a new paradigm cannot *compel* assent. There is no precise point at which resistance to the change of paradigm becomes illogical.[2] Proponents of the new paradigm and defenders of the old one may each be able to lay claim to be acting "rationally"; the fact that neither side can persuade the other does not undermine the claim each can make to have good reasons for what they assert. "The point I have been trying to make," Kuhn says in the Postscript to *SSR*, "is a simple one, long familiar in philosophy of science. Debates over theory-choice cannot be cast in a form that fully resembles logical or mathematical proof. . . . Nothing about that relatively familiar thesis implies either that there are no good reasons for being persuaded or that those reasons are not ultimately decisive for the group. Nor does it even imply that the reasons for choice are different from those usually listed by philosophers of science: accuracy, simplicity, fruitfulness, and the like. What it should suggest, however, is that such reasons function as values and that they can thus be differently applied, individually and collectively, by men who concur in honoring them."[3]

It is with the implications of this thesis that I will be mainly concerned in this essay. The values a good theory is expected to embody enable comparisons to be made, even when the rival theories are incommensurable. Kuhn makes it clear that "incommensurable" for him does not imply "incomparable." *SSR*, he notes, "includes many explicit examples of comparisons between successive theories. I have never doubted either that they were possible or that they were essential at times of theory choice."[4] What he wanted to emphasize, he says, is that "successive theories are incommensurable (which is not the same as incomparable) in the sense that the referents of some of the terms which occur in both are a function of the theory within which those terms appear," and hence that there is no neutral language available for purposes of comparison. Nonetheless, translation is in principle possible.[5] But to translate another's theory is still not to make it one's own. "For that one must go native, discover that one is thinking and working in, not simply translating out of, a language that was previously foreign."[6] And that transition cannot simply be willed, he maintained, however strong the reasons for it may be. This is what enabled him to maintain his most characteristic claim, even after the qualifiers he inserted in the Postscript: "The conversion experience that I have likened to a gestalt switch remains, therefore, at the heart of the revolutionary process. Good reasons for choice provide motives for conversion and a climate in which it is more likely to occur. Translation may, in addition, provide points of entry for the neural reprogramming that, however inscrutable at this time, must underlie conversion. But neither good reasons nor translation constitute conversion, and it is that process we must explicate in order to understand an essential sort of scientific change."[7]

How is the transition to be explicated? Kuhn has only some hints to offer: "With respect to divergences of this sort, no set of choice criteria yet proposed is of any use. One can explain, as the historian characteristically does, why particular men made particular choices at particular times. But for that purpose one must go beyond the list of shared criteria to characteristics of the individuals who make the choice. One must, that is, deal with characteristics which vary from one scientist to another without thereby in the least jeopardizing their adherence to the canons that make science scientific."[8]

And he mentions such characteristics as previous experience as a scientist, philosophical views, personality differences. In the years since *SSR* appeared, sociologists of science have made much of these factors, often in ways that Kuhn himself would disavow. It was his stress on the role of these factors, he later remarked, that led critics to dub his views "subjectivist." They forgot his stress on the "shared criteria" that guide (but do not dictate) theory choice.[9] I will take him at his word here, assuming that the rationality of theory choice in his account rests on the persistence of these criteria that enable theories to be compared and evaluated, relatively to one another, even when they are incommensurable.

2 How Deep Do Revolutions Go?

Here we immediately encounter a difficulty. *Do* these criteria persist? Can they bridge paradigm differences? How deep, in short, do revolutions go? There is an ambiguity in Kuhn's response to this question. In a celebrated paragraph in *SSR,* he describes paradigm change as follows: "Like the choice between competing political institutions, that between competing paradigms proves to be a choice between incompatible modes of community life. Because it has that character, the choice is not and cannot be determined merely by the valuative procedures characteristic of normal science, for these depend in part upon a particular paradigm, and that paradigm is at issue. When paradigms enter, as they must, into a debate about paradigm choice, their role is necessarily circular. Each group uses its own paradigm to argue in that paradigm's defense."[10]

Since the evaluative procedures depend on the paradigm, and the paradigm itself is in question, there can be no agreed-upon way to adjudicate the choice between rival paradigms. Though he goes on to say that the resulting circularity does not *necessarily* undercut the arguments used, he concludes that the status of such arguments can at best be only that of persuasion. They "cannot be made logically or even probabilistically compelling for those who refuse to step into the circle. The premises and values shared by the two parties to a debate over paradigms are not sufficiently extensive for that."[11]

What prevents the rival parties from agreeing as to which paradigm is the better, then, is in part the fact that the norms in terms of which this debate could be carried on are themselves part of the paradigm, so that there is no neutral methodological ground, or at least not enough to enable agreement to be reached. How important is this sort of "circularity" to Kuhn's account of the inability of either side in a paradigm debate to muster an entirely cogent argument in its own behalf? If a circularity in regard to evaluative procedures were to hold in general in such cases, then scientific revolutions *would* indeed seem to be the irrational, or at least minimally rational, affairs that Kuhn's critics take him to be saying they are. One way to find out is to direct attention to the examples he gives of scientific revolutions and ask what paradigm change amounts to in each of these cases.

When the question is put in this way, it is clear that there is a striking difference in the depth of the different changes classified by Kuhn as "revolutions." At one end of the spectrum is the Copernican revolution, the charting of which led him to the writing of *SSR* in the first place. At the other end would be, for example, the discovery of X rays. Somewhere in the middle might come the discovery of the oxygen theory of combustion.[12] We have a choice in some cases, it would seem, between saying that only a small part of the paradigm changed and saying that an entire paradigm changed but that the "paradigm" in this case comprised only a fraction of the beliefs, procedures, and so forth, of the scientists involved.

Take the case of X rays. Kuhn insists that their discovery did accomplish a revolution in his sense. Yet he recognizes that at first sight this episode scarcely seems to qualify. After all, no fundamental change of theory occurred. No troublesome anomalies were noted in advance. There was no prior crisis to signal that a revolution might be at hand. Why then, he asks, can we not regard the discovery of X rays as a simple extension of the range of electromagnetic phenomena? Because, he responds, it "violated deeply entrenched expectations . . . implicit in the design and interpretation of established laboratory procedures."[13] The use of a particular apparatus "carries with it the assumption that only certain sorts of circumstances will arise." Roentgen's discovery "denied previously paradigmatic types of instrumentation their right to that title." That was sufficient, in his view, to constitute it a "revolution" in the sense in which he is proposing to use that term.

I will call this a shallow revolution because so much was left untouched by it. Electromagnetic theory was not replaced or even altered in any significant way. There were no challenges to accepted ways of assessing theory or to what counts as proper explanation. The textbooks, the sets of approved problem solutions, did not change much. What changed were the experimental procedures used in working with cathode-ray equipment and the expected outcomes of such work. And, of course, there were some important long-range implications for theory (as we now know). Such "revolutions" ought, it would seem, to be fairly frequent. Much would depend on how literally one should take the criteria Kuhn specifies as being the symptoms of impending revolution: previous awareness of anomaly and a resistance to a threatened change in procedures or categories.[14]

We are much more likely to think in terms of "revolution" in cases where one large-scale theory replaces another. Kuhn's favorite example is the replacement of phlogiston theory by the oxygen theory of combustion.[15] It meant a reformulation of the entire field of chemistry, a new conceptual framework, a new set of problems. Another example he gives of this sort of intermediate revolution, as we might call it, is the discovery of the Leyden jar and the resulting emergence of "the first full paradigm for electricity."[16] Prior to this discovery, Kuhn remarks, no single paradigm governed electrical research. A number of partial theories were applied, none of them entirely successful. The new conceptual framework enabled normal science to get under way, even though one-fluid and two-fluid theories were still in competition.

These changes involved the formulation of a new and more comprehensive theory. But they left more or less unchanged the epistemic principles governing the paradigm debate itself. Both sides would have agreed as to what counts as evidence, as to how claims should be tested. Or more accurately, to the extent that the scientists involved would have disagreed on these issues, their disagreements would not have been paradigm-dependent to any significant extent. So far as we can tell, Priestley and Lavoisier applied the same sorts of criteria to the assessment of theory, though they might not have attached the same weight to each criterion.

In Kuhn's favorite example of a scientific revolution, the Copernican one, this was, of course, not the case. This was a revolution of a much more fundamental sort because it involved a change in what counted as a good theory, in the procedures of justification themselves. It was not abrupt; indeed, it took a century and a half, from Copernicus's *De revolutionibus* to Newton's *Principia,* to consummate. And what made it revolutionary was not just the separation of Newtonian cosmology or Newtonian mechanics from their Aristotelian counterparts but the gradual transformation in the very idea of what constitutes valid evidence for a claim about the natural world, as well as in people's beliefs about how that world is ordered at the most fundamental level.[17]

It can thus be called a *deep* revolution, by contrast with the others described above. The Aristotelians and the Galileans totally disagreed as to how agreement itself should be brought about. So did the Cartesians and the Newtonians. The Galileans made use of idealization, of measurement, of mathematics, in ways the Aristotelians believed were illegitimate. The Newtonians allowed a form of explanation that the Cartesians were quite sure was improper. The shift in paradigm here meant a radical shift in the methodology of paradigm debate itself. Paradigm replacement means something much more thoroughgoing in such a case.

Have there been other deep revolutions in the more recent history of natural science? Newton's success means the success of a methodology which is still roughly the methodology of natural science today. Perhaps only one deep revolution was needed to get us to what Kuhn calls "mature" science. The two major revolutions in the physics of our own century did not run quite so deep. But they *did* involve principles of natural order, that is, shared assumptions as to what count as acceptable ways of articulating physical process at its most basic level. In the quantum revolution, what separated Bohr and Einstein was not just a difference in theoretical perspective but a disagreement as to what counted as good science and why. Quantum theory, in its Copenhagen interpretation, came much closer to a deep paradigm replacement than it would have done in Einstein's way of taking it.

In the Postscript to *SSR,* Kuhn addressed the ambiguity of the notion of paradigm and proposed a new label. A disciplinary matrix is the answer to the question, "What does [a community of specialists] share that accounts for the relative fullness of their professional communication and the relative unanimity of their professional judgments?"[18] Some of its principal components, he says, are symbolic generalizations, models of the underlying ontology of the field under investigation, concrete problem solutions, and the values governing theory appraisal.

It is clear, then, that for there to be a revolution in Kuhn's sense of the term this last component does not have to be at issue. Only in a deep revolution does one side challenge the other in regard to the appropriate methodology of theory assessment. When X rays were discovered, there was no dispute as to how their reality should be tested. When a Kuhnian revolution takes place, it is evidently not necessary that the entire paradigm should change. Only a part of the disciplinary matrix need be affected for there to be a sufficient change in worldview to qualify as "revolutionary." What 'revolutionary' means in practice is a change that falls outside the normal range of puzzle-solving techniques and whose resolution cannot, therefore, be brought about by the ordinary resources of the paradigm.

The implicit contrast is between puzzle solving, with its definitive ways of deciding whether a puzzle really *is* solved, and paradigm debate, where no such means of ready resolution exists. Whether so sharp a contrast is warranted by the actual practice of science may well be questioned. Decision between rival theories is an everyday affair in any active part of science. There may be an accepted general framework within which problems are formulated, but new data constantly pose challenges to older subtheories within that framework. This was the main issue dividing Kuhn and his Popperian critics in the late 1960s. It is clear in retrospect that there was merit on both sides of that dispute but that each was focusing on a particular aspect of scientific change to the exclusion of others.

The appraisal of rival theories within a paradigm is not a simple matter of puzzle solving. The history of high-energy physics over the past thirty years, for example, has seen one theory dispute after another. The notorious divisions at the moment among paleontologists about the causes of the Cretaceous extinction or between planetary physicists about the origin of the moon are only two of the more obvious reminders of the fact that deep-seated disagreement about the merits of alternative theories is a routine feature of science at its most "normal." As we have seen, Kuhn traced the roots of paradigm disagreement to two different sources: an "incommensurability" of a complex sort between two ways of looking at the world and a set of criteria for theory choice that function as values to be maximized rather than as an effective logic of decision. But this latter source of difference characterizes theory disputes generally and not just the more intractable ones that Kuhn terms paradigm disagreements. What we have here, I suspect, is a spectrum of different levels of intractability, not just a sharp dichotomy between revolutions and puzzle solutions. Nevertheless, Kuhn's dichotomy, though rather idealized, did serve to bring out in a forceful and dramatic way how complex, and how far from a simple matter of demonstration, the choice between theoretical alternatives ordinarily is.

3   The Virtues of a Good Theory

What makes this choice a *rational* one for Kuhn, as we have seen, is the fact that scientists are guided by what they would regard as the virtues of a good theory. And there has been a certain constancy in that regard, according to him, across all but perhaps the deepest of revolutions: "I have implicitly assumed that, whatever their initial source, the criteria or values deployed in theory choice are fixed once and for all, unaffected by their transitions from one theory to another. Roughly speaking, but only roughly speaking, I take that to be the case. If the list of relevant values be kept short (I have mentioned five, not all independent) and if their specification is left vague, then such values as accuracy, scope, and fruitfulness are permanent attributes of science."[19]

This is a strong assertion indeed. Ironically, it is stronger than that now made by some of those who, like Laudan and Shapere, have chided Kuhn in the past for his subjectivism.[20] They argue that the values involved in theory choice are in no sense fixed; Shapere objects to any such claim as an objectionable form of essentialism. According to Laudan and Shapere, these values themselves change gradually as theories change or are replaced. They change for *reasons,* they insist, these reasons functioning as some sort of higher-level arbitration. But there is no limit in principle as to how *much* they might change over time. To put this in a more direct way, there is no constraint on how different the criteria of a good theory might be in the science of the far future from those we rely on today, unlikely though a radical shift might be.[21] In the original text of *SSR,* Kuhn proposed what sounds like a rather different view:

When paradigms change, there are usually significant shifts in the criteria determining the legitimacy both of problems and of proposed solutions. . . . [This is] why the choice between competing paradigms regularly raises questions that cannot be resolved by the criteria of normal science. To the extent, as significant as it is incomplete, that two scientific schools disagree about what is a problem and what a solution, they will inevitably talk through each other when debating the relative merits of their respective paradigms. In the partially circular arguments that regularly result, each paradigm will be shown to satisfy more or less the criteria that it dictates for itself and to fall short of a few of those dictated by its opponent.[22]

The criteria governing theory choice are described here as strongly paradigm-dependent and thus as suffering "significant shifts" from one paradigm to the next. The resulting partial circularity in paradigm assessment leads rival scientists to "talk through each other." This was the theme, of course, that Paul Feyerabend picked up on. One can see how severely it limits the notion that there are "good reasons" for paradigm change. Here, then, is a clear instance of how Kuhn's later construals soften the radical overtones of the earlier work.

Kuhn does not hesitate to speak of the values involved in theory appraisal as "permanent attributes of science." He allows that the manner in which these values are understood and the relative weights attached to them have changed in the past and may change again in the future. But he wants to emphasize that these changes at the metalevel tend to be slower and smaller in scale than the changes that can occur at the level of theory:

If such value changes had occurred as rapidly or been as complete as the theory changes to which they related, then theory choice would be value choice, and neither could provide justification for the other. But, historically, value change is ordinarily a belated and largely unconscious concomitant of theory choice, and the former's magnitude is regularly smaller than the latter's. For the functions I have here ascribed to values, such relative stability provides a sufficient basis. The existence of a feedback loop through which theory change affects the values which led to that change does not make the decision process circular in any damaging sense.[23]

One would need, however, to know just how and why changes in theory bring about changes at the metalevel of theory assessment in order to judge how large these latter changes might become without undermining the claim that a rational choice is being made. Is the "relative stability" of the criteria governing theory choice a contingent historical finding, or is it a necessary feature of any activity claiming the title of science? There are suggestions of both views in the passage I have just quoted. Historically, these values have in fact been stable,

Rationality and Paradigm Change in Science

Kuhn remarks. But he adds that if they were not, if one had to choose the criteria of choice themselves in the act of choosing between theories, there would be no fulcrum. The process would lack justification; it would be circular in a way that would be damaging to its claim to qualify as science.

The presumption appears to be that *really* deep revolutions do not occur, that is, revolutions where there is *no* sharing of epistemic values between one paradigm and the other. Kuhn allows that large-scale theory change may involve smaller-scale changes in the values believed to be appropriate to theory appraisal. In such cases, adoption of the new paradigm carries with it adoption of a somewhat different "rationality" at the metalevel. The advantages of the new theory are so marked, in terms of a minimal level of shared values, that a shift in the values themselves is ultimately taken to be warranted. This, it can be argued, is what happened in the seventeenth century as the balance shifted between Aristotelians and Galileans. Galileo set out to undermine Aristotle's physics in its *own* terms first and then to present an alternative that, in terms of consistency, empirical adequacy, and future potential, could claim a definite advantage, even in terms of criteria the Aristotelian might be brought to admit. That, at any rate, would be the grounds, in Kuhn's perspective, for regarding the Scientific Revolution as a "rational" shift in the way in which natural science was carried on.

In a recent essay Kuhn argues that we learn to use the term 'science' in conjunction with a cluster of other terms like 'art', 'medicine', 'philosophy'. To know what science is, is to know how it relates to these other activities.[24] Identifying an activity as scientific is to single out "such dimensions as accuracy, beauty, predictive power, normativeness, generality, and so on. Though a given sample of activity can be referred to under many descriptions, only those cast in this vocabulary of disciplinary characteristics permit its identification as, say, science; for that vocabulary alone can locate the activity close to other scientific disciplines and at a distance from disciplines other than science. That position, in turn, is a necessary property of all referents of the modern term, 'science.'"[25]

He immediately qualifies this last very strong claim by noting that not every activity that qualifies as "scientific" need be predictive, not all need be experimental, and so forth. And there is no sharp line of demarcation between science and nonscience. Nonetheless, there is a well-defined cluster of values whose pursuit marks off scientific from other activities in a relatively unambiguous way and that gives the term 'science' the position it occupies in the "semantic field." This marking off is not a mere matter of convention. The taxonomy of disciplines has developed in an empirical way; a real learning has taken place. If someone were to deny the rationality of learning from experience, we would not know what he or she is trying to say. One cannot, he maintains, further *justify* the norms for rational theory choice. He cites C. G. Hempel to the effect that this inability is a testimony to our continuing failure to solve the classical problem of induction.[26]

Kuhn rests his case, then, both for the rationality of science and for its distinctiveness as a human activity mainly on the values governing theory choice in science. But he does not chronicle their history, disentangle them from one another except in a cursory way, or inquire in any detail into how and why they have changed in the ways they have. Many of these variations, he remarks, "have been associated with particular changes in scientific theory. Though the experience of scientists provides no philosophical justification for the values they deploy (such justification would solve the problem of induction), those values are in part learned from that experience, and they evolve with it."[27]

But what justification other than the experience of scientists is *needed* to justify the values they deploy? Kuhn has, I suspect, altogether too lofty a view of what "philosophical" justification might amount to. And he has too readily allowed himself to be intimidated by that most dire of philosophers' threats: "That *can't* be right: if it were, it would solve the problem of induction." My own guess is that attention to the role of values in theory appraisal might well dissolve the problem Hume bequeathed us about the grounds for inductive inference. But whether that be true or not, the criteria employed by scientists in theory evaluation enjoy whatever sanction is appropriate to something learned in, and tested by, experience.

4 How Might Epistemic Values Be Validated?

Suppose a scientist were to doubt whether a particular value, say simplicity, is really a desideratum in a practical situation of theory choice facing him or her. The rationality of the choice depends, presumably, on what sort of answer can be given to this kind of question. Two different sorts of answers suggest themselves. One is to look at the track record and decide how good a guide simplicity has proved to be in the past. (There are obvious problems about how the criterion itself is to be understood, but I will bracket these for the moment.) A quite different sort of response would be that simplicity is clearly a desideratum of theory because ____, where we fill the blank with a reason why on the face of it, a simple theory is more likely to be a good theory (if indeed one *can* find a convincing reason). Both of these responses would, of course, need further clarification before they could begin to carry any conviction.

First, what does it mean to ask how good a guide simplicity has been in the past? Guide to what? Some kind of ordering of means and ends is clearly needed here. Some of the values we have been talking about seem to function as goals of the scientific enterprise itself: predictive accuracy (empirical adequacy) and explanatory power are the most obvious candidates. One can trace each of these goals back a very long way in human history. In some sense, they may be as old as humanity itself. The story of how they developed in the ancient world, how the skills of prediction came to be prized in many domains, how explanatory accounts of natural process came to be constructed, is a familiar one. Less familiar is the realization that these goals were not linked together in any organic way at the beginning. Indeed, they were long considered antithetical in the domain of astronomy, the most highly developed part of the knowledge of nature in early times. One of the consequences, perhaps the most important consequence, of the Copernican revolution was to show that they *are* compatible, that they can be successfully blended. This was an empirical discovery about the sort of universe we live in. It was something we *learned* and that now we *know*.

Each of these goals has come to be considered valuable in its own right, an end in itself.[28] An activity that gives us accurate knowledge of the world we live in and consequently power over its processes can come to seem worthwhile for all sorts of reasons. An activity that allows us to understand natural process, that allows our imaginations to reach out to realms inaccessible to our senses, holds immediate attraction. What it is to understand will, of course, shift as the principles of natural order themselves shift. So this goal of explaining lacks the definiteness of the goal of predicting; as theory changes, so will the contours of what counts as explaining.

Much more would have to be said about all this, but I am going to press on to make my main point.[29] Other epistemic values serve as *means* to these ends; they help to identify theories more likely to predict well or to explain. Some of these are quite general and would apply to any epistemic activity. Logical consistency (absence of contradiction) and compatibility with other accepted knowledge claims would be among these. They are obviously not goals in themselves; they would not motivate us to carry on an activity in the first place. But we have found that these values are worth taking seriously as *means*. Or should I say, it has always been obvious that we must not neglect them, if it is knowledge we are seeking?

Other values are more specific to science, for example, fertility, unifying power, and coherence (i.e., absence of ad hoc features). Once again, these are clearly not primary goals. They are not so much deliberately aimed at as esteemed when present. And they are esteemed not in themselves but because they have proved to be the marks of a "good" theory, a theory that will serve well in prediction and explanation. A long story could be told about this, beginning with Kepler, Boyle, and Huygens and working through Herschel, Whewell, and a legion of others who have drawn attention to the significance of these three virtues.

Once again, the story is an ambiguous one: it can be told in two quite different ways. According to one way of telling it, these values can be shown to have played a positive historical role in theory choice; we have gradually learned to trust them as clues. According to the other, a series of acute thinkers (some of the most prominent of them listed above) have realized that these values *ought* to serve as indicators of a good theory. These are what one would *expect* a priori from a theory that purported to predict accurately and explain correctly. When Kepler and Boyle drew attention to the importance of such criteria, it was not to point to their efficacy in the earlier history of natural philosophy but to recommend them on general epistemic grounds.[30]

The question of how to validate the values that customarily guide scientific theory choice can now be addressed more directly. The goals of predictive accuracy (empirical adequacy) and explanatory power serve to define the activity of science itself, in part at least. If, as Kuhn notes, one relinquishes the goal of producing an accurate account of natural regularity, the activity one is engaged in may be worthwhile, but it is not science.[31] The notion of epistemic justification does not directly apply to the goals themselves. One might ask, of course, whether the pursuit of these goals is justifiable on *moral* grounds. Or one might ask, as a means of determining whether effort expended on them is worthwhile, whether the goals are in fact attainable. We have learned that in general they *are* attainable. This is something one could not have known a priori. And we have learned much about the *methods* that have to be followed for theory construction to get under way, methods of experiment, of conceptual idealization, of mathematical formulation, and the rest. All of this had to be *learned,* and no doubt there is still much to discover in this regard.

The other values, being instrumental, are justified when it is shown that they serve as means to the ends defined by the primary goals. And this, as we have seen, can be done in two ways: by an appeal to what we have learned from the actual practice of science or by an analysis in epistemological terms of the aims of theory and what, in consequence, the marks of a good theory should be. Ideally, both ways need to be followed, each serving as check for the other. The appeal to historical practice works not so much as a testimony to what values have actually guided scientists in their theory choices but as a finding that reliance on certain values has *in fact* served the primary goals of science. Might it cease to?

This is the Humean echo that seems to worry Kuhn so much. One might respond, as he does, that learning from experience is part of what it is to be rational. We cannot demonstrate that experience will continue to serve as a reliable guide. But demonstration is not what is called for. Kuhn has done more than anyone else, perhaps, to show that rational theory choice does not require the cogency of demonstration. We know that the predictive powers of natural science have enormously increased, and we know something of the theory characteristics that have served to promote this expansion. No future development could, so far as I can see, lead us to deny these knowledge claims, which rest not just on a perception of past regularities but on an understanding, partial at least, of why these regularities took the course they did. We can, and almost surely will, learn more about what to look for in a good theory. But no further evidence seems to be needed to show that coherence in a theory is a value to be sought, so that, other things being equal, a more coherent theory is to be preferred to a less coherent one.

5 Rationality without Realism?

Over the years since *SSR* appeared, Kuhn has, as we have seen, become more and more explicit about the basic rationality that underlies theory choice in science. It is a complex rationality with many components, allowing much latitude for difference among the defenders of different theories. But it has remained relatively invariant since the deep revolution that brought it into clear focus in the seventeenth century. One might almost speak of a *convergence* here. Kuhn clearly believes that scientists have a pretty good grip on the values that *ought* to guide the appraisal of rival theories, and that this grip has improved as it has been tested against a wider and wider variety of circumstances.

But he has not softened his stance in regard to the truth character of theories in the least. In a well-known passage in the Postscript, he insists that the only sort of progress that science exhibits is in puzzle solving: later theories solve more puzzles than earlier ones, or (to put this in a different idiom) they predict better. But there is, he insists, "no coherent direction of ontological development"; there is no reason to think that successive theories approximate more and more closely to the truth.[32] "The notion of a match between the ontology of a theory and its 'real' counterpart in nature now seems to me illusive in principle."[33] Kuhn thus rejects in a most emphatic way the traditional realist view that the explanatory success of a theory gives reason to believe that entities like those postulated by the theory exist, i.e., that the theory is at least approximately true.

He does not argue for this position in SSR, aside from a remark about Einstein's physics being closer in some respects to Aristotle's than to Newton's. But it is clear what the grounds for it are in his mind: the incommensurability of successive paradigms implies a discontinuity between their ontologies. By separating the issues of comparability and commensurability, he believes he can retain a more or less traditional view in regard to the former while adopting an instrumentalist one in regard to the latter. The radical challenge of *SSR* is directed not at rationality but at realism. The implications of the familiar Kuhnian themes of holism and paradigm replacement are now seen to be more significant for the debate about realism than for the issue of scientific rationality, on which they had so great an initial impact.

Kuhn's influence on the burgeoning antirealism of the last two decades can scarcely be overestimated. His views on theory change, on problems about the continuity of reference, are reflected in the work of such notable critics of realism as Arthur Fine, Bas van Fraassen, and especially Larry Laudan.[34] Kuhn's own emphasis on science as a puzzle-solving enterprise would lead one to interpret him in an instrumentalist manner. At this point I am obviously not going to open a full-scale debate on realism versus instrumentalism.[35] But I would like to pull out one thread from that notorious tangle. Kuhn's way of securing scientific rationality by focusing on the values proper to theory choice might well have led him (I argue) to a more sympathetic appreciation of realism. I am not saying that rationality and realism are all of a piece, that to defend one is to commit oneself to the other. Most of the current critics of realism would be emphatic in their defense of the overall rationality of scientific change. But a closer study of the values to which Kuhn so effectively drew attention should, to my mind, raise a serious question about the adequacy of an instrumentalist construal of the puzzle-solving metaphor. If such a construal is adopted, it is hard to make sense of those many episodes in the history of science where values other than mere predictive accuracy played a decisive role in the choice between theories.

To show this, I will focus on a case history from Kuhn's own earlier work, *The Copernican Revolution*. At issue are the relative merits of the Ptolemaic and the Copernican systems prior to Galileo's work. Kuhn points out that there was little to choose between the two on the score of predictive accuracy. "Judged on purely practical grounds," he concludes, "the Copernican system was a failure; it was neither more accurate nor significantly simpler than its Ptolemaic predecessors."[36] Yet it persuaded some of the best astronomers of the time. And it was they who ultimately produced the "simple and accurate" account that carried the day. How *did* it persuade them?

In Kuhn's view, "The real appeal of sun-centered astronomy was aesthetic rather than pragmatic. To astronomers the initial choice between Copernicus' system and Ptolemy's could only be a matter of taste, and matters of taste are the most difficult of all to define or debate."[37]

But such matters cannot be regarded as unimportant, he goes on, as the success of the Copernican Revolution itself testifies. Whatever it was that persuaded so many of those most skilled in astronomy to make what we would now regard as the right step obviously must be looked at with care. Those who were equipped "to discern geometric harmonies" obviously found "a new neatness and harmony" in the heliocentric system. What Copernicus offered was "a new and aesthetic harmony" that somehow carried conviction in the right quarters.

But now let us see how Copernicus's own argument went, in the crucial chapter 10 of book 1 of *De revolutionibus.* He points to two different sorts of clues. First, the heliocentric model allows one to specify the order of the planets outward from the central body in an unequivocal way, which Ptolemy's model could not do. Furthermore, the Copernican model has the planetary periods increase as one moves outward from the sun, just as one would expect. What Copernicus claims to discover in the new way of ordering the planets is a "clear bond of harmony," "an admirable symmetry." But why should this carry conviction, especially since (as Kuhn emphasizes) Copernicus in the end had to retain an inelegant and far from harmonious-seeming tangle of epicycles?

He had stronger arguments. The heliocentric model could *explain,* that is, provide the *cause* of, a whole series of features of the planetary motions that Ptolemy simply had to postulate as given, as inexplicable in their own right. For example, even in ancient times it had been suggested that Venus and Mercury appear to have the sun as their center of rotation, since, unlike the other planets, they accompany the sun in its motion across our sky. Or again, it had long been noted that the superior planets (Mars, Jupiter, Saturn) are at their brightest when in opposition (rising together in the evening or setting together in the morning). Assuming that brightness is a measure of relative distance, this is explained if we are viewing the planetary motions from a body that itself is orbiting the sun as center. This "proves,"

Copernicus somewhat optimistically concludes, that the center of motion of the superior planets is the same as that of the inferior planets, namely the sun.

Kuhn comments that it does "not actually prove a thing. The Ptolemaic system explains these phenomena as completely as the Copernican," although the latter can be said to be "more natural."[38] Here I must disagree. The Ptolemaic system does not *explain* the phenomena mentioned above at all. Ptolemy is forced to postulate that the center of the epicycle for both Venus and Mercury always lies on the line joining the earth and sun. Kuhn says that in this way Ptolemy "accounts for" this feature of their motions. But this is surely not *accounting for* in the sense of explaining. Kuhn evidently equates prediction and explanation in these passages, not an unusual assumption at the time his book was written.

But he allows that Copernicus gives a "far more natural" account than does Ptolemy. Why? And what does 'natural' mean in the lexicon of an instrumentalist? Ptolemy's restriction on the deferent radii swept out by Venus and Mercury "is an 'extra' device, an *ad hoc* addition,"[39] one that Copernicus can discard. Kuhn is surely on the right track here. But this is *not* an aesthetic argument, an appeal to taste. Copernicus himself makes the genre to which it belongs quite clear. He says that Copernicus is able to assign the *cause* of these features of the planetary motions, whereas Ptolemy is not. There is no reason in Ptolemy's system for them, other than the mere need to get the predictions right. They are, as Kuhn himself says, ad hoc.

Copernicus gives another set of arguments based on the retrograde motions. Their relative size and frequency from one planet to another and the lack of any such motions on the part of the sun and moon are exactly what one would be led to expect in a system where we are observing the motions from the third planet and the moon is not a true planet but a satellite of earth. Later, in the *Mysterium cosmographicum*, Kepler developed these arguments more fully and added some of his own, for example, the striking fact that in the Ptolemaic model, the period of rotation for each planet on either the deferent or the epicycle circle is *exactly* one year, something which seemed like an extraordinary piece of adjustment, especially since Ptolemy took the planets to be dynamically independent of one another. Kepler is clear that the issue here is one of causal explanation; one of the systems can provide such an explanation, the other cannot. He is also clear that the criterion of prediction alone will not be enough to decide in all cases between two rival accounts of the planetary motions and thus that a different genre of argument (he calls it "physical") is needed.[40] This he urged as a refutation of the instrumentalism of his opponent, Ursus.

The competition may have been neutral between Ptolemy and Copernicus where *prediction* of planetary motions was concerned, but the two systems were quite unequal as *explanation*. No better illustration could be found of the distinction between these two concepts, and of the consequent importance of criteria of theory appraisal other than that of predictive or descriptive accuracy. Copernicus's criterion of "naturalness," the elimination of ad hoc features, the virtue that might today be called coherence, is not aesthetic; it is epistemic. He is not just appealing to his reader's taste, or sense of elegance. He is not assuming that the simpler, the more beautiful, models are more likely to be true. He is saying that a theory that makes causal sense of a whole series of features of the planetary motions is more likely to be true than one that leaves these features unexplained.

Copernicus and those who followed him believed that they had good arguments for the reality of the earth's motion around the sun. They sometimes overstated the force of those arguments, to be sure, using terms like 'proof' and 'demonstration'. The natural philosophers of the day were not yet accustomed to the weaker notions of likelihood and probability. Galileo found, to his cost, that he had to speak in terms of demonstration if his claims for the Copernican system were to be taken seriously. He did not have a demonstration, but from our perspective, he called effectively on the criterion of coherence in his critique of the geostatic alternative, just as Copernicus had earlier done.

As we look back on those debates, we are ready to allow that the coherence arguments of Copernicus and Galileo *did* carry force, that they *did* give a motive for accepting the new heliocentric model as true. And their force came from something other than predictive advantage. Kuhn's point in regard to theory assessment, one that became clearer in his successive formulations of it, was that the different theory values were not reducible to one another, and hence that no simple algorithm, no logic of confirmation such as the logical positivists had sought, underlay real-life theory decision. What I have tried to do here is to carry this insight further and to note the special epistemic weight carried by certain of these values. Besides coherence, one could make similar cases for fertility and unifying power. It is hard to make sense of the role played by these values if one adopts the instrumentalist standpoint that Kuhn feels compelled to advocate.

The case for scientific realism rests in large part on these "super-empirical" values. That is, when we ask about a particular theory, how likely is it that it is true (correlatively, how likely is it that something like the explanatory entities it postulates actually exist), it is to these virtues that we are inclined to turn. To say that a theory simply "saves the phenomena," though this carries *some* epistemic weight, leaves open the suspicion of its being ad hoc. If a theory be thought of simply as an hypothetico-deductive device, it would seem plausible to suppose that other devices might account as well or better for the phenomena to be explained. It is only when the *temporal* dimension is added, when a theory is evaluated in a historical context, when its success in unifying domains over time or in predicting new sorts of phenomena are taken into account, that conviction begins to emerge. Theories are not assessed simply as predictors; they are not confirmed purely by the enumeration of consequences.

My conclusion is that the diversity of the expectations scientists hold up for their theories argues not only for the tentative character of theory choice, Kuhn's original point, but also for its properly epistemic character. This leaves us, of course, with a problem: how can the difficulties in regard to incommensurability be reconciled with the epistemic force of such arguments as that of Copernicus? Kuhn emphasized the discontinuities of language across theory change so strongly that he left no room for the possibility of convergence, for the possibility that the theories of the paleontologists of today, for example, not only solve more puzzles than those of yesteryear but also tell us, with high degree of likelihood, what actually happened at distant epochs in the earth's past.

The Kuhnian heritage is thus a curiously divided one. Kuhn wanted to maintain the rational character of theory choice in science while denying the epistemic character of the theory chosen. The consequent tensions are, of course, familiar to every reader of current philosophy of science. Thirty years later, *The Structure of Scientific Revolutions* still leaves us with an agenda.

Notes

1. See *The Essential Tension (ET)*, pp. 320–339. In his effort to ward off the charge of subjectivism, Kuhn might also have pointed to "The Function of Measurement in Modern Physical Science" (*Isis* 52 [1961]: 161–190; reprinted in *ET*, pp. 178–224), which appeared before *SSR* and whose theme was that "measurement can be an immensely powerful weapon in the battle between two theories" (*ET*, p. 211), that "the comparison of numerical predictions . . . has proved particularly successful in bringing scientific controversies to a close" (*ET*, p. 213). Or he could have recalled an even earlier paper, "The Essential Tension" (*The Third University of Utah Research Conference on the Identification of Scientific Talent*, ed. C. W. Taylor [Salt Lake City: University of Utah Press, 1959], 162–174; reprinted in *ET*, pp. 225–239), whose title referred to the opposition between the themes of tradition and innovation in science and which argued that it is the very effort to work within a tightly construed tradition that leads eventually to the recognition of anomalies that in turn prepares the way for revolution (*ET*, p. 234). One further paper that Kuhn might have called on was "A Function for Thought Experiments" (*L'Aventure de la science*, ed. Mélanges Alexandre Koyré [Paris: Hermann, 1964], vol. 2, pp. 307–334; reprinted in *ET*, pp. 240–265), which describes how failures of expectation induce the crisis that is the usual prelude to paradigm change (*ET*, p. 263).

2. "Objectivity," p. 30.

3. *The Structure of Scientific Revolutions*, 2nd ed. (Chicago: University of Chicago Press, 1970), p. 199.

4. "Metaphor in Science," in *Metaphor and Thought*, ed. Andrew Ortony (Cambridge: Cambridge University Press, 1979), 409–419; see p. 416.

5. In a recent essay Kuhn distinguishes between translation and interpretation and shows how communication can occur even where languages are incommensurable ("Commensurability, Comparability, Communicability," *PSA 1982* [Philosophy of Science Association], 1983: 669–688). In a comment Philip Kitcher remarks that Kuhn, in his later readings of *SSR*, has progressively weakened the dramatic doctrine of the original work in ways, be it said, of which Kitcher approves ("Implications of Incommensurability," *PSA 1982*, 1983: 689–703).

6. *SSR*, p. 204.

7. *SSR*, p. 204.

8. "Objectivity," p. 324.

9. "Objectivity," p. 325.

10. *SSR*, p. 94.

11. *SSR*, p. 94.

12. In *SSR* Kuhn himself distinguishes between "major paradigm changes, like those attributable to Copernicus and Lavoisier," and "the far smaller ones associated with the assimilation of a new sort of phenomenon, like oxygen or X-rays" (p. 92).

13. *SSR*, p. 58.

14. *SSR*, p. 62. It is not clear to me that the discovery of X rays satisfies either of these criteria in any other than a minimal way.

15. *SSR*, p. 199.

16. *SSR*, p. 62. There might be some question as to whether, in fact, a single theory of electricity did emerge at this time. But that is not to the point of my inquiry.

17. I have worked out this theme in some detail in my "Conceptions of Science in the Scientific Revolution," in *Reappraisals of the Scientific Revolution*, ed. David Lindberg and Robert Westman (Cambridge: Cambridge University Press, 1990).

18. *SSR*, p. 182.

19. "Objectivity," p. 335.

20. See Larry Laudan, *Science and Values* (Berkeley: University of California Press, 1984); Dudley Shapere, *Reason and the Search for Knowledge* (Dordrecht: Reidel, 1984). I have discussed the ironies of this particular divergence more fully in "The Shaping of Scientific Rationality," in *Construction and Constraint*, ed. E. McMullin (Notre Dame: University of Notre Dame Press, 1988), pp. 1–47.

21. Nicholas Rescher defends a somewhat similar position in regard to how different from ours the "science" carried on by the inhabitants of a distant planet might be: "Science *as we have it*—the only 'science' that we ourselves know—is a specifically human artifact that must be expected to reflect in significant degree the particular characteristics of its makers. Consequently, the prospect that an alien 'science'-possessing civilizaiton has a *science* that we could acknowledge (if sufficiently informed) as representing the same general line of inquiry as that in which we ourselves are engaged seems extremely implausible" ("Extraterrestrial Science," *Philosophia Naturalis* 21 [1984]: 400–424; see p. 413.

22. *SSR*, pp. 109–110.

23. "Objectivity," p. 336.

24. "Rationality and Theory Choice," *Journal of Philosophy* 80 (1983): 563–570; see p. 567.

25. Ibid., p. 568.

26. C. G. Hempel, "Valuation and Objectivity in Science," in *Physics, Philosophy, and Psychoanalysis*, ed. R. S. Cohen and L. Laudan (Dordrecht: Reidel, 1983), 73–100.

27. "Objectivity," p. 335.

28. See my "Values in Science," *PSA 1982*, 1983: 3–25.

Ernan McMullin

29. The story sketched so lightly here is told in much more detail in my "Goals of Natural Science," *Proc. American Philosophical Association* 58 (1984): 37–64.

30. For a fuller historical treatment, see my "Conceptions of Science in the Scientific Revolution."

31. "Rationality and Theory Choice," p. 569.

32. *SSR,* p. 206.

33. *SSR,* p. 206.

34. Laudan's much-quoted essay, "A Confutation of Convergent Realism," in *Scientific Realism,* ed. J. Leplin (Berkeley: University of California Press, 1984), 218–249, presents in detail the sort of arguments that Kuhn would need to support his own rejection of convergence.

35. See my "Case for Scientific Realism," in *Scientific Realism,* ed. J. Leplin, 8–40, and "Selective Anti-realism," *Philosophical Studies* 61 (1991): 97–108.

36. *The Copernican Revolution* (New York: Random House, 1957), p. 171.

37. Ibid., p. 172.

38. Ibid., p. 178.

39. Ibid., p. 172.

40. Kepler's clearest treatment of this issue will be found in the *Apologia Tychonis contra Ursum* (1600). See Nicholas Jardine's translation of this work in *The Birth of History and Philosophy of Science* (Cambridge: Cambridge University Press, 1984). Michael Gardner extracts a "Kepler principle" to the effect that it counts in favor of the realistic acceptance of a theory if it explains facts that competing theories merely postulate. See "Realism and Instrumentalism in Pre-Newtonian Astronomy," in *Testing Scientific Theories,* ed. John Earman (Minneapolis: University of Minnesota Press, 1983), 201–265; p. 256.

The Historians Look

A Mathematicians' Mutiny, with Morals

J. L. Heilbron

In 1784 a small party of mathematicians threatened to secede from the Royal Society of London. Their mutiny, soon put down, makes an instructive story for both the historian and the philosopher of science. The episode took place at the beginning of the rapid quantification of the physical sciences, which is supposed to have brought physicists (or natural philosophers) and mathematicians closer together. The further reason may be advanced that T. S. Kuhn published an important paper about this rapprochement and its antecedents that has received wide attention and acceptance.[1] The morals to be drawn from the mutiny touch the propriety of Kuhn's approach and, more generally, the equivocal use of words like "physics," "chemistry," and "mathematics" and their derivatives in the historiography of science.

By 1780 several subjects now belonging to physics but then classed as "mixed mathematics" had attained a high level of abstraction and quantification: mechanics, hydrostatics including pneumatics, and optics. Other subjects of mixed mathematics not now belonging to physics, positional astronomy and geodesy, in particular, had reached a similar or higher level. Still other subjects, which became central parts of physics during the eighteenth century, like electricity, magnetism, and thermal phenomena, were quantified during the period from 1780 to 1820. "Quantification" here signifies exact and reproduceable measurement successfully linked mathematically to theoretical entities applicable to a wide range of experimental arrangements. Coulomb's demonstration of the "laws" of interaction of particles of the hypothetical electrical fluids by means of a torsion balance is a familiar example of quantification in physics in the 1780s, although his experimental results fell well below the accuracy then expected in other rapidly quantifying subjects, like thermal effects in gases.

How did it come about that previously qualitative branches of natural philosophy became quantified around 1800 and at the same time joined with parts of mixed mathematics and chemistry to constitute classical physics? One answer credits the school of Laplace with the first successful mathematizing of the previously qualitative branches: "The first mathematizers were uniformly French."[2] (In Kuhn's parlance, "mathematizers" impose mathematical theory and "quantifiers" make exact measurements; in this chapter, "quantification" refers to the joint result of both activities.) The French mathematizers, proficient in continental calculus and celestial mechanics, tried to model accounts of electricity, magnetism, and heat on the gravitational theory of planetary motions. As Kuhn observed, they did not create the theoretical entities that made quantification possible or, it must be added, perfect the necessary instruments of measure. Those who did typically were not French: Alessandro Volta, J. C. Wilcke, H. B. de Saussure, and J. A. Deluc, for example, and, in England, Henry Cavendish, G. A. Shuckburgh, Thomas Young, and, to reach back a little further, Joseph Black and Benjamin Franklin.

The Royal Society's mutineers of 1784 identified themselves as mathematicians; the main membership of the Society conceded them the title. They devoted themselves assiduously to mixed mathematics, but none of them, with the partial though instructive exception of Shuckburgh, worked at the quantification of physics. To be sure, the case is partial and parochial. But it suggests that we should not look upon the momentous turn in the history of physics that brought about its quantification as the result of the accretion of subjects mathematized seriatim by mathematicians. Rather, we should consider it the invention of people who did not regard themselves as mathematicians but who embodied that esprit geometrique, that quantitative spirit, that can be followed through many aspects of the life and thought of the late Enlightenment.[3]

This chapter has three parts: an account of the mutiny, an indication of the range of mathematics in England during the late eighteenth century as implied by the work of the mutineers, and historiographical observations suggested by the story.

1   A Mutiny of Mathematicians

The mutiny

In 1778 Sir John Pringle, a distinguished physician and a scientific and political crony of Benjamin Franklin's, indicated his intention not to seek reelection as president of the Royal Society of London. Pringle was close to some of the mutineers; his successor, Joseph Banks, whose election he did not favor, was close to the King.[4] Banks strove to preserve order and honor in the Society and interested himself particularly in the choice of his officers and the election of fellows. He preferred the gentleman to the journeyman of science and the rarities of natural history to the generalities of mathematics. A lampoon published during the mutiny had Banks, "the man who travell'd far and near / to show the world what fools breed here," demonstrate his learning by exhibiting "a Toe-nail of Robin Hood, preserved in a Tobacco paper of Little John's."[5]

The new president's predilections, both as administrator and collector, showed themselves to disadvantage early in 1781, when he solicited votes in opposition to two candidates recommended for their mathematics. The stronger of the two, Henry Clerke, a teacher in Manchester, had written three books that demonstrated his command of geometry, algebra, and the fluxional calculus; he had the support of the Astronomer Royal, Nevil Maskelyne, and of the professor of mathematics at the Royal Military Academy, Charles Hutton, to whom he had dedicated one of his books. This dedication tells much about the frustrations of an ambitious provincial mathematician: Hutton's candor, impartiality, and encouragement of others, says Clerke, "will raise him a far more noble Monument to Posterity, than that conspicuous Station in the Mathematic World he now so deservedly possesses; and which even those pitiful, groveling Sons of Envy who have so long been carping must allow, to be only a just Reward for Superior Merit."[6] The other disappointed candidate, Major Joseph

Frederick Wallet Desbarres, who had studied under the Bernoullis, taught Captain Cook, and "surveyed the coasts of North America," had the support of respectable fellows, but nonetheless could not pass through the eye of Banks' needle.[7]

To strengthen his hand against those dissatisfied with his performance in office, Banks moved toward replacing officers he inherited with men more congenial to him. In the fall of 1783, rumors circulated that Paul Matthew Maty, one of the Society's secretaries and an under librarian at the British Museum who marched to his own capricious drummer, would be replaced; that came to pass as a result of the mutiny. Maskelyne, who had opposed Banks over the acceptance of papers for the *Philosophical Transactions,* was dropped from the Council at the annual elections held in December 1783.[8] And the previous November, Banks had contrived to discharge Hutton from his post under Maty as an assistant secretary for foreign correspondence.

The complaint against Hutton came down to negligence in thanking foreigners for gifts and letters to the society.[9] The charge did not seem colorable to his friends. Hutton had risen from the Newcastle collieries, which employed his father and where he labored briefly in the pits, by sheer hard work, which brought him, via the trade of school master, to the professorship in mathematics at the Woolwich military academy in 1773. Those who chose him for this position knew he was not lazy or neglectful. Two of the electors—Astronomer Royal Maskelyne and Samuel Horsley, editor of Newton's mathematical papers—became the leaders of the mutineers.[10]

The year that Banks won office, Hutton published two valuable papers in the Society's *Transactions.* One, on the ballistics of cannon balls, earned him the Society's most prestigious award for contributions to knowledge, the Copley medal; the other, a calculation that was "more laborious, and at the same time called for more ingenuity, than has probably been brought into action in any computation undertaken by a single person since the preparation of logarithmic tables," deduced a value for the density of the earth from measurements made by Maskelyne on the behavior of a pendulum near a mountain in Perthshire. As a mathematician, Hutton was expert but not creative; as a correspondent, overworked but not dilatory; as a person, mild and equable, with what his colleague Olinthus Gregory described as a "constitutional and . . . unconscious aversion to the pedantry and parade of science."[11]

Since Hutton lived in Woolwich, he could not collect foreign correspondence daily in London. Nor had he time to dance attendance on Banks, who expected obeisance as a baronet (he became Sir Joseph in 1781) as well as as a president.[12] In November 1783 the Council, except for Maskelyne and Maty, decided that the person responsible for foreign correspondence should reside in London. Hutton resigned. His friends objected to the proceeding and, at a stormy meeting on 11 December 1783, called for a vote of thanks to Hutton for his services. Banks opposed the motion, which nonetheless carried, 30 to 25. Thus emboldened, Hutton delivered a defense of his conduct at the next weekly meeting, which 45 of the 60 fellows present thought persuasive.[13]

Maskelyne and others then began canvassing against Banks. "They give out that it is a struggle of the men of science against the Maccaronis of the Society," Banks's agent, Charles Blagden, reported. "Your last council is held as a matter of derision." The renegades were succeeding "chiefly among the inferior members of the Society, where the language that all the members are alike, and that they ought no longer to be rode by your Maccaroni gentlemen, would certainly be acceptable. One of the greatest articles of complaint against you is keeping out deserving men [like Henry Clerke], because they are not of a certain rank." Sober and senior fellows, like Henry Cavendish and the physician William Heberden, when prodded by Blagden, agreed to support a motion of confidence in Banks but nothing that would censure the opposition, for, as Cavendish observed, Banks had put nonentities on his council. This strategy, characteristic, according to a recent analysis, of the mode of action of great whig magnates like Cavendish, prevailed at the meeting of the full society on 8 January 1784.[14]

Banks's party succeeded by bringing in fainéant fellows favorable to him, some 70 or even 100 more people than ordinarily attended meetings. The victory did not end the dissension, however. Banks reckoned the residual hard-core opposition to him at 42; that agrees with a vote taken on 12 February on a motion to reinstate Hutton, which lost 47 to 85. According to Horsley, the usual attendance at the meetings, some 60 fellows, consituted the "true effective members"; if so, his claim that five sixths of the effective membership opposed Banks early in 1784 was not farther out than the rhetorical requirements of the moment demanded.[15]

On 26 February 1784 Horsley, Francis Masères (a wealthy lawyer and tedious mathematician), and other mutineers spelled out the charges for which Hutton had become the stalking horse. The rejoinder by Banks's ally, Thomas Anguish, the Accountant-General of Britain, exposed a fundamental division within the Republic of Letters. "It was not sufficient to be a Mathematician, to be a Fellow of [the] Society; there were other qualities requisite, social qualities, the lack of which might make a man unfit, however competent he might be in learning."[16] In Anguish's opinion, no one who deserved admission had been denied. On the contrary, "if he had one fault to find with the administration it was that improper people had got among us who disgraced the Society by involving them in frivolous dissensions."[17]

The charge that Banks had packed the council with men ignorant of mathematics opened extensive rhetorical opportunities to Horsley's group. The council had to approve papers for publication in the Society's *Philosophical Transactions*. For this task, a well-rounded intelligence alone did not suffice. "What is this to professional skill? to that accuracy of science which arises from having been employed only about one object? to that acquaintance, in short with the *minutiae*, and, if we may so call them, the finesses of those dry studies which mostly occupy us in the times of our meetings, and without which no man is fit to judge of philosophical inventions?" Or, as Lieutenant James Glenie of the army engineers, a rigorous geometer and mutineer, would have put it, had his philosophical fellows not shouted him down at the February meeting, a Banksian council would be "incapable of examining or even perusing the various papers on mathematical, mechanical, astronomical, optical, and chemical subjects, etc., that may come before them . . . , [and under such a council] this house, instead of being the resort of philosophers, [would] become a cabinet of trifling curiosities, and degenerate into a virtuoso's closet decorated with plants and shells."[18]

The tumultuous meetings heard often from Horsley. In a momentarily famous speech delivered at the January meeting packed with Banks's men, he threatened that he and his band of true scholars—

"the scientific part of the Society"—would secede. "When the hour of secession comes, the President will be left with his train of feeble *Amateurs,* and that Toy upon the table [the Society's mace, a gift from the Royal founder, Charles II], the Ghost of the Society in which Philosophy once reigned and Newton presided as her minister."[19] In reviewing *An Authentic Narrative* in a journal he edited, Maty singled out this exordium as the equal of Cicero and as a full refutation of the *opprobrium mathematicum,* the supposed incapacity of mathematicians to express themselves energetically. *"Toy upon the Table! Ghost of the Society where Philosophy once reigned and Newton presided as her minister!* What imagery! *Feeble Amateurs!* What a substantive! What an adjective to couple with it! The strength of all modern languages united, could not have brought together two others to express the same idea."[20] "Who can produce from the pages of Demosthenes, of Tully, or Rousseau, an effusion of eloquence more apposite to its particular purpose, or breathing a loftier tone of indignant vehemence and sublimity?"[21]

Banks's unpunctuated notes on this bombast are instructive. "Those gentlemen [Horsley and company] might easily be informed that howsoever respectable mathematics as a science might be it by no means can pretend to monopolize the praise due to learning it is indeed little more than a tool with which other sciences are hewd into form. Sir Isaac Newton demonstrated it is true the discoveries which made him immortal by the help of mathematics but he owes his immortality to his discoveries in Natural Philosophy not mathematics."[22] Hutton's year-long reduction of Maskelyne's data was, to Banks, a typical piece of mathematics: a work of sheer drudgery to obtain a result, a number for the average density of the earth, of no earthly interest.

As controversy continued during the early spring of 1784, Blagden and Cavendish proposed that some men of independent judgment should form a committee to make resolutions that would bring peace.[23] They did not have to bother. Horsley and Maty became increasingly shrill and alienated their support; they sustained an important defeat on 25 March, when Maty presented the Society with a copy of *An Authentic Account* (the mutineers' version of the dispute) on behalf of its anonymous compiler Horsley. Horsley moved thanks, seconded by Hutton and Maty. Banks opposed the motion on the ground that the *Account* contained much that did not conduce to the honor of the Society. Several members urged that the motion be withdrawn. Banks decided not to put it to a vote. Maty refused to continue his duties as secretary to the meeting unless Banks called the vote. Banks did not budge. Maty resigned on the spot. Hutton ran for the post vacated by his ally but was roundly defeated (139 to 39) by Banks's man Blagden.[24]

The substantive defense against the mutiny mounted by Banks's side turned on the desirability of having people with large views and broad educations in charge of the society's affairs. Their principal polemic, *Canons of Criticism, Extracted from the Beauties of "Maty's Review," and the Oratorical Powers of Dr Horsley*, observed that the Society had been formed to promote knowledge, not to multiply details or to make a living: "The Man who, though he may not have confined his Studies to any Single Branch of Science, in his large Grasp embraces Knowledge of every kind; courting it, not for a pitiful Subsistence, but through real Love . . . ; he, Sir, is an Honour to the Royal Society. . . . A profound Knowledge of the Mathematics is not the Qualification most requisite in the Secretary of the Royal Society . . . ; a general Acquaintance with the Sciences and classical Learning are of much more Consequence."[25]

Kippis divided the Society's membership into three categories: "real philosophers," "men of general literature," and "the nobility and gentlemen of rank and fortune." The literary men function as a filter. Unlike the real philosophers, who easily succomb to their private and picayune enthusiasms, the men of general literature have little difficulty forming "a right opinion concerning the general value of the philosophical observations and experiments which are produced at the Society's meetings."[26] As for the rich and aristocratic, they are necessary for prudent management. "The Society at large is greatly indebted to the Noblemen and Gentlemen, who compose the Council, and who take the trouble of conducting the business of the Society."[27] Had it not been for the staunch and generous intervention of the literary men, the noblemen, and the gentlemen, the Society might have succumbed to the mathematicians and made the great mistake of taking Horsley as its president.[28]

A Mathematicians' Mutiny, with Morals

The mutineers

The mutiny has been variously interpreted: as a power play by Horsley; as a fight between "professors" or "mathematical practitioners" and "gentlemen"; as a ripple of the great political events of the time; and, as the mutineers themselves represented it, as a struggle between mathematicians and natural historians.[29] The politics of Pringle's resignation, the lowly origins of Hutton, the religious polemics of Horsley, and Banks's opposition to "the spurious philosophy of the theorists [and] the atheists"[30] might indicate that something other than the merits of mathematics moved the mutineers. To assay these possibilities, more mutineers are needed than the leaders Horsley, Hutton, Masères, Maskelyne, Maty, and Glenie.

A few might be dragged from the poor poetry of John Wolcott, alias Peter Pindar, an apothecary, physician, and minister who made his living as a satirist. Banks was the butt of several of his lampoons. One of them names mutineers:[31]

Peter: Think of the men, who[m] SCIENCE so reveres!
Horsley, and Wilson, Maskelyne, Masères
Landen, and Hornsby, Atwood, Glenie, Hutton—

Sir Joseph: Blockheads! for whom I do not care a button!
Fools, who to mathematics would confine us
And bother all our ears with plus and minus.

From these limp verses we infer that Sir John Wilson (senior wrangler at Cambridge, 1761; fellow of the Royal Society, 1783; a judge), John Landen (self-taught mathematician and surveyor; FRS, 1766), Thomas Hornsby (Savilian Professor of Astronomy at Oxford; FRS, 1763), and George Atwood (third wrangler, 1769; FRS, 1776; a Cambridge don) supported the mutiny and enjoyed the disapprobation of Banks. George Shuckburgh-Evelyn (FRS, 1774) also can be enrolled in the mutiny on the strength of a report from Blagden and a letter from one of Banks's secretaries.[32]

The social backgrounds of the mutineers and fellow travelers ranged from below modest (Hutton's father was a collier, Landen's a yeoman) through the upper middle professional class (Glenie's father was an army officer, Maskelyne's a senior secretary to government officials, Horsley's a minister of the Church, Masères's a physician, Maty's the Librarian of the British Museum), to the wealthy and aristocratic (Wilson was the son of a "man of property," Shuckburgh of a baronet). One attained high preferment in the church (Horsley became a bishop); another was knighted for professional achievement (Judge Wilson); a third ranked as the highest official mathematical practitioner in the country (Astronomer Royal Maskelyne); a fourth succeeded to the family baronetcy (Shuckburgh). Their mathematical education had a similar variety: Hutton and Landen were self-taught; Glenie went through the Royal Military Academy at Woolwich; Atwood, Masères, Maskelyne, and Wilson were Cambridge wranglers (recipients of a B.A. with first-class honors in mathematics); Horsley and Maty also attended Cambridge; Hornsby and Shuckburgh managed at Oxford. Those at the lower end of the social scale—Hutton and Landen—took their first steps toward recognition by publishing mathematical puzzles in the *Ladies' Diary,* of which Hutton became editor, and thereby a patron of struggling mathematicians.[33]

Hornsby, Horsley, Maskelyne, and Maty were ordained ministers of the Church of England. Their mathematics and their ministry did not cause them to agree about doctrine, however. Horsley made a reputation by attacking the religious beliefs of Joseph Priestley, who claimed, among other things, that the primitive church had not been Trinitarian.[34] Priestley held with Banks during the mutiny.[35] Maty shared this Socinianism and became so perplexed with other doubts about the 39 articles that he gave up preaching.[36] Maskelyne and Hornsby appear to have been unproblematic churchmen.

Three generalizations may be drawn from this prosopography. For one, although the leading mutineers and fellow travelers did not come from a single social stratum, all of them worked for a living. For several of them—Atwood, Glenie, Hutton, Landen, Maskelyne, and perhaps Horsley—ability at mathematics was a means of upward mobility. Mathematics had about it an aroma of trade and tedium that did not recomend it to gentlemen.

The second generalization is that, except for Maty, the leading mutineers were proper mathematicians, in the eighteenth-century meaning of the word. The third is that for a time the mutiny had the allegiance of the majority of fellows familiar with mathematics. An anonymous reviewer of *An Authentic Account,* trying to refute

Horsley's claim to the support of the largest part of the productive members of the Society, could name only one mathematical fellow opposed to the mutineers. That was Henry Cavendish, "who, so far from being a *feeble amateur,* may alone be considered as a learned society."[37] The others mentioned were the chemist Priestley, another chemist (Richard Kirwan), two physicians, a geologist, an engineer who liked numbers (John Smeaton), and an astronomer who did not (William Herschel). Astronomy had its Banksian, or natural-histori- cal, side in random telescopic observations. Peter Pindar picked upon poor Herschel as a typical Banks man. "When on the moon he first began to peep / The wond'ring world pronounced the gazer, deep / But, wiser now th' unwond'ring world, alas! / Gives all poor Herschel's glory to his glass."[38]

2   Mathematics of the Mutineers

Piety or faithfulness to Newton's concerns marked the mutineers' approach. Horsley made an early contribution by editing, in an "el- egant monument of typography," five thick volumes of Newton's works. This monument was not a critical success. Lord Brougham, in his kindly way, rated it "as signal a failure as any on record in the history of science."[39] We have some geometrical theorems from Hors ley, in the ancient mode, written out in Latin, "in which language indeed works of science ought to be composed." They do little to disconfirm Brougham's judgment that as a mathematician Horsley was "nearly, if not altogether insignificant."[40] The respect for the monuments of the past that inspired Horsley's Newton appears also in his editions of Greek mathematicians; in Hutton's essays on the history of mathematics;[42] and in Masères's six stout volumes of re- prints, translations, and original papers of people he called scriptores logarithmici, most of whom he collected from a history of logarithms by Hutton.[43]

The volumes of the *Scriptores logarithmici* contain many notes on geometry and analysis by Landen, Glenie, and Masères himself. Lan- den, the self-taught surveyor, was the most creative of the three. His first sustained effort, a set of *Lucubrations* composed "at my evening hours" and published in 1755, remained geometrical and Newtonian, although Landen solved algebraically the problems he posed geometrically. He devoted lucubrations to the summation of series and to the solution of hard definite integrals, or "whole fluents" in Newtonian parlance, a line he continued vigorously.[44] One of his results, the rectification of a hyperbolic arc in terms of two elliptic arcs, gave rise to the Landen transformation, which has some importance in the theory of elliptic functions.[45]

Landen departed from Newton in trying to devise a formulation of the principles of the calculus that did not involve an appeal to velocity. His "residual analysis" operates with the ratio $[f(x) - f(x')] / (x - x')$ when $x = x'$, an approach to the concept of the derivative as a limit of a ratio developed further by Lagrange.[46] In another set of memoirs, Landen took up the notion of a rigid body around an axis passing through its center of gravity. The treatment never loses sight of the physical, or rather geometrical, problem: Landen locates points on the spinning body where the effects of the angular velocities impressed by the forces acting on it cancel and thus identifies the instantaneous axis of rotation in a way more intuitive than Euler's.[47] Generalizing his approach, Landen arrived at equations that differed from those obtained by Euler and applied by d'Alembert. The disagreement did not alter his conviction, however, and he vindicated his ideas about spinning tops by correcting Newton's calculation of the rate of precession of the equinoxes.[48]

James Glenie was throughout rigorously geometrical and old-fashioned. His forte was constructing ratios, to various powers, entirely geometrically.[49] He tried to base even arithmetic on geometry by rewriting fractions or products as ratios: 5·6 must be understood as shorthand for the compound ratio (5:1)·(6:1). Such expressions, he observed, "are not only geometrical but universally metrical." They are "so natural, so scientific, and so beautiful, that they cannot fail to furnish the mind with the highest pleasure, satisfaction, and delight." His method produced interesting exercises in dimensional analysis and an idiosynchratic formulation of the differential calculus.[50] Glenie ended his career as a geometer with demonstrations of the impossibility of squaring the circle and of the difficulty of making a good living as a tutor of mathematical subjects. He died in poverty.[51]

The most conservative of the mathematicians, Masères, had trouble advancing beyond negative numbers. He strove to make algebra as luminous as geometry, in which a minus magnitude has no meaning, and wrote a large book to show how to solve equations without resorting to negative numbers. Instead of clarity and simplicity, however, he produced the confusion of a multitude of special cases, which he set out in great prolixity.[52] He later summed some series, offered a proof of the binomial theorem, and computed other things, seldom by the straight path.[53] His logorrhea over his logarithms won him the reputation of an expert in the higher mathematics among his fellow lawyers and of an erudite and generous patron among aspiring mathematicians.[54]

Masères's interest in making available mathematical tracts grew with his income from his law practice and his sinecure as cursitor baron of the Exchequer. The fourth volume of *Scriptores logarithmici* is given over largely to navigation, the fifth to compound interest. These departures from the logarithmic path opened the wide prospect of mixed mathematics. Masères floated the notion of a series of *scriptores* in addition to *nautici: statici* (writers on simple machines, hydrostatics, and the catenary), *phoronomici* (kinematics, including Newton's "most profound, but very difficult work," the *Principia*), *centrobarici* (centers of gravity), and *optici*.[55] Many of the mutineers worked at these subjects and another associated with them since ancient times, positional astronomy.

Hornsby, Horsley, and Maskelyne were the astronomers of the group. Maskelyne's obligations as the King's astronomer kept his eyes glued to the telescope; he published four volumes of observations and four dozen editions of the *Nautical almanac*. He played a major part in observations of the transits of Venus, in the testing of Mayer's and Harrison's solutions to the longitude problem, and in finding the earth's shape and the local value of terrestrial gravity.[56] Hornsby left his observations, the evidence of 25 years of sight-seeing at the Radcliffe Observatory in Oxford, to posterity, which printed them in 1932. He did not get around to it himself because he had had to labor for decades with "assiduity and unremitted diligence" at the pious and useful task of publishing, in three large folio volumes, the observations of his predecessor in the chair of astronomy at Oxford, the Astronomer Royal James Bradley.[57] Hornsby observed the Venus transits of the 1760s and deduced the excellent value of 8″.78 for the solar parallax (about 0.25 percent off the current value). He thought that he had reached the limit of the humanly possible. "The learned of the present time may congratulate themselves on obtaining as accurate a determination of the sun's distance, as perhaps the nature of the subject will admit."[58]

The only *scriptor opticus* among the mutineers was Horsley, who took it upon himself to defend Newton's theories against doubts thrown up by the irreverent Bemjamin Franklin. In a letter read to the Royal Society in 1752, Franklin joked that received wisdom left him in the dark about light. If, as Newton's followers taught, light consists of particles shot from the sun, the sun should diminish in size and the particles, discharged with "a force exceeding a twenty-four pounder," should drive planets and dust alike before them. Nonsense, replied Horsley. Let particles of light have diameters of a million-millionths of an inch, and a density three times that of iron. Their velocity can be calculated using the newly found value of the solar parallax and the time of their transit to the earth. Horsley made out, correctly, that the force with which one such particle strikes the earth is to the force with which an iron ball ¼ of an inch in diameter moving at 1,000 yards/sec hits as 7.317×10^{-29} is to 1. Hence the momentum of a Horsley light particle is equal to that of a ¼ inch iron ball that takes 455XXI seconds (Horsley wrote powers of ten as Roman numerals) or more than 144,000 million million Egyptian years to go a foot. The argument showed that the force of Franklin's objections diminished with the size of the light particles. To eliminate them entirely, one had only to take the particles to be vanishingly small. Why not? "I cannot apprehend, from any quarter, so unphilosophical an objection, as that the extreme minuteness of the particles of light . . . is an argument against their existence. Size is a mere accident. . . . One Being only is absolutely Great. . . . In respect of Him, all things that are, are little."[59]

Mechanics had its main exponent in George Atwood, who, like Landen, has achieved the status of an eponym. "Atwood's machine," which Atwood introduced into his lectures in Cambridge to measure the velocity and acceleration of falling bodies, remained a standard teaching device until well into our century. Atwood proceeded analytically, and felt obliged to apologize for doing so; his treatise, he said, should "be considered as auxilliary and subservient to . . . authors who have written geometrically on the principles of motion."[60] The machine figured not only in a course of mixed mathematics but also in one on natural philosophy, which comprised—besides mechanics, hydrostatics, optics, and astronomy—the core qualitative physical subjects of electricity and magnetism. Atwood might therefore appear to have blurred the distinction between the mutineers' material and experimental physics. In fact, his treatment rather confirms than confuses the distinction, since the course was entirely nonmathematical apart from some exercises in proportionality.[61]

When playing the mathematician rather than the pedagogue, Atwood analyzed idealized practical problems, like the oscillation of watch springs, supposed subject to restoring forces proportional to any positive power of the angle of twist, or the stability of floating bodies against rotations around horizontal axes through their centers of gravity. Atwood explicitly placed this last work among the most important subjects treated by mathematicians and, as if still feeling guilty about his recourse to analysis, quoted Archimedes at length. "The construction of Archimedes . . . is here inserted, in order that the agreement between the solutions by analytical investigation, and geometrical construction, may appear in the most satisfactory point of view."[62] For his investigations of the rotations of floating bodies, Atwood received the Copley medal for 1796.

The most elaborate exercise in practical analytical mechanics undertaken by the mutineers was the cooperation of Maskelyne and Hutton on the deduction of the density of the earth from measurements made on a symmetrical hill in Scotland called Mount Schehallien. Under Maskelyne's direction, surveyors mapped the hill and its immediate neighborhood while he measured the angular distance of a certain star from the zenith as seen from temporary observatories north and south of the hill. Let the observed zenith distances be P and Q, and assume that the plumb bob is drawn off the vertical by a small angle ϵ in opposite directions at each station. Then $\epsilon = (P - Q)/2 - \Delta\lambda$, where $\Delta\lambda$ represents the difference in latitude between the two stations. Maskelyne measured P and Q astronomically and $\Delta\lambda$ geodetically.[63]

Hutton worked up the survey results. He divided the hill into 1,000 cylindrical pillars to fit the surveyed profile and computed the sum of their gravitational attractions on each plumb bob. Now ϵ is proportional to the ratio of the densities of the hill and the earth. Putting in the numbers, Hutton estimated the earth's density to be 4.5 times that of water. Newton had guessed the factor to be five or six. Close. "So much justness was there even in the surmises of this wonderful man!"[64] For the weighing of Mount Schehallien, Maskelyne received the Copley medal for 1775.

Shuckburgh's hypsometry ranks with Maskelyne's surveys of Schehallien as the geodetic masterpiece of the mutineers. Shuckburgh worked in the mountains of Savoy, to which he carried an excellent equatorial telescope, two barometers, and a few thermometers of the best English manufacture. He laid out a baseline with a 55-foot chain and, aiming from its end points, took the angular height of an Alp from a station near Geneva. It remained to climb to the spot so triangulated with a barometer to measure the pressure and thermometers to correct the barometer, all to check a rule proposed by Jean Antoine Deluc for deducing heights from barometer readings. It was heavy work for a baronet. On one ascent he had to stay the night with two shepherdesses and two cows in a "rather too artless habitation." The next morning, eager to leave the hut, he rushed to the summit, some 6,000 feet high. "I was imprudently the first of the company: the surprize was perfect horror, and two steps further would have sent me headlong from the rock." But he got what he wanted. His triangulated heights always exceeded his barometric ones in a way that allowed him to improve minutely upon Deluc's rule.[65]

Hutton had been asked to reduce the measurements of Schehallien because he suffered more than any other mutineer from a lust for calculation. He amused himself by recomputing, and discovering errors, in other peoples' tables. He kept track of all the reciprocals and square roots he ever worked out ("the results preserved of many years' occasional and accidental calculations in various subjects") to such effect that in 1775 he could publish tables of reciprocals to seven decimal places and of square roots to ten places for all integers from 1 to 1,000.[66] He tried to organize a group of calculators to compute trigonometric tables in terms of an angular unit that might well be called the hutton: 157,079.63267948966 huttons make up a fourth of a circle (a hutton is 10^{-5} radian). During the last year of his life, when he may have done his minimum, he published the fifteenth edition of his *Arithmetic,* the eighth of his *Compendious Measurer,* and the sixth of his *Mathematical Tables.*[67] On this scale, Maskelyne "did not publish much," or so Hutton evaluated his friend's literary output of 8 books, 49 nautical almanacs, and 30 contributions to the Royal Society's *Transactions*.[68]

Hutton's award-winning work on the velocity of cannon balls built on the pioneering inquiries into "this part of mixed mathematics" made by Benjamin Robins in the 1740s. Robins, who invented the ballistic pendulum, worked only with bullets.[69] Hutton took on cannon balls. He obtained a fair agreement for measured values of the muzzle velocity v with the formula $v = (P/W)^{1/2}$, P being the weight of the charge and W that of the shot. He fancied that his principles were "sufficient for answering all the enquiries of the speculative philosopher, as well as those of the practical artillerist."[70] In this last claim he certainly went too far, since he had not determined the effect of air resistance. The practical artillerist, represented by Glenie, would have to rely on "a sort of random or guess-work" until the problem of resisted motion, "the most difficult part of mixed mathematics," was subdued.[71] Hutton tried to subdue it via experiments with the ballistic pendulum that lasted many years. In the end, however, he had to content himself with exhibiting his results in tables and graphs that eluded analytical formulation.[72]

Civil, military, and naval architecture also belonged to mixed mathematics. The mutineers made contributions to them all. At the outset of his career, Hutton advertised himself and his school in Newcastle by writing a book about the principles of bridge construction just after a flood had washed out several local bridges; as an indication of his competence to undertake the analysis, he styled himself "mathematician" on his title page. Atwood also studied bridges. In 1801, prompted by plans for a new crossing of the Thames, he published an instructive booklet, written *more geometrico*, on the design of arches to bear various loads. Hutton returned to the subject as a consultant on the proposed Thames bridge and, for good measure, reprinted his earlier opportunistic tract.[73] Military architecture had its representative in Glenie's neat geometrical construction of lines of defense applicable to cases where the ground or other constraints prevented the classical solution in which the lines make equal angles with the ends of the sections of external walls.[74] Atwood's investigation of the stability of ships represents naval architecture. Navigation received attention in Masères's *Scriptores logarithmici* and in Atwood's general theory of the sextant, which permitted estimates of errors arising from imperfect alignment of the mirrors.[75]

A staple of mixed mathematicians was business arithmetic. Masères outdid himself rendering the principles of annuities in a quarto volume of over 750 pages. He had a strong philanthropic interest in establishing annuities on a sound actuarial basis. A bill to allow parishes to set them up in accordance with his calculations and guaranteed by the poor rate passed the House of Commons in 1772 but was rejected by the Lords. His huge book on the subject represented a second attempt, which he recognized as overdone: "I expect that many of my readers will wholly pass over many large parts of [it]."[76]

Atwood also served up business mathematics in great detail and for a practical purpose in a study of the price of bread in Britain from the time of King John to that of George III. He learned that the official profit margin allowed bakers by the regulations had remained constant at around 13 percent for 492 years. His practical problem was to recommend whether the hoary system needed change. Just before Atwood worried this subject, Shuckburgh worked up the price of wheat and animals from 1050 to 1795 to discover the rate of depreciation of money. He found a 20-fold inflation over the period.[77] We might also include as business mathematics Glenie's calculation that the cost of fortifying naval bases as proposed by the Board of Ordnance amounted to more than the replacement cost of the entire British navy.[78]

There remain mensuration, metrology, and chronology. Hutton wrote at length and eloquently about mensuration, "a subject of the greatest use and importance, both in affairs that are absolutely necessary in human life, and in every branch of mathematics." He mentioned estimating the capacities of vessels, gauging liquors, constructing buildings, measuring timber, surveying lands, checking the work of masons and carpenters, and making machines. Maskelyne and Horsley contributed equally to mensuration and metrology by translating Deluc's hypsometric rule into English measure.[79] Metrology further claimed the attention of Horsley and Shuckburgh, who respectively rendered the Hellenic stadion and the Roman foot into English feet and inches.[80] In a much more serious exercise in the same line, Shuckburgh proposed a new scheme of weights and measures to be based on a length defined by a pendulum and on a weight of distilled water contained in a unit volume fixed by the unit length. He defined the unit length *a* as the interval of the difference between the points of suspension of a pendulum when beating 42 and when beating 84 times a second at the latitude of London. With the most refined comparator he could procure, Shuckburgh determined *a* to be equal to 59.89358 inches of the existing Parliamentary standard. The proposal might appear to owe something to the metric reforms, which were just coming into force when Shuckburgh made it. On the contrary: he had ruminated about it for two decades but had not brought it forward for want of sufficiently precise instruments to work to the ten-thousandth part of an inch.[81]

Chronology belonged to Horsley. Like Newton, he tried to settle a disputed date in ancient history by calculating the times of astronomical events in Greek calendars. The subject was sufficiently obscure to allow Horsley to hold opinions different from the master's. In his edition of Newton's writings, Horsley protested against Newton's insistence on taking Chiron the Centaur and his daughter Hippo as reliable astronomers. In Horsley's opinion, a centaur could amount to no more than a doctor, or perhaps an astrologer.[82]

3 The Morals

In general

One of the first lessons to collect from the story is the wide meaning of "mathematics" in early modern times. It ran from the higher reaches of analysis (as practised by Landen) through the ancient mixed sciences of astronomy (Maskelyne), optics (Horsley), and mechanics (Atwood) down to gunnery (Glenie, Hutton), architecture (Atwood, Hutton), surveying (Maskeyline, Shuckburgh), chronology (Horsley), metrology (Shuckburgh), and business arithmetic (Maseres). An even wider variety of meanings perplexed "physics." Around 1700 the word indicated all qualitative natural science, from the concepts of space and time through natural history. By 1800 "physics" had come to signify primarily the branches of experimental natural philosophy, from mechanics to magnetism, but traces of the old meaning survived, and a new one, akin to "juggling," had come in to capture the popular application of physics to parlor tricks.[83]

A second lesson is that the conservatism of English mathematicians in respect of foundations did not inhibit application. They might insist on geometrizing arithmetic or negativing the minus sign, but they would also calculate annuities or loxodromes as if they had full confidence in the underpinnings of algebra. Their conservatism affected not so much the manner but the range of their applications. In general, they stayed well within the precinct of traditional mixed mathematics. There they could anticipate the pleasure of finishing much tedious work without much competition from physicists.

To be able to describe these phenomena, the historian needs to be able to distinguish at least three usages of the names of sciences. One is application by contemporaries, which might be placed in quotation marks. The other two fall within the discretion of the historian: a synchronic significance, used of a science or sciences within a narrowly restricted period of time, which might be indicated with plain roman type; and a diachronic significance, used of a science over a longer span, often with an implied reference to its modern form, which might be shown with italics. With this way of writing, the riddle whether Lavoisier compassed a revolution *in* or *into* chemistry, which historians agitated a few years ago, may be easily dissolved. The proposition behind the first preposition was that the revolution occurred within an established discipline; that behind the second, that the revolution subjected whatever chemists did (there were then people called "chemists") to the techniques and standards of another discipline, physics, and so established a new discipline.[84] Lavoisier did bring techniques using experimental physics (synchronic meaning, in roman), like his calorimeter and gazometer, into chemistry; the result may have been a revolution in *chemistry* (diachronic meaning, in italics); he himself liked to say that his theory of combustion transformed both "physics" and "chemistry" (Lavoisier's usage, in quotation marks).[85]

A similar riddle, with a similar solution, has been posed about Siméon-Denis Poisson. One expert points to Poisson's election to the class of *physique generale* of the Paris Academy of Sciences in 1812, just after the completion of his analysis of the distribution of electricity on the surfaces of conductors, as proof that the "oft-proclaimed unification of mathematics and physics was [then] no longer a dream." Another expert counters that Poisson was a mathematician of the late eighteenth century, not a modern mathematical physicist, or better, that he was "an example of a practitioner of a discipline that shares only a name in common with its twentieth-century descendant."[86] According to the first interpretation, Poisson might be labeled a mathematical physicist vintage 1810 by virtue of his membership in the Academy's physics class and his expertise as a géo-metre. According to the second, he was not a *mathematical physicist* in a sense recognized today. In this orthography, both parties may be right, although the second statement may not hold any interest for the historian.

In the period from 1780 to 1820, physics struggled to define its borders between a newly powerful chemistry on one side and an old multifaceted mathematics on the other. The confusion on the chemical border derived largely from the discovery of the various types of gases (which historians close to the event reckoned as a turning point in "physics" as well as in "chemistry"[87]) and from advances in understanding heat, especially the roughly simultaneous inventions of the concepts of latent and specific heat by Joseph Black (usually taken as a *chemist*) and Johann Carl Wilcke (usually taken as a *physicist*). A further cause for overlap was the standard distinction between "physics" as the study of sensible motions and "chemistry" as the study of insensible ones.[88] That required literal-minded textbook writers to subordinate "electricity" and "magnetism" to "chemistry," apparently because the particles of the fluids supposed responsible for electrical and magnetic phenomena acted on one another through undetectable forces and motions.[89] There is much that appears to the modern reader as *physics* in the *Annales de chimie* and much that appears as *chemistry* in the *Annalen der Physik*. Their editors felt a difficulty too, which they acknowledged by adding "et physique" to the title of the first in 1816 and "und Chemie" to that of the second in 1816.

German dictionaries and textbooks toward the end of the century announced a serious redrawing of the boundaries between physics and mathematics: "the objects of applied mathematics . . . belong in themselves to physics"; "applied mathematics is made up really only of individual parts of physics"; up-to-date physics will consist of "the choicest parts" of applied mathematics together with "the most necessary topics of chemistry."[90] The acquisition would not come easy or in full. F. A. C. Gren, an early editor of the *Annalen der Physik*, recognized that he would have to learn more mathematics to keep his textbooks current and to understand all the papers submitted to his journal. But "although he devoted himself with the most creditable zeal to the study of mathematics . . . , his work showed that he had not made himself at home with the rigor of mathematical discourse."[91] Also toward the end of his career, Pierre Prevost, a late representative of the Genevan school, whose compulsion for hypsometry had raised the general level of accuracy of physical instruments, found the exertion of mastering "the choicest parts of applied mathematics" unwelcome and wasteful. "It is a shame that by piling up so many calculations (hardly necessary in my opinion), [Fourier] has given me so much useless trouble. He could and should have set forth these interesting results [on radiant heat] more simply."[92]

Thomas Young, one of the most creative British physicists of the period, suffered from a similar allergy to extensive computations and analyses. Young found little to admire in what Laplace's protégé, J. B. Biot, touted as "the only example of attraction in terrestrial phenomena [so far] successfully quantified," that is, a prolix and intricate memoir by Laplace on capillarity. According to Young, it was an "ostentatious parade," an exercise in "difficulty, obscurity, and perplexity," one of several works calculated to "discourage, at the same time that they astonish, the student."[93] In Young's opinion, the Laplacians, having missed the natural boundaries of the physical and mathematical sciences, were busy gerrymandering them.

Three meanings of the "mathematizing of physics" may be distinguished: the capture by *physics* of branches previously belonging to *mathematics,* the quantification of new branches, and the application of mathematics to correct and manipulate the results of measurement.[94] In mathematizing by capture, physics took on the lower levels of analytic mechanics, including unperturbed motion in Keplerian ellipses, but eschewed the higher reaches, like the formulations of Lagrange or the perturbation methods of Laplace. The work of Atwood and Landen on angular motion represents the most demanding of the mathematics of captured mechanics around 1800. Lagrange's and Hamilton's formulations eventually entered mainstream *physics,* in connection with electromagnetism and statistical mechanics. Perturbation methods followed in connection with the old quantum theory. *Physics* did not acquire astronomy or its eighteenth-century fellow traveller geodesy. As for optics, the years around 1800 saw a most dramatic transformation. Apart from geometrical optics or ray tracing and the technical improvement of lenses, the study of light had not produced much during the eighteenth century that qualified it as a quantitative science. Horsley's arithmetic nonsense about the momentum of light particles suggests the main difficulty: the basic elements of the theory, although easily quantified, could not be brought successfully into confrontation with experiment.[95] The great mathematicians working with the Newtonian theory on the Continent did little better than Horsley, as Laplace's handling of the problem of refraction shows. Like Horsley's computations, Laplace's only placed a mathematical fig leaf over the main points of physical interest.

Laplace tried to bring the interaction of light particles and common matter into line with his conceptions of affinity, cohesion, and capillarity by supposing a force between the interacting particles that diminishes as some power n of the distance between them. The problem in quantification became the deduction of the value of n from the phenomena. Knowing n, the physicist would be able to investigate other situations in which the same or similar forces occur, distinguish them from one another, make predictions, and do indicated experiments, in analogy to the application of laws found to govern gravitation, electricity, and magnetism. In fact, Laplace never learned anything about n or the forces he claimed to be studying. He set up problems so that all expressions involving unknown short-distance forces occurred within integrals running to infinity. These definite integrals appeared as opaque constants, not further analyzable, in subsequent calculations; their values had to be taken from experiment. Laplace worked such analyses for capillarity (which prompted the courtly comments of Young quoted earlier) as well as for refraction, and some of his disciples, for example, Etienne Malus, did the same for optical polarization. In contrast to Laplace, Malus was an exact experimentalist as well as a mathematician. But when he tried to quantify theory, he did so as a calculating workhorse, as "[if] set on mathematic grounds to prance / show all his paces, not a step advance."[96]

The second meaning of mathematizing physics was the quantification of previously qualitative branches, like electricity, magnetism, and heat. Here the essential step was to discern quantifiable theoretical entities that could be brought into fruitful confrontation with experiment. Successfully quantified sciences, like *mechanics* and *astronomy*, had worked with quantifiable entities—distance, angle, and time—literally incorporated in the instruments that measured them: standard rulers, graduated circles, and clocks. Velocity was immediately defined as the quotient of numbers obtained from direct measurements. The introduction of mass as a measure of the strength of gravitational pull, and thus of weight, led to the first general quantitative theory based on a fundamental entity other than space and time. It took four generations before people felt comfortable enough with Newton's technique to extend it to other sets of phenomena. That occurred in the 1780s with the introduction of electric and magnetic fluids sui generis, whose elements interacted by the same law of force that regulated the attractions between gravitating particles.

The construction of mathematical theories in both cases required more quantifiable entities than did the theory of gravity, since bodies differ considerably in their electrical and magnetic properties, whereas all matter gravitates. Hence the need for, and the provision of, the concepts of electrical tension and capacity. At the same time, during the 1770s and 1780s, the theory of heat acquired quantifiable concepts closely analogous to terms in the theory of electricity: heat capacity (with an obvious analogy in electrical theory) and latent heat, the heat that does not affect a thermometer (analogous to an uncharged body's electrical fluid, which does not register on an electrometer). To complete the parallel, heat was represented as a fluid, caloric, whose elements repelled one another according to a law of distance that no one ever discovered.[97]

Very few mathematicians, as defined by the mutineers' range, contributed to the creation of quantifiable concepts in imitation of the theory of gravitation. Concepts effective for electricity came from Alessandro Volta, who has been accused, unfairly, of not understanding arithmetic; C. A. Coulomb, who had the mathematics of a well-trained engineer; and Henry Cavendish, a Banks man, at home with fluxions, to be sure, but celebrated for his precise experiments with electricity, heat, and gases. Neither Joseph Black nor Johann Carl Wilcke, who did the parallel work in the theory of heat, was a mathematician. Laplace's study of heat may be counted an exception. He began with calorimetric measurements in collaboration with Lavoisier and ended with his brilliant and well-known deduction of the speed of sound in air by applying the calculus to the theory of caloric. Fortunately, Laplace's analysis required only the concept of a conserved mass of caloric, not information about the repulsion between caloric elements.[98]

The third meaning of the mathematizing of physics was probably the one most often intended by contemporaries who noticed the increasing quantification of the subject. The great and rapid improvement of the instruments of physical measurement—the thermometer, barometer, hygrometer, and electrometer—made everyone who used them sensitive to the value of number. Meteorological instruments occupied the vanguard. During the 1770s the thermometer and barometer became precise, intercomparable, and reliable tools. Contemporaries remarked at the suddenness of "the singular success with which this age and nation has introduced a mathematical precision, hitherto unheard of, into the construction of philosophical instruments." Shuckburgh wrote these lines in 1779, referring to the labors of a committee of the Royal Society that two years earlier had established procedures for reliably setting the fixed points of thermometers.[99]

As for the barometer, it owed the first steps in its perfection to Deluc, who in 1772 published two wordy volumes about the ills and cures of meteorological instruments. Like Shuckburgh, whom he inspired, he had a passion to measure mountains. To do so accurately required that the barometer's mercury column be purged of air and water, that it be bent into a siphon, and that its readings be corrected for changes in temperature according to a formula Deluc devised.[100] The height of the mountain could be obtained from the cleared barometric readings by a maneuver with logarithms, which probably taxed Deluc's mathematics to their limit. It was the rule involving logarithms that Horsley and Maskelyne turned into English and Shuckburgh improved. Laplace later had a hand in reworking it. Many people have learned logarithms expressly to use some form of Deluc's rule, thereby glimpsing "the numberless [!] conveniences logarithms can effect in the most ordinary business of civil life."[101]

It was in this third sense of the mathematizing of physics that mathematicians participated. The committee of the Royal Society impaneled to calibrate thermometers had Cavendish as chair and a perfect balance of mathematicians and nonmathematicians as members: Alexandre Aubert (an amateur but exact astronomer and Pringle's choice as his successor as president of the Royal Society), Horsley, and Maskelyne, in the first party, and Deluc, William Heberden (a physician), and Joseph Planta (Maty's colleague as secretary of the Society and under-librarian at the British Museum) in the second.[102] Shuckburgh also made measurements relative to the perfection of instruments, for example, the rates of expansion of mercury and air with temperature, which helped him improve on Deluc. He further checked Deluc on the depression of the boiling point of water with decrease in pressure, that is, with increase of the height of the mountain on which the boiling occurred.[103]

Shuckburgh, as we know, went with the mutineers, although his work just described did not belong to traditional mixed mathematics. His ambiguous position on a productive boundary between disciplines may be further illustrated, and even symbolized, by the fact that his colleague and counterpart Major William Roy adhered weakly to Banks. Like Shuckburgh, Roy made precise measurements of the expansion of air and the relations between readings of air and mercury thermometers. But again, like Shuckburgh, he devoted himself primarily to a main line in eighteenth-century mixed mathematics, geodesy. His preference for geodetic over astronomical determinations of latitude and longitude caused a rift between himself and Maskelyne, while his position as head of trigonometric surveys conducted by the Ordnance department and his interest in Roman antiquities tilted him toward authority in the form of Banks.[104]

The work of the Royal Society's committee on thermometers and of Deluc, Roy, and Shuckburgh represented the desire, even the compulsion, typical of the late eighteenth century, to make physical instruments as precise as astronomical ones. The quest obscured the division between mixed mathematics and natural philosophy. The study and perfection of instruments, especially the thermometer and barometer, not the requirements of high theory or (in England at least) the initiatives of mathematicians, provoked the quantification of the physical sciences during the two decades on either side of 1800.[105]

Re Kuhn

These conclusions extend the schema that Kuhn proposed in his essay "*Mathematical* versus *Experimental* Traditions in the Development of *Physical Science*." Italics have been added to his title in accordance with the convention proposed earlier. The convention, or rather recognition that something like it is necessary, may also be regarded as an extension of Kuhn's work, in this case *The Structure of Scientific Revolutions*. His argument that theories on either side of a revolution are incommensurable because the meanings of terms change so radically applies also, and perhaps more plausibly, to the names of disciplines. We are told that relativists and Newtonian mechanists were unable to talk coherently together because "mass" did not mean the same thing in the two theories. A fortiori, people who write about the history of physics over a long span need to keep in mind that the mismatch between the modern meaning and earlier ones may subvert their efforts entirely if they allow the meanings to mingle.

In his paper on the two traditions, Kuhn works with two sets of convenient fictions. One set divides those who think the sciences are many (the multiplists) from those who think they are one (the singlists). Multiplists take the modern division of the sciences for granted and work up detailed and technical accounts of each in isolation from the others. They work diachronically and should write largely in italics, if at all. Singlists invoke global factors—like economics, social forces, and religion—to account for all the scientific thought of an age. They perform synchronically. Kuhn's other set of opposites divides sciences into "classical" and "Baconian." These are, respectively, the mathematical and the experimental "traditions." The classical sciences—comprising astronomy, optics, statics, and harmonics— reached an advanced stage in antiquity. The Baconian sciences, which depended upon the deployment of physical instruments, began life in the seventeenth century as qualitative and largely untheoretical. Their opposition, or essential tension, dissolved with the quantification of the Baconian sciences, which began in the late eighteenth

century and continued, with a "remarkably rapid and full mathe-
matization," during the first quarter of the nineteenth.[106]

This outline of the development of *physics,* in which junior branches
of the subject, like electricity and magnetism, joined senior branches,
like optics and mechanics, under the yoke of *mathematics* shortly before 1800, is a useful one even though the "traditions" supposed to
be its protagonists did not exist. The classical sciences during most
of their long history were not protophysics, but mathematics. They
belonged to a wider set of subjects that the greatest ancient writer of
texts on mathematics, Claudius Ptolemy, considered to be all of a
piece. To make up Ptolemy's canon, we must add to the four Kuhn
mentions—astronomy, optics, statics, and harmonics—the mathematical parts of astrology and geography. These additional subjects satisfy the condition that Kuhn sets on classical sciences: they required
technical knowledge not easily accessible and they gave rise to an
ongoing tradition. They do not satisfy the condition of survival into
physics. That is why their addition to the group is essential. Thus
expanded, Kuhn's "classical sciences," a retrospective classification,
become the material of "mixed mathematics," a classification found
in Aristotle and current in the time of Ptolemy.[107]

The expanded classical sciences defined implicitly by the Ptolemaic
texts suffered and prospered variously during the Middle Ages. In
the Latin West, the tradition almost expired; yet it grew on the edges,
in calendry and chronology, or, to use the term then employed,
"computing," sciences very necessary to regulating the observances
of the Roman Catholic Church. In the Islamic countries the tradition
developed and deepened, partly also in consequence of practical
requirements: geography and navigation (for crossing deserts as well
as seas) for a great empire; spherical trigonometry (for orienting
mosques) and gnomonics (for determining times of prayer) for a
demanding religion; exact astronomy and astrology for a fatalistic
people; and improved mathematics, tables, and instruments for the
calculators.

This material came west in the twelfth and thirteenth centuries,
where it formed the core of mixed mathematics. The Renaissance
added surveying, fortification, and civil architecture, and much enlarged navigation and perspective. (Kuhn mentions these subjects
and their connection with the classical sciences [pp. 55–56, 61] but

does not notice the difficulty they present for his historiography.) A late Renaissance text, John Dee's preface to the first English printing of Euclid, sets out no fewer than two dozen branches of mixed mathematics, some traditional, some Dee's own fantastic creations. We find astrology and astronomy, architecture and navigation, perspective and horology, etc. But we also find Thaumaturgike, "the Art Mathematicall, which giveth certaine order to make Straunge workes, of the sense to be perceived, and of men greatly to be wondered at," and Archemastrie, which "teacheth to bryng to actuell experience sensible, all worthy conclusions by all the Artes Mathematicall proposed, and by true Naturall Philosophie concluded." Kuhn's statement that "the cluster of mathematical sciences . . . reconstituted [during the Renaissance] closely resembled its Hellenic progenitor" (p. 39) is true only in the sense that Switzerland was part of the Roman Empire.

Two more data points, at 100 year intervals, will indicate the persistence of this generous concept of applied *mathematics*. A curriculum for undergraduates, drawn up by Isaac Newton, calls for a lecturer in natural philosophy and for another in mathematics. The natural philosopher would teach the usual introductory general concepts— time, space, body, place, motion, force, mechanical power, gravity, hydrostatics, and projectiles—before turning to matters specific to his subject: the system of the world, the constitution of the earth "and the things therein," meteors, elements, minerals, vegetables, animals, and, "if he have skill therein," anatomy. The mathematician would offer some "easy and useful practical things" as an inducement; then Euclid, spherical geometry, trigonometry, cartography, optics, astronomy, music, geography.[109]

The second datum, from the first edition of the *Encyclopedia Britannica*, divides mathematics into "pure" and "mixed," the latter including, among other things, astronomy, optics, geography, hydrostatics, mechanics, fortification, and navigation.[110] This division was standard and stationary during the later eighteenth century. It occurs in Chambers's *Dictionary* of 1750 and recurs in Hutton's *Philosophical and Mathematical Dictionary* (1795, 1815). It figures in the advertisement of Hutton's first school at Newcastle in 1760 and in his *Course of Mathematics* of 1800 (which has the significant addition of "the measures of altitudes by the barometer and thermometer").[111]

The most elaborate bibliography of mathematics of the eighteenth century, F. W. A. Murhard's *Litteratur der mathematischen Naturwissenschaften*, the first volume of which was published in 1797, includes among "applied mathematics" the mechanical, optical, and astronomical sciences. "But you will look in vain here for everything that belongs more to physics than to mathematics."[112] He would have judged perplexing, if not unintelligible, a knowledgeable modern historian's assessment that "around 1800, the branches of physics [that is, *physics*] which had received substantial mathematical treatment were terrestrial and celestial mechanics, and some aspects of optics."[113] Murhard cites literature on all branches of cameral science, machines, navigation, and so on, but tiring after five volumes, he did not complete his original design to cover astronomy, gnomonics, chronology, and music, nor did he do justice to architecture, fortification, or gunnery.

One more point about the anachronistic construction of the classical sciences needs comment. Kuhn says that they all underwent radical reconstruction during the sixteenth and seventeenth centuries, "and in physical science [of which they were then not a part] such transformations occurred nowhere else" (p. 40). He lists these reconstructed sciences as algebra, analytic geometry, and calculus; heliocentric astronomy and noncircular planetary orbits; kinematics; "the first acceptable solution to the classical problem of refraction, and a drastically altered theory of colors;" and pneumatics (the revolution in hydrostatics). This list errs by excess and by defect: by excess in enrolling theories of color and of the constitution of the atmosphere, and in claiming "the newly central sun" (p. 41), under the rubric of "classical sciences"; by defect in omitting surveying, cartography, and navigation, which also sustained very important advances during the "scientific revolution." A counterthesis seems to come closer to the record: the classical sciences, that is, *mixed mathematics,* advanced rapidly in scope and technique during the sixteenth and seventeenth centuries; some of this technique served *natural philosophy,* which simultaneously underwent a revolution in qualitative theories about the nature and constitution of the world.

The story of the Baconian sciences is more quickly told. These are introduced (p. 41) as "research fields" that employed the array of instruments newly available: the barometer, thermometer, hygrometer, air pump, and, later, the electrical machine. They did not undergo a revolution during the Scientific Revolution because, having no substantial previous history, they had nothing to reconstruct (pp. 52–53). The Baconian sciences initially busied themselves with exploration and compilation; their general theories, like corpuscularism, stood too far from their business to direct laboratory work (pp. 43–44). But during the last third of the eighteenth century, Baconian sciences like electricity, magnetism, and the study of heat reached a state "very like the classical sciences in antiquity" (p. 47). They became "developed sciences"; they "at last came of age" (pp. 47, 48). Their quantifiable theories of the 1780s provided grist for the powerful mathematical mill of Laplace and his followers. The barriers between the classical and the Baconian sciences fell. Classical physics was born (pp. 61–63).

These aperçus have the merit of raising the question how physics came to be quantified during the years 1780 to 1820. An outline of an answer has been suggested in connection with the episode of the mutineers. But it begs the question. It points to a new fascination with the accuracy of "philosophical instruments" around 1780, but it does not seek the origins of the novelty. The answer to this larger puzzle cannot be found in multiplist histories. The singlist approach across sciences and nonsciences therefore must be attempted. The contributors to a recent volume entitled *The Quantifying Spirit in the Eighteenth Century* have made a start by tracing their quarry from astronomy through meteorology to geodesy and also into chemistry, botany, technology, forestry, cameral science, classificatory systems, and encyclopedias.[114] They find a rapid acceleration in the use of mathematics in all these subjects during the 1780s. A good explanation of the quantification of the physical sciences will have to make reference to the needs of bureaucratic states and the military, the felt compulsion to control rapidly multiplying information about the world, the influence of instrumentalist philosophies, and no doubt much else.

Kuhn's scheme, which calls for a fuller explanation of the type indicated, is characteristic of his mode of analysis and indicative of his strength as a teacher. His clarity of thought, resourcefulness in argument, and enthusiasm for ideas inspired his students to fill in the blanks. His persuasiveness derives from a powerful dialectic often made more powerful still by a resonance between his personal experiences and his historiographical ideas. The dialectic is most familiar in *The Structure of Scientific Revolutions,* in which normal scientific practice (the thesis) generates anomalies that prompt counter theories and procedures (the antithesis), which, after a time of troubles, give rise to a new normal practice (the synthesis). The autobiographical resonance is perhaps clearest in the story of Kuhn's discovery that Aristotle's physics was not bad modern science but good old philosophy.[115] The involvement of self while retaining respect for the historical actor, the technique of scrutinizing texts not for what sounds familiar but for what seems bizarre, and the reliance on a clear and simple schema as a first approximation to a historical reconstruction were lessons of great value. So too was Kuhn's passionate interest in the work of his graduate students. Eager to know how our research results would fit his general ideas, he gave us to understand that we were engaged in an intellectual adventure of great moment. Some of us think we still are.

Notes

See the bibliography for abbreviations used in the notes.

1. Kuhn, *Essential Tension* (1977), pp. 31–65.

2. Kuhn, *Essential Tension* (1977), pp. 61–63.

3. It has been treated recently in Frängsmyr et al., *Quantifying Spirit* (1990).

4. Howse, *Maskelyne* (1987), p. 159; Gregory, *PM* 56 (1820): 243; O'Brian, *Banks* (1987), pp. 194–195; Aubert, *N&R* 9 (1951): 82–83.

5. Snip, *Puppet Show* (1663 [1784?]), pp. 23, 31.

6. Certificate for Henry Clerke, rejected Feb. 1781, signed by Thos. Percival, Thos. Henry, C. White, Thos. Bayley, Maskelyne, and Hutton (RS). Clerke, *Dissertation* (1779), p. iii.

7. Certificate for Desbarres, rejected 18 Jan. 1781, signed by R. P. Jodrell, John Grant, G. Fordyce, and R. Richardson (RS). Horsley in Horsley et al., *An Authentic Narrative* (1784), pp. 56–57.

8. Blagden to Banks, 16 Oct. 1783 (DTC 3:125–128).

9. Kippis, *Observations* (1784), pp. 58–60; Brougham, *Men of Letters* (1845), vol. 2, pp. 365–366, and *Philosophers* (1855), pp. 359–361.

10. Bruce, *Memoir of Charles Hutton* (1823), pp. 3–6, 10–16; Gregory, *Imperial Magazine* 5 (1823): 202.

11. Hutton, *PT* 68 (1778): 50–85, 689–778. Quotes from Gregory, *Imperial Magazine* 5 (1823): 211, 219–220, respectively.

12. Kippis, *Observations* (1784), pp. 66–67, 74–76; anon., *Appeal* (1784), pp. 29–31; Gregory, *PM* 56 (1820): 244.

13. Weld, *History* (1848), vol. 2, pp. 155–160; Lyons, *Royal Society* (1944), pp. 212–213; Howse, *Maskelyne* (1989), pp. 159–160. The motions are given in Kippis, *Observations* (1784), pp. 7–22.

14. Blagden to Banks, 22, 23, 24, and 27 Dec. 1783 (DTC 3: 171, 174); Banks to Blagden, 26 Dec. 1783 (RS, B.25); McCormmach in Garber, *Beyond History of Science* (1990), pp. 37–41.

15. Banks, "Notes," RS, M.M. 1, f. 30, p. 8; Horsley et al., *An Authentic Narrative* (1784), pp. 3, 23, 45, 77, 84; anon., *Supplement* (1784), pp. 9–14.

16. Horsley et al., *An Athentic Narrative* (1784), pp. 56–57, 109; anon., *History of the Instances of Exclusion* (1794), pp. 5–7, attributed to Maty by Kippis, *Observations* (1784), p. 144.

17. Banks, "Notes," RS, M.M. 1, f. 31, pp. 5–6.

18. Quotes from, respectively, anon., *History* (1784), p. 17, and Horsley et al., *An Authentic Narrative* (1784), p. 73; O'Brian, *Banks* (1787), pp. 210–211; Kippis, *Observations* (1784), pp. 101–106.

19. Horsley et al., *An Authentic Narrative* (1784), pp. 65, 66 (quote). *History of the Exclusions* (1784), p. 21: "Expect the creation of a new Society, a real Academy of Sciences."

20. Maty, notice in *New Review* 5 (1784): 210–216, quote on p. 214.

21. Anon., "Banks" (1801), p. 393, a piece favorable to Banks and probably intended to ridicule Maty.

22. Banks, "Notes," RS, M.M. 1, 46a.

23. Blagden to Banks, 5 and 6 Apr. 1784 (DTC 4: 20, 22); McCormmach in Garber, *Beyond History of Science* (1990), pp. 44–45.

24. Kippis, *Observations* (1784), pp. 145–146, 153–155; Weld, *History* (1848), vol. 2, pp. 164–165; O'Brian, *Banks* (1987), pp. 209–211.

25. Anon., *Canons*, 2nd ed. (1785?), pp. 51–52, 68.

26. Kippis, *Observations* (1784), pp. 95–96, 105, respectively.

27. "An old member," letter to the editor, *The Morning Post*, 2 Jan. 1784, in RS, M.M. 1, 45. Compare Archenholtz, *England* (1784), vol. 2, pp. 214–215: "The nobility [in the Royal Society and the Royal Society of Antiquities] do not in general contrib-

J. L. Heilbron

ute by means of their writing to the splendour of letters, and the progress of science; they willingly, however, empty their riches in defraying the expenses of these establishments."

28. The opinion of Brougham, *Men of Letters* (1845), vol. 2, p. 370, and *Philosophers* (1855), p. 364.

29. Respectively, Carter, *Banks* (1987), pp. 194–202; Miller, *BJHS* 16 (1783): 5, 10; McCormmach in Garber, *Beyond History of Science* (1990), p. 37; Jebb, *Horsley* (1909), pp. 54–55; Cameron, *Banks* (1952), p. 131; and Hall, *All Scientists Now* (1984), p. 2.

30. Anon., "Banks" (1801), p. 391.

31. Pindar, "Peter's Prophecy" (1789), in *Works* (1794), vol. 2, pp. 111–112; Vales, *Pindar* (1973), pp. 13–19, 62–67.

32. Blagden to Banks, 30 Dec. 1783 (DTC 3:185); Robert Brown to Edward Knatchbull, 15 Mar. 1846, in O'Brian, *Banks* (1988), p. 210. Compare Howse, *Maskelyne* (1989), p. 209. Shuckburgh later trimmed and asked Banks, who refused, to support the election of a friend of his (Shuckburgh to Banks, and reply, 27 Mar. 1784 [DTC 4:18–19]).

33. Gregory, *Imperial Magazine* 5 (1823): 205–207.

34. Weld, *History* (1848), vol. 2, pp. 168. Horsley opened fire on Priestley in 1778. The squabble, which produced many tracts on both sides, peaked during the year of the mutiny. See Jebb, *Horsley* (1909), pp. 45–46, and Nichols, *Literary Anecdotes* (1812), vol. 4, pp. 678–679.

35. Priestley to Banks, 25 Apr. 1790 ("thinking, as I have always professed to do, that the Society is honoured by your being its President"), quoted in Cameron, *Banks* (1952), p. 160. Griffin (*N&R* 38 [1983]: 3, 7) identifies Priestley with Banks and Kippis.

36. Maty's public declaration of antifaith appears in *GM* 47 (1777): 467.

37. *Monthly Review*, Apr. 1784, p. 301.

38. Pindar, "Peter's Prophecy" (1789), in Pindar, *Works* (1794), vol. 2, p. 124. Pindar adds, "As for [Herschel's] mathematical abilities, they can scarcely be called the shadows of science."

39. Nichols, *Literary Anecdotes* (1812), vol. 4, p. 683 ("elegant . . . typography"); Brougham, *Men of Science* (1845), vol. 2, pp. 371–372, and *Philosophers* (1855), pp. 364–365; Playfair, *Edinburgh Review* 4 (1804): 258–259. Whiteside (*Papers* [1967], vol. 1, pp. xxvi–xxvii; vol. 4, pp. 336, 421; vol. 7, p. 305) is not so harsh as Brougham.

40. Horsley, *PT* 65 (1775): 301–310. Quotes from Horsley, *Elementary Treatises* (1801), p. xii, and Brougham, *Men of Science* (1845), vol. 2, p. 366 (*Philosophers* [1855], p. 360).

41. Horsley brought out editions of Euclid's *Elements* and *Data* in 1802 and 1803, which earned him faint praise from the stingy *Edinburgh Review* 4 (1804): 272.

42. Hutton wrote histories of trigonometric tables, logarithms, and algebra in *Mathematical and Philosophical Subjects* (1812), vol. 1, pp. 278–306, 306–454, and vol. 2, pp.

143–305, resp., and he worked hard to include historical and biographical information in his *Dictionary*.

43. Masères, in *SL*, vol. 1 (1791), p. i. Compare Babbage in *Scriptores optici* (1823), p. 3; Hutton, *Mathematical and Philosophical Subjects* (1812), pp. 306–454.

44. Landen, *Lucubrations* (1755), preface, p. 5, and *PT* 60 (1770): 442; Landen, *Memoirs*, vol. 2 (1789), contains an appendix of over 150 pages of integrals.

45. Landen, *PT* 65 (1775): 283–289, and *Memoirs* (1780), vol. 1, pp. iv, 23–36, vol. 2, pp. 24–25; Richelot, *Die Landensche Transformation* (1868), pp. 55–60.

46. Landen, *Residual Analyses* (1764); Boyer, *Concepts of the Calculus* (1939), pp. 236–237; Cajori, *Mathematical Notations* (1929), vol. 2, pp. 206, 214; Guicciardini, *Newtonian Calculus* (1989), pp. 85–88, 128.

47. Landen, *PT* 67 (1777): 266–270, and *Memoirs* (1780), vol. 2, pp. 65–82.

48. Landen, *PT* 75 (1785): 315–326, and on the continental mathematicians, pp. 312–313, 327–332; Landen, *Memoirs* (1780), vol. 1, p. iv; vol. 2, pp. 83–89; Hutton, *Dictionary* (1815), vol. 1, pp. 711–712; Landen, *Memoirs* (1780), vol. 2, pp. 50–57, correcting *Principia*, bk. 3, prop. 39.

49. Glenie, *PT* 67 (1777): 452, 457.

50. Glenie, *Doctrine of Universal Comparison* (1789), pp. 3–4, 31 (quote), and *Antecedental Calculus* (1793), which starts with *PT* 67 (1777): 450–457; Guicciardini, *Newtonian Calculus* (1989), p. 104. Compare Horsley, *Practical Mathematics* (1801), p. xli.

51. Glenie, in Masères, *SL*, vol. 4, pp. 335–336; *PT* 66 (1776): 73–91.

52. Masères, *SL*, vol. 4, p. xi (on Glenie), and *Dissertation on the Use of the Negative Sign* (1758), pp. i–iii, 29–34; *Dictionary of Scientific Biography*, vol. 10, pp. 158–159.

53. Masères, *PT* 67 (1777): 187–230, and *SL*, vol. 1, pp. 233–344.

54. Wallace, *Masères* (1919), p. 32, and Babbage in *Scriptores optici* (1823), p. 3.

55. Masères, *SL*, vol. 4, pp. ix–x; *Gentleman's Magazine* 72, no. 2 (1801): 998.

56. Howse, *Maskelyne* (1989), pp. 23–25, 33–35, 74–85, 110–112, 129–141.

57. Hornsby, *Observations . . . 1774 to 1798* (1932), and *Astronomical Observations* (1798), vol. 1, pp. i–ii (quote).

58. Hornsby, *PT* 61 (1771): 574 (quote); Nichols, *Literary Anecdotes* (1812), vol. 3, p. 703, vol. 8, pp. 232, 260; *Dictionary of Scientific Biography* 6:512–513, and *Dictionary of National Biography* 9:1266–1267. Compare Horsley, *PT* 59 (1769): 185, and *PT* 57 (1767): 183–184.

59. Horsley, *PT* 60 (1770): 418–423, 434–435 (quote); *PT* 61 (1771): 547–558. Compare Cantor (*Optics* [1983], pp. 53–59, 143), who may take Horsley more seriously than he deserves.

60. Atwood, *Treatise* (1784), pp. xiii (quote), 298.

61. Atwood, *Description* (1776), with a little arithmetic about radians (pp. 89 f.), and *Analysis* (1784), which sometimes uses the rule of three.

62. Atwood, *PT* 84 (1794): 129–136 (watches); *PT* 86 (1796): 56, 103, 105 (quote), 108–109, 124–125.

63. Maskelyne, *PT* 65 (1775): 500–542; Howse, *Maskelyne* (1989), pp. 134–141.

64. Hutton, *PT* 68 (1778): 690, 748–758, 779–783 (quote), reprinted in Hutton, *Mathematical and Philosophical Subjects* (1812), vol. 2, pp. 1–68. Of course, Hutton knew the earth is not a perfect sphere and may, like Horsley (*Remarks* [1774], p. 12), have doubted that it is an ellipsoid of revolution.

65. Shuckburgh, *PT* 67 (1777): 513–515, 518, 533–537 (quote), 547, 562–569. He reduced his rule to tables (pp. 571–591) and "to the capacity of such persons as are but little conversant with mathematical computations."

66. Hutton, in Masères, *SL*, vol. 1 (1791), pp. ii–iii, and in Masères, *Permutations and Combinations* (1795), p. 605 (quote); reprinted in Hutton, *Mathematical and Philosophical Subjects* (1812), vol. 1, pp. 459–485.

67. Hutton, *PT* 74 (1784): 21–34, and *Mathematical and Philosophical Subjects* (1812), vol. 2, pp. 121–133; Gregory, *Imperial Magazine* 5 (1823): 221.

68. Hutton, *Dictionary* (1815), vol. 2, pp. 22–23.

69. Pringle, *Discourses* (1783), pp. 268, 270 (quote); Wilson, in Robins, *Gunnery* (1805), p. vii.

70. Hutton, *PT* 68 (1778): 50–52, 75, 83, 85 (quote).

71. Glenie, *History of Gunnery* (1786), pp. vi (quote), 32–37, 43–59.

72. Hutton, *Mathematical and Philosophical Subjects* (1812), vol. 2, pp. 306–384, vol. 3, pp. 1–153.

73. Atwood, *Construction and Properties of Arches* (1801); Hutton, *Principles of Bridges* (1772), p. iv, in Hutton, *Mathematical and Philosophical Subjects* (1812), vol. 1, pp. 1–115, together with his consultative opinion (pp. 127–144).

74. Glenie, *Concise Observations* (1793), pp. 5–6.

75. Atwood, *PT* 71 (1781): 397, 414, 421–425.

76. Masères, *Proposal* (1783), pp. 33–68, and *Principles* (1783), pp. ix (quote), 1–90 (the principles of annuity calculations), 605–626 (text of bill).

77. Atwood, *Review* (1801), vol. 5, pp. 22–26; Shuckburgh, *PT* 88 (1798): 176.

78. Glenie, *Short Essay* (1785), pp. 40–45, and *Reply* (1785), p. 55; Lennox, *Answer* (1785), p. 41.

79. Hutton, *Mensuration* (1788), p. v; Maskelyne, *PT* 64 (1774): 158–170; Horsley, *PT* 64 (1774): 214–216.

80. Horsley, in Vincent, *Voyage* (1797), pp. 523–525; Shuckburgh, *PT* 88 (1798): 169.

81. Shuckburgh, *PT* 88 (1798): 133–139.

82. Horsley, in Vincent, *Voyage of Nearchus* (1797), pp. 505–522; Horsley, in Newton, *Opera* (1785), vol. 5, pp. 63–65; Manuel, *Newton* (1963), pp. 75, 88, 188.

83. Heilbron, *Electricity* (1979), pp. 9 17.

84. Donovan, *Osiris* 4 (1988): 215, 221–222, 226–228; Perrin, *Osiris* 4 (1988): 55–64, and *Isis* 81 (1990): 262, 265–267; Melhado, in Freudenthal, *Études* (1989), p. 116.

85. Lavoisier to Franklin, Feb. 1790, quoted by Donovan, *Osiris* 4 (1988): 227–228. Compare Levere, in Levere and Shea, *Nature* (1990), pp. 210, 218–219; Lundgren, in Frängsmyr, *Quantifying Spirit* (1990), pp. 245–246, 257–263.

86. Respectively, Home, *BJHS* 16 (1983): 251–259 (quote), and Garber, in Garber, *Beyond History of Science* (1990), pp. 159, 161 (quote).

87. For example, J. C. Fischer, Antoine Libes, and Friedrich Murhard. See Heilbron, in Rousseau and Porter, *Ferment of Knowledge* (1980), pp. 364–365.

88. Nicholson, *Introduction to Natural Philosophy* (1790), vol. 2, p. 112, repeated by Thomson, *System of Chemistry* (1802), vol. 1, pp. 2–3.

89. Nicholson, *Natural Philosophy* (1790), vol. 2, pp. 288–366. For what it's worth, Nicholson dedicated his work to Banks, as "president of that respectable body of men, among whom the true philosophy had its origin, and to whom it owes a great part of its improvements."

90. Gehler, *Handwörterbuch* (1799), s.v. "Physik," and Lichtenberg, in Erxleben, *Anfangsgründe*, 6th ed., quoted in Gilles, *Erxleben* (1978), pp. 12, 15; Fischer, *Physikalisches Wörterbuch*, vol. 3 (1800), pp. 889–890.

91. Fischer, in Gren, *Grundriss* (1808), p. xv.

92. Prevost, "Journal in Time," 9 June 1825, in Weiss, *Prevost* (1988), p. 89.

93. Young, *Quarterly Review*, Feb. 1809, pp. 109, 110; Biot to Blagden, 19 Jan. 1806 (RS, B.152): "C'est le seul exemple d'attractions soumises au calcul, dans les phénomènes terrestres."

94. See Lorenz, *NTM* 27, no. 1 (1990): 25.

95. See Cantor, *Optics* (1983), pp. 50–51, 76–77, 86–87.

96. The couplet is Pope's, "Duncaid," 4:265–266, with "mathematic" for "metaphysic." The technique of the mathematical fig leaf is laid bare in Heilbron, *Some Quantitative Science* (1992), chap. 3.

97. For details and literature, see Heilbron, *Some Quantitative Science* (1992), chap. 1.

98. Heilbron, *Some Quantitative Science* (1992), chap. 3; T. S. Kuhn, *Isis* 49 (1958): 142–150.

99. Shuckburgh, *PT* 69 (1779): 362 (quote), 374 ("these days of precision"), and *PT* 67 (1777): 513; Cavendish et al., *PT* 67 (1777): 816–857.

100. Feldman, *HSPS* 15, no. 2 (1985): 164–177; Maskelyne, *PT* 64 (1774): 159.

101. Dhombres-Firmas, *Journal de physique* 75 (1812): 275. The most advanced mathematics in Nicholson, *Natural Philosophy* (1790), is the logarithmic rule of hypsometry (vol. 2, pp. 36–48).

102. See Middleton, *Thermometer* (1966), pp. 127–129, and Aubert, *N&R* 9 (1951): 82–83.

103. Shuckburgh, *PT* 68 (1778): 682–683, and *PT* 69 (1779): 366, resp. Compare Shuckburgh, *PT* 88 (1798): 180–181, a very elaborate analysis of the probable errors in the graduation of a linear rule made by Troughton.

104. Roy, *PT* 67 (1777), esp. 689–695. Roy's allegiance during the discussions appears in a letter from Blagden to Banks, 23 Dec. 1783 (DTC, 3, 175); his relations with Maskelyne, in one from Widmalm, in Frängsmyr et al., *Quantifying Spirit* (1990), pp. 185–189.

105. See Feldman, *HSPS* 15, no. 2 (1985): 127, 195, also in Frängsmyr et al., *Quantifying Spirit* (1990), pp. 156–157; and Garber, in Garber, *Beyond History of Science* (1990), pp. 163, 165–166, 173.

106. Kuhn, *Essential Tension* (1977), pp. 31–36, 41–42, 61 (quote). Further references to this book will be given by page numbers inserted in the text.

107. Aristotle, *Physics*, II.2, 194a7 ("the more physical branches of mathematics, such as optics, harmonics, and astronomy"), to which *Metaphysics*, B.2, 997b26, and M.3, 1078a15, add geodesy and mechanics, respectively. These "mathematical sciences" (M.3, 1077a33) are intermediate between mathematics and physics. In his gloss in *Physics*, II.2, Aquinas calls them "intermediate sciences," in contrast to "purely mathematical sciences" (Aquinas, *Commentary* [1963], pp. 80, 164).

108. Dee, in Billingsley, *Euclid* (1570), A.i, A.iii.

109. Newton, in Ball, *Cambridge Papers* (1918), p. 245, and in Hall and Hall, *Unpublished Papers* (1962), p. 370.

110. *Encyclopedia Britannica* (1771), vol. 3, p. 30.

111. Hutton, "Advertisement," 14 Apr. 1760, in Bruce, *Hutton* (1823), p. 50, and *Course of Mathematics* (1880), vol. 2, p. 244 (quote). Compare Playfair, *Lectures on Some of the Practical Parts of Mathematics* (1793): astronomy, geography, geodesy, gnomonics, navigation, gunnery, fortification.

112. Murhard, *Litteratur* (1797), vol. 1, p. xii (quote).

113. Grattan-Guiness, in Jahnke and Otte, *Problems* (1981), p. 349.

114. Frängsmyr et al., *Quantifying Spirit* (1990).

115. Kuhn, *Essential Tension* (1977), pp. xi–xiii, where Kuhn talks of the beginning of his "enlightenment" in 1947.

Bibliography

The following abbreviations are used here and in the notes:

BJHS: *British Journal for the History of Science.*
DTC: Dawson Turner Copies (of Banks's correspondence), Botany Library, British Museum (Natural History). Cited by bound volume and page.
HSPS: *Historical Studies in the Physical and Biological Sciences.*
N&R: *Notes and Records, Royal Society.*
PM: *Philosophical Magazine.*
PT: *Philosophical Transactions, Royal Society.*
RS: Royal Society of London. "RS, M.M." signifies "Miscellaneous manuscripts"; "RS, B.*x*," item B.*x* in the Charles Blagden Papers; both series held at the RS.
SL: Francis Maseres, *Scriptores logarithmici*, 6 vols. (London: J. Davis, 1791–1807).

Anon. 1701. *An Essay on the Usefulness of Mathematical Learning, in a Letter from a Gentleman in the City to His Friend in Oxford.* Oxford: A. Peisley.

Anon. 1784. *An Appeal to the Fellows of the Royal Society, Concerning the Measures Taken by Sir Joseph Banks, Their President, to Compel Dr. Hutton to Resign the Office of Secretary to the Society for Their Foreign Correspondence.* London: J. Debrett.

Anon. 1784. *An History of the Instances of Exclusion from the Royal Society, Which Were Not Suffered to Be Argued in the Course of the Late Debates, with Strictures on the Formation of the Council, and Other Instances of the Despotism of Sir Joseph Banks, the Present President, and of His Incapacity for His High Office, by Some Members of the Minority.* 2nd ed. London: J. Debrett.

Anon. 1784. *Supplement to the Appeal to the Fellows of the Royal Society, Being Letters Taken from the Public Advertiser and Morning Post.* London: J. Debrett.

Anon. [1785?] *Canons of Criticism, Extracted from the Beauties of Maty's Review, and the Oratorical Powers of Dr. Horsley.* New ed. London: J. Ridgeway.

Anon. 1801. "Sir Joseph Banks." In *Public Characters of 1800–1801*, pp. 370–401. London: R. Phillips.

Anon. 1823. "Memoir of Olinthus Gregory, LL.D., Professor of Mathematics, Royal Military Academy, Woolwich." *Imperial Magazine* 5, cols. 777–792.

Aquinas, Thomas. 1963. *Commentary on Aristotle's Physics.* Trans. Richard J. Blackwell et al. London: Routledge and Kegan Paul.

Archenholtz, J. von. 1784. *A Picture of England.* 2 vols. London: E. Jeffrey.

Atwood, George. 1776. *A Description of the Experiments, Intended to Illustrate a Course of Lectures, on the Principles of Natural Philosophy, Read in the Observatory at Trinity College, Cambridge*. London.

Atwood, George. 1781. "A General Theory for the Mensuration of the Angle Subtended by Two Objects, of Which One Is Observed by Rays after Two Reflections from Plane Surfaces, and the Other by Rays Coming Directly to the Spectator's Eye." *PT* 71:395–434.

Atwood, George. 1784. *An Analysis of a Course of Lectures on the Principles of Natural Philosophy, Read in the University of Cambridge*. London: T. Cadell.

Atwood, George. 1784. *A Treatise on the Rectilinear Motion and Rotation of Bodies*. Cambridge: J. & J. Merrill.

Atwood, George. 1794. "Investigations, Founded on the Theory of Motion, for Determining the Times of Vibration of Watch Balances." *PT* 84:119–168.

Atwood, George. 1796. "The Construction and Analysis of Geometrical Propositions, Determining the Positions Assumed by Homogeneal Bodies Which Float Freely, and at Rest, on a Fluid's Surface; Also Determining the Stability of Ships and of Other Floating Bodies." *PT* 86:46–130.

Atwood, George. 1801. *A Dissertation on the Construction and Properties of Arches*. London: W. Bullmer.

Atwood, George. 1801. *Review of the Statutes and Ordinances of Assize, Which Have Been Established in England from the Fourth Year of King John, 1202, to the Thirty-Seventh of His Present Majesty*. London: T. Egerton et al.

Aubert, Théodore. 1951. "Alexander Aubert, F.R.S., Astronome, 1730–1785." *N&R* 9:79–95.

Babbage, Charles, ed. 1823. *Scriptores optici; or, A Collection of Tracts Relating to Optics*. London: R. Wilks.

Ball, W. W. Rouse. 1918. "Isaac Newton on University Studies." In W. W. R. Ball, *Cambridge Papers*, pp. 244–251. London: Macmillan.

Boyer, Carl. 1939. *The Concepts of the Calculus: A Critical and Historical Discussion of the Derivative and the Integral*. New York: Columbia University Press.

Brougham, Henry. 1845–1846. *Lives of Men of Letters and Science Who Flourished in the Time of George III*. 2 vols. London: Colburn.

Brougham, Henry. 1855. *Lives of Philosophers of the Time of George III*. London and Glasgow: Griffin.

Bruce, John. 1823. *A Memoir of Charles Hutton*. Newcastle: T. and J. Hodgson.

Cajori, Florian. 1929. *A History of Mathematical Notations*. 2 vols. Chicago: Open Court.

Cameron, H. C. 1952. *Sir Joseph Banks, K.B., P.R.S.: The Autocrat of the Philosophes*. London: Batchworth.

Cantor, Geoffrey. 1983. *Optics after Newton. Theories of Light in Britain and Ireland, 1704–1840*. Manchester: Manchester University Press.

Carter, Harold B. 1987. *Sir Joseph Banks (1743–1820): A Guide to Biographical and Bibliographical Sources*. London: St. Paul's Bibliographies.

Cavendish, Henry, et al. 1777. "Report of the Committee Appointed by the Royal Society to Consider the Best Method of Adjusting the Fixed Points of Thermometers." *PT* 67:816–857.

Clerke, Henry. 1779. *A Dissertation on the Summation of Infinite Converging Series with Algebraic Divisors . . . , Translated from the Latin of A. M. Lorgna . . . , to Which is Added an Appendix Containing All the Most Elegant and Useful Formulae Which Have Been Investigated for the Summing of the Different Orders of Series*. London: for the author.

Crosland, Maurice. 1980. "Chemistry and the Chemical Revolution." In Rousseau and Porter, *Ferment of Knowledge*, pp. 389–418 (1980).

[Dawson, J.] 1769. *Four Propositions, Etc., Shewing, Not Only, That the Distance of the Sun, as Determined from the Theory of Gravity, by a Late Author, Is, Upon His Own Principles, Erroneous, but Also, That It is More Than Probable This Capital Question Can Never Be Satisfactorily Answered bu Any Calculus of the Kind*. Newcastle: W. Charnley.

Dawson, Warren R. 1958. *The Banks Letters: A Calendar of the Manuscript Correspondence of Sir Joseph Banks*. London: British Museum.

Dee, John. 1570. "Mathematicall Preface." In Henry Billingsley, trans., *The Elements of Geometry of the Most Auncient Philosopher Euclide of Megara*. London: I. Daye.

Deluc, Jean André. 1772. *Recherches sur les modifications de l'atmosphère*. 2 vols. Geneva.

Dhombres-Firmas, L. A. 1812. "Extrait [des Memoires sur la formule barometrique, par L. Ramond]." *Journal de physique* 75:253–275.

Donovan, Arthur. 1988. "Lavoisier and the Origins of Modern Chemistry." *Osiris* 4:214–231.

Erxleben, J. C. P. 1794. *Anfansgründe der Naturlehre*. 6th ed. Ed. G. C. Lichtenberg. Göttingen. 1st ed., 1772.

Feldman, Theodore S. 1985. "Applied Mathematics and the Quantification of Experimental Physics: The Example of Barometric Hypsometry." *HSPS* 15, no. 2: 127–197.

Feldman, Theodore S. 1990. "Late Enlightenment Meteorology." In Frängsmyr et al., *Quantifying Spirit*, pp. 143–177.

Fischer, J. C. 1798–1806. *Physikalisches Wörterbuch, oder Erklärung der vornehmsten zur Physik gehörigen Begriffe und Kunstwörter so wohl nach atomistischer als auch nach dynamischer Lehrart betrachtet*. 7 vols. Göttingen: J. C. Dietrich.

Fox, Robert. 1971. *The Caloric Theory of Gases from Lavoisier to Regnault*. Oxford: Oxford University Press.

Frängsmyr, Tore, J. L. Heilbron, and Robin E. Rider, eds. 1990. *The Quantifying Spirit in the Eighteenth Century*. Berkeley: University of California Press.

Garber, Elizabeth, ed. 1990. *Beyond History of Science: Essays in Honor of Robert E. Schofield*. Bethlehem: Lehigh University Press.

Garber, Elizabeth. 1990. "Siméon-Denis Poisson: Mathematics versus Physics in Early Nineteenth-Century France." In Garber, *Beyond History of Science*, pp. 156–176.

Gehler, Johann Samuel Traugott. 1799–1801. *Physikalisches Wörterbuch*. 6 vols. Leipzig: Schwickert.

Gilles, Bernhard. 1978. *J. Ch. P. Erxlebens "Anfansgründe der Naturlehre" als Spiegelbild der physikalischen Wissenschaft im letzten Viertel des 18. Jahrhunderts*. Thesis, University of Mainz.

Glenie, James. 1766. "Propositions Selected from a Paper on the Division of Right Lines, Surfaces, and Solids." *PT* 66, no. 1: 73–91.

Glenie, James. 1777. "The General Mathematical Laws Which Regulate and Extend Proportion Universally; or, A Method of Comparing Magnitudes of Any Kind Together, in all the Possible Degrees of Increase and Decrease." *PT* 67 (1777): 450–457.

Glenie, James. 1785. *A Reply to the Answer to a Short Essay on the Modes of Defence Best Adapted to the Situation and Circumstances of This Island, Etc., in a Letter to His Grace the Duke of Richmond*. London: G. and T. Wilkie.

[Glenie, James.] 1785. *A Short Essay on the Modes of Defence Best Adapted to the Situation and Circumstances of This Island*. 2nd ed. London: G. and T. Wilkie.

Glenie, James. 1786. *The History of Gunnery, with a New Method of Deriving the Theory of Projectiles in Vacuo from the Properties of the Square and Rhombus*. Edinburgh: J. Balfour; London: T. Cadell and J. Nourse.

Glenie, James. 1789. *The Doctrine of Universal Comparison, or general proportion*. . . . London: G. and J. Robinson.

Glenie, James. 1791. "A Problem Concerning the Construction of a Triangle, by Means of a Circle Only, though the Properties of Its Sides Leads to an Equation of the Tenth Order." In Maseres, *SL*, vol. 4, pp. 335–338.

Glenie, James. 1793. *The Antecedental Calculus; or, A Geometrical Method of Reasoning, without Any Consideration of Motion or Velocity Applicable to Every Purpose, to Which Fluxions Have Been or Can Be Applied*. London: G. and J. Robinson.

Glenie, James. 1793. *A Few Concise Observations on Construction*. London.

Grattan-Guiness, Ivor. 1981. "Mathematical Physics in France, 1800–1835." In H. N. Jahnke and M. Otte, eds., *Epistemological and Social Problems of the Sciences in the Early Nineteenth Century*, pp. 349–370. Dordrecht: Reidel.

Green, H. G., and H. J. J. Winter. 1944. "John Landen, F.R.S." *Isis* 35:6–10.

Gregory, Olinthus. 1823. "Brief Memoir of the Life and Writings of Charles Hutton, LL.D." *Imperial Magazine* 5, cols. 201–227.

[Gregory, Olinthus.] 1820. "A Review of Some Leading Points in the Official Character and Proceedings of the Late President of the Royal Society." *PM* 56: 161–174, 241–257.

Gren, F. A. C. 1808. *Grundriss der Naturlehre*. 5th ed. Ed. E. G. Fischer. Halle: Hemmerde und Schwetschke.

Griffin, William P. 1983. "Priestley in London." *N&R* 38:1–16.

Guicciardini, Niccolò. 1989. *The Development of the Newtonian Calculus in Britain, 1700–1800*. Cambridge: Cambridge University Press.

Hall, A. R., and Marie Boas Hall. 1962. *Unpublished Scientific Papers of Isaac Newton*. Cambridge: Cambridge University Press.

Hall, Marie Boas. 1984. *All Scientists Now: The Royal Society in the Nineteenth Century*. Cambridge: Cambridge University Press.

Heilbron, J. L. 1979. *Electricity in the Seventeenth and Eighteenth centuries*. Berkeley: University of California Press.

Heilbron, J. L. 1980. "Experimental Natural Philosophy." In Rousseau and Porter, *Ferment of Knowledge*, pp. 357–387.

Heilbron, J. L. 1992. *Some Quantitative Science around 1800*. Berkeley: Office for History of Science and Technology, University of California, Berkeley.

Home, R. W. 1983. "Poisson's Memoirs on Electricity: Academic Politics and a New Style in Physics." *British Journal for the History of Science* 16:239–259.

Hornsby, Thomas. 1771. *A proposal to purchase astronomical instruments*. Oxford.

Hornsby, Thomas. 1771. "The Quantity of the Sun's Parallax, as Deduced from the Observations of the Transits of Venus, on June 3, 1769." *PT* 61:574–579.

[Hornsby, Thomas, ed.] 1798. *Astronomical Observations, Made at the Royal Observatory at Greenwich, from the year MDCCL to the year MDCCLXII, by the Rev. James Bradley*. 3 vols. in 1. Oxford: Oxford University Press.

Hornsby, Thomas. 1932. *Observations Made with the Transit Instrument and Quadrant at the Radcliffe Observatory . . . , 1774–1798*. London: H. Milford and Oxford University Press.

Horsley, Samuel. 1767. "A Computation of the Distance of the Sun from the Earth." *PT* 57:179–185.

Horsley, Samuel. 1768. *The True Notion of Centripetal and Centrifugal Forces*. n.p.

Horsley, Samuel. 1769. "Venus Observed upon the Sun at Oxford, June 3, 1769." *PT* 59:183–188.

Horsley, Samuel. 1770. "Difficulties in the Newtonian Theory of Light, Considered and Removed." *PT* 60:417–440.

J. L. Heilbron

Horsley, Samuel. 1771. "A Supplement to a Former Paper Concerning Difficulties in the Newtonian Theory of Light." *PT* 61:547–558.

Horsley, Samuel. 1774. "M. de Luc's Rules for the Measurement of Heights by the Barometer, Compared with Theory and Reduced to English Measures of Length and Adapted to Fahrenheit's Scale of the Thermometer; with Tables and Precepts for Expediting the Practical Application of Them . . . , addressed to Sir John Pringle." *PT* 64:214–301.

Horsley, Samuel. 1774. *Remarks on the Observations Made in the Last Voyage Towards the North Pole for Discovering the Acceleration of the Pendulum in Latitude 79°50'.* London: B. White and P. Elmsley.

Horsley, Samuel. 1775. "De polygonis areâ vel perimetro maximis et minimis, inscriptis circulo, vel circulum circumscribentibus." *PT* 65:301–310.

Horsley, Samuel. 1791. *Speech . . . on the Second Reading of the Bill for the Relief of Papists.* London.

Horsley, Samuel. 1797. "On the Rising of Constellations." In Vincent, *Voyage of Nearchus*, pp. 505–522.

Horsley, Samuel. 1797. "On the Small Stadium of Aristotle." In Vincent, *Voyage of Nearchus*, pp. 523–525.

Horsley, Samuel. 1801. *Elementary Treatises on the Fundamental Principles of Practical Mathematics.* Oxford: Oxford University Press.

Horsley, Samuel. 1813. *The Speeches in Parliament.* Dundee: J. Chalmers et al.

Horsley, Samuel, et al. 1784. *An Authentic Narrative of the Discussions and Debates in the Royal Society, Containing the Speeches at Large of Dr. Horsley, Dr. Maskelyne, Mr. Maseres, Mr. Poore, Mr. Glenie, Mr. Watson, and Mr. Maty.* London: J. Debrett.

Howse, Derek. 1989. *Nevil Maskelyne, the Seaman's Astronomer.* Cambridge: Cambridge University Press.

Hutton, Charles. 1772. *The Principles of Bridges, Containing the Mathematical Demonstrations of the Properties of Arches, the Thickness of the Piers, the Force of the Water against Them, Etc.* Newcastle: T. Saint.

Hutton, Charles. 1775. *Miscellanea mathematica: Consisting of a Large Collection of Curious Mathematical Problems and Their Solutions.* London: G. Robinson and R. Baldwin.

Hutton, Charles. 1778. "An Account of Calculations Made from the Survey Measures Taken at Schehallien, in Order to Ascertain the Mean Density of the Earth." *PT* 68:689–788.

Hutton, Charles. 1778. "The Forces of Fired Gunpowder and the Initial Velocities of Cannon Balls, Determined by Experiments, from Which Is Also Deduced the Relation of the Initial Velocity to the Weight of the Shot and the Quantity of the Powder." *PT* 68:50–85.

Hutton, Charles. 1784. "Project for a New Division of the Quadrant." *PT* 74:21–34.

Hutton, Charles. 1786. *Tracts Mathematical and Philosophical*. London: G. and J. Robinson.

Hutton, Charles. 1787. *Elements of Conic Sections, with Select Exercises in Various Branches of Mathematics and Philosophy*. London: J. Davis.

Hutton, Charles. 1788. *A Treatise on Mensuration, both in Theory and Practice*. 2nd ed. London: Robinson et al.

Hutton, Charles. 1791. "Preface and Introduction to His Mathematical Tables." In Maseres, *SL* 1 (1791): i–cxxi.

Hutton, Charles. 1795–1796. *A Mathematical and Philosphical Dictionary*. 2 vols. London: J. Davis.

Hutton, Charles. 1798, 1811. *A Course of Mathematics . . . Compared, and More Especially Designed, for the Use of the Gentlemen Cadets in the Royal Military Academy at Woolwich*. 3 vols. Vols. 1–2: London: G. and J. Robinson; vol. 3: London: Rivington et al.

Hutton Charles. 1801. *A Complete Treatise on Practical Arithmetic and Bookkeeping*. 6th ed. London: G. Robinson and R. Baldwin.

Hutton, Charles. 1812. *Tracts on Mathematical and Philosophical Subjects*. 3 vols. London: Rivington et al.

Hutton, Charles. 1815. *Philosophical and Mathematical Dictionary*. 2nd ed., 2 vols. London: the author. 1st ed., 1795.

[Hutton, Charles.] 1816. *A Catalogue of the Entire, Extensive, and Very Rare Mathematical Library of C. Hutton . . . , Together with Some Very Curious Mathematical Instruments, Formerly the Property of the Celebrated Dr. Franklin, Which Will Be Sold at Auction by Leigh and Sotheby, June 11th, 1816*. London: Leigh and Sotheby.

Jebb, Heneage Horsley. 1909. *A Great Bishop of One Hundred Years Ago, Being a Sketch of the Life of Samuel Horsley, LL.D*. London: Arnold.

Jebb, John. 1787. *Works*. Ed. John Disney. 3 vols. London: T. Cadell et al.

Kippis, Andrew. 1784. *Observations on the Late Contests in the Royal Society*. London: G. Robinson.

Kuhn, Thomas S. 1958. "The Caloric Theory of Adiabatic Compression." *Isis* 49:132–140.

Kuhn, Thomas S. 1977. *The Essential Tension: Selected Studies in Scientific Tradition and Change*. Chicago: University of Chicago Press.

Landen, John. 1755. *Mathematical Lucubrations, Containing New Improvements in Various Branches of the Mathematics*. London: J. Nourse.

Landen, John. 1764. *Residual Analysis*. London: the author.

Landen, John. 1770. "Some New Theorems for Computing the Areas of Certain Curved Lines." *PT* 60:441–444.

Landen, John. 1771. *Animadversions on Dr. Stewart's Computation of the Sun's Distance from the Earth.* London: for the author.

Landen, John. 1775. "An Investigation of a General Theorem for Finding the Length of Any Arc of Any Conic Hyperbola by Means of Two Elliptic Arcs." *PT* 65:283–289.

Landen, John. 1777. "A New Theory of the Rotatory Motion of Bodies Affected by Forces Disturbing Such Motion." *PT* 67:266–295.

Landen, John. 1780, 1789. *Mathematical Memoirs Respecting a Variety of Subjects.* 2 vols. London: for the author.

Landen, John. 1785. "On the Rotatory Motion of a Body of Any Form Whatever, Revolving, without Restraint, about Any Axis Through Its Center of Gravity." *PT* 75:311–332.

La Rochefoucauld, F. F. de. 1933. *A Frenchman in England, 1784.* Trans. S. C. Roberts. Cambridge: Cambridge University Press.

Lemay, Pierre, and Ralph E. Oesper. 1948. "Pierre Louis Dulong: His Life and Work." *Chymia* 1:171–190.

[Lennox, Charles.] 1785. *An Answer to "A Short Essay on the Modes of Defence Best Adapted to the Situation and Circumstances of This Island."* London: J. Almon.

Levere, Trevor. H. 1990. "Lavoisier: Language. Instruments, and the Chemical Revolution." In T. H. Levere and W. B. Shea, eds., *Nature, Experiment, and the Sciences*, pp. 207–223. Dordrecht, Boston: Kluwer.

Lorenz, Martina. 1990. "Rezeption und Bewertung des Newtonschen Gravitationgezetzes in Physikbüchern im Zeitalter der Aufklärung in Deutschland." *NTM: Schriftenreihe für Geschichte der Naturwissenschaften, Technik, und Medizin* 27, no. 1: 25–39.

Lundgren, Anders. 1990. "The Changing Role of Numbers in Eighteenth-Century Chemistry." In Frängsmyr et al., *Quantifying Spirit*, pp. 245–266 (1990).

Lyons, Henry. 1944. *The Royal Society, 1660–1940: A History of Its Administration under Its Charters.* Cambridge: Cambridge University Press.

Lysaght, Averil M. 1971. *Joseph Banks in Newfoundland and Labrador, 1766: His Diary, Manuscripts and Collections.* Berkeley: University of California Press.

McCormmach, Russell. 1990. "Henry Cavendish on the Proper Method of Rectifying Abuses." In Garber, *Beyond History of Science*, pp. 35–51.

Manuel, Frank E. 1963. *Isaac Newton, Historian.* Cambridge: Harvard University Press.

Masères, Francis. 1758. *A Dissertation on the Use of the Negative Sign in Algebra . . . Shewing How Quadratic and Cubic Equations May Be Explained, without the Consideration of Negative Roots.* London: Samuel Richardson.

Masères, Francis. 1777. "A Method of Finding the Value of an Infinite Series of Decreasing Quantities of a Certain Form, When It Converges Too Slowly to Be Summed in the Common Way." *PT* 67:187–230.

A Mathematicians' Mutiny, with Morals

Masères, Francis. 1783. *The Principles of the Doctrine of Life-Annuities, Explained in a Familiar Manner So As to Be Intelligible to Persons Not Acquainted with the Doctrine of Chances and Accompanied with a Variety of New Tables . . . Accurately Compiled from Observations.* London: B. White.

Masères, Francis. 1783. *A Proposal for Establishing Life-Annuities in Parishes for the Benefit of the Industrious Poor.* London: B. White.

[Masères, Francis.] 1791. *The Moderate Reformer; or, A Proposal to Correct Some Abuses in the Present Establishment of the Church of England.* London: G. Stafford.

Masères, Francis, comp. 1791–1807. *Scriptores logarithmici; or, A Collection of Several Curious Tracts on the Nature and Construction of Logarithms, Mentioned in Dr. Hutton's Historical Introduction to His New Edition of Sherwin's Mathematical Tables.* 6 vols. London: J. Davis.

Masères, Francis, comp. 1795. *The Doctrine of Permutations and Combinations, . . . Together with Some Other Useful Mathematical Tracts.* London: B. & J. White.

Maskelyne, Nevil. 1774. "M. de Luc's Rule for Measuring Heights by the Barometer, Reduced to the English Measure of Length and Adapted to Fahrenheit's Thermometer and Other Scales of Heat and Reduced to a More Convenient Expression." *PT* 64:158–170.

Maskelyne, Nevil. 1775. "An Account of Observations Made on the Mountain Schehallien for Finding Its Attraction." *PT* 65:500–542.

Maty, Paul Henry. 1777. Letter to *Gentleman's Magazine*, 22 Oct. *Gentleman's Magazine* 47:466–468.

Maty, Paul Henry. 1788. *Sermons Preached in the British Ambassador's Chapel at Paris in the Years 1774, 1775, 1776.* London: T. Cadell et al.

Melhado, Evan M. 1989. "Metzger, Kuhn, and Eighteenth-Century Disciplinary History." In Gad Freudenthal, ed., *Études sur Hélène Metzger*, pp. 111–134. Paris: Fayard.

Middleton, W. E. Knowles. 1966. *A History of the Thermometer and Its Use in Meteorology.* Baltimore: Johns Hopkins University Press.

Miller, John Philip. 1983. "Between Hostile Camps: Sir Humphry Davy's Presidency of the Royal Society of London, 1820–1827." *British Journal for the History of Science* 16:1–47.

Murhard, F. W. A. 1797–1805. *Litteratur der mathematischen Wissenschaften: Bibliotheca mathematica.* 5 vols. Leipzig: Breitkopf and Härtel.

Newton, Isaac. 1779–1785. *Opera quae extant omnia; commentariis illustrabat Samuel Horsley.* 5 vols. London: J. Nichols.

Nichols, John. 1812. *Literary Anecdotes of the Eighteenth Century.* 6 vols. London: Nichols.

Nicholson, William. 1790. *An Introduction to Natural Philosophy.* 3rd ed., 2 vols. London: J. Johnson.

O'Brian, Patrick. 1987. *Joseph Banks: A Life*. London: Collins Harvill.

Perrin, Carleton E. 1988. "Research Traditions, Lavoisier, and the Chemical Revolution." *Osiris* 4:53–81.

Perrin, Carleton E. 1990. "Chemistry as Peer of Physics: A Response to Donovan and Melhado on Lavoisier." *Isis* 81:259–270.

Pindar, Peter. 1789. *Peter's Prophecy; or, The President and the Poet; or, An Important Epistle to Sir Jos. Banks, on the Approaching Election of a President of the Royal Society*. Dublin: Colles et al.

Pindar, Peter. 1794. *Works*. 3 vols. London: John Walker.

Playfair, John. 1793. *Prospectus of a Course of Lectures on Some of the Practical Parts of the Mathematics*. Edinburgh.

[Playfair, John.] 1804. Review of Samuel Horsley's *Èuclidis elementorum libri priores XII* (Oxford: 1802) and *Euclidis datorum liber* (Oxford: Oxford University Press, 1803). *Edinburgh Review* 4:257–272.

Playfair, John. 1812, 1814. *Outlines of Natural Philosophy, Being Heads of Lectures Delivered in the University of Edinburgh*. 2 vols. Edinburgh: A. Constable; London: Longman et al.

Pringle, John. 1783. *Six Discourses, Delivered . . . on Occasion of Six Annual Assignments of George Copley's Medal, to Which Is Prefaced the Life of the Author, by Andrew Kippis*. London: W. Strahan and T. Cadell.

Richelot, F. J. 1868. *Die Landensche Transformation in ihrer Anwendung auf die Entwicklung der elliptischen Funktionen*. Königsberg: Hübner and Matz.

Robins, Benjamin. 1805. *New Principles of Gunnery, Containing the Determination of the Force of Gunpowder*. Ed. Charles Hutton. London: F. Windgrave.

Rouse, William. 1816. *An Investigation of the Errors of All Writers on Annuities . . . , Including Those of Sir Isaac Newton, Demoivre, Dr. Price, Mr. Morgan, Dr. Hutton, Etc., Etc.* London: for the author.

Rousseau, G. S., and Roy Porter, eds. 1980. *The Ferment Of Knowledge: Studies in the Historiography of Eighteenth-Century Science*. Cambridge: Cambridge University Press.

Roy, William. 1777. "Experiments and Observations Made in Britain in Order to Obtain a Rule for Measuring Heights with the Barometer." *PT* 67:653–788.

Shuckburgh-Evelyn, G. A. W. 1777. "Observations Made in Savoy in Order to Ascertain the Height of Mountains by Means of the Barometer, Being an Examination of Mr. de Luc's Rules, Delivered in His *Recherches sur les modifications de l'atmosphère*." *PT* 67:513–597.

Shuckburgh-Evelyn, G. A. W. 1778. "Comparison between George Shuckburgh's and Colonel Roy's Rules for the Measurements of Heights with the Barometer." *PT* 68:681–688.

Shuckburgh-Evelyn, G. A. W. 1779. "On the Variation of the Temperature of Boiling Water." *PT* 69:362–375.

Shuckburgh-Evelyn, G. A. W. 1793. "Account of the Equatorial Instrument." *PT* 83:67–128.

Shuckburgh-Evelyn, G. A. W. 1798. "An Account of Some Endeavors to Ascertain a Standard of Weight and Measures." *PT* 88:133–182.

Snip, Simon, pseud. 1663 [1784?]. *The Philosophical Puppet Show, or Snip's Inauguration to the President's Choir, Addressed to Sir J____ B____, Celebrated Connoisseur of Chickweed, Caterpillars, Black Beetles, Butterflies, and Cockle-Shells.* London: Wm. Green.

Stewart, Matthew. 1763. *The Distance of the Sun from the Earth Determined by the Theory of Gravity.* Edinburgh: A. Kincaid and J. Bell; London: A. Millar et al.

Thomson, Thomas. 1802. *A System of Chemistry.* 4 vols. Edinburgh: Bell and Bradfute et al.

Vales, Robert L. 1973. *Peter Pindar (John Wolcot).* New York: Twayne.

Vincent, William. 1797. *The Voyage of Nearchus from the Indes to the Euphrates . . . , to Which Are Added [Two] Dissertations . . . by Samuel Horsley.* London: T. Cadell.

Wallace, W. Stewart, ed. 1919. *The Masères Letters, 1766–1768.* Toronto: University of Toronto Library. Also in *University of Toronto Studies: History and Economics* 3, no. 2 (1919).

Weld, Charles Richard. 1848. *A History of the Royal Society, with Memoirs of the Presidents.* 2 vols. London: Parker.

Whiteside, D. T., ed. 1967–1981. *The Mathematical Papers of Isaac Newton.* 8 vols. Cambridge: Cambridge University Press.

Widmalm, Sven. 1990. "Accuracy, Rhetoric, and Technology: The Paris-Greenwich Triangulation, 1784–88." In Frängsmyr et al., *Quantifying spirit*, pp. 179–206.

Wilson, James. 1805. "Preface." In Robins, *Principles of Gunnery*, pp. v–xlvi.

Windischmann, Karl Joseph. 1802. "Grundzüge zu einer Darstellung des Begriffs der Physik und der Verhältnisse dieser Wissenschaft zur gegenwärtigen Lage der Natur- kunde." *Neue Zeitschrift für spekulative Physik* 1, no. 1. 78–160.

Winter, H. J. J. 1944. *John Landen, F. R. S., Mathematician and Land Agent.* Peterborough: Natural Historical, Scientific, and Archeological Society.

Science and Humanism in the Renaissance: Regiomontanus's Oration on the Dignity and Utility of the Mathematical Sciences

N. M. Swerdlow

The theorems of Euclid have the same certainty today as a thousand years ago. The discoveries of Archimedes will instill no less admiration in men to come after a thousand centuries than the delight instilled by our own reading.

In April of 1464 Johannes Müller of Königsberg (1436–1476) in Franconia, generally known by the Latin form of his name as Regiomontanus, gave a series of lectures at the University of Padua on *De scientia stellarum,* a brief introduction to astronomy by the ninth-century Arabic writer al-Farghānī that had been translated into Latin in the twelfth century. This in itself was an ambitious undertaking because Farghānī's treatise was far more comprehensive and required far more elucidation than the elementary *De sphaera* of Sacrobosco or the *Theorica planetarum,* attributed to Gerard of Cremona, that made up the common fare of university lecturing on astronomy. Indeed, in selecting this text, Regiomontanus showed that he was prepared to give a course encompassing chronology, geography, spherical astronomy, planetary theory, eclipses, and some of the astronomical parts of astrology. But he was well prepared for this as only a year or so earlier he had completed the *Epitome of the Almagest,* a thorough and masterful exposition of Ptolemy's astronomy that for the first time made the difficulties of the work comprehensible to anyone in Europe who wished to study astronomy in a serious way. And although only twenty-seven years old at the time of the lectures, he was already the most learned and proficient astronomer and mathematician of his age.

Regiomontanus's lectures are unfortunately lost except for an inaugural oration on the mathematical sciences in general, on their dignity and utility. Such inaugural orations before a course of lectures were standard fare among humanists in the universities of Italy in the fifteenth century, and had antecedents in the preceding two centuries. They consisted in essence of frequently ostentatious advertisements for the dignity of the subject of the lectures and of its superiority to other branches of learning. In principle they were attended by the entire faculty of the university, or at least the faculty of arts and medicine, and a large number of students, and after the professor had eloquently extolled his subject, those that were sufficiently enticed could return for the remaining lectures while the rest would go their own way. Thus, if the course were on, say, Horace, the virtues of poetry would be praised, if on Cicero, the virtues of rhetoric, if on Lucan or Caesar, the virtues of history, and in each case it would be shown that this particular subject above all was the best guide to eloquence or good conduct or political acumen or wisdom or whatever admirable virtue an educated gentleman should aspire to attain. But mathematics was something else. Although humanists were beginning to make an important contribution to mathematics in the discovery and translation of ancient texts, it was not really a subject associated with the *studia humanitatis* in the universities, and could not be said to lead to the virtues instilled by grammar, rhetoric, poetry, history, and moral philosophy. Rather, in the university it was more a relic of medieval instruction, treating for the most part very elementary texts, and characterized by a rather tedious, scholastic sort of exposition, that is, distinctions of parts, definitions of terms, and such, which it must be confessed was also not entirely absent from humanist teaching.

Regiomontanus wished to do something different, something new and important, and his intentions are set out in the inaugural oration. He wished, first of all, to bring the mathematical sciences into, or at least have them recognized as equal—if not superior—to humanistic studies, and to show that they too could contribute to the most highly sought virtue in every field of study: philosophy, the sciences, medicine, law, and theology. Indeed, he makes no concession in proclaiming mathematics in every way a more certain path to truth than scholastic philosophy, which he appears to despise, and a requisite part of the education of physicians, canon and civil lawyers, and theologians, not to mention architects and military engineers. Just as more than a century later Bacon looked to the sciences in general, so Regiomontanus looked to the mathematical sciences as nothing less than the principal instrument of progress in all knowledge and likewise in technology. And the oration also displays that most celebrated aspect of humanistic studies, namely, the public announcement of newly discovered texts, of Archimedes, Apollonius, and Diophantus, all of whom Regiomontanus was one of the first Europeans to read, and almost certainly the first to read with real understanding, for his excitement here is at least equal to that engendered in the humanist scholars of his age by the discovery of Greek historians, poets, and philosophers. Further, it is evident, both from the oration and from Regiomontanus's other writings, that his appreciation of the mathematical sciences was profound in a way far beyond that of his contemporaries, in a way that one associates more with the late sixteenth and early seventeenth centuries—Viète and Fermat, Galileo and Kepler come to mind—and I would not hesitate to say that he was the first European to display, although within the limitations of his more restricted knowledge, the intellectual and aesthetic sensibility of modern mathematics.

Finally, a very important point made in the oration is that one of the mathematical sciences is superior to all the rest—to geometry, arithmetic, music, optics, and several others—requires the greatest dedication to learn, and offers the greatest rewards, intellectual, spiritual, even material, and that is astronomy, particularly its judicial part, the continuation of religion by other means. "Through this angelic science," he says, "we are brought near to immortal God no less than through the other arts we are set apart from wild beasts." Perhaps this is to be expected since the following lectures are to be on an astronomical text also pertinent to astrology. Yet Regiomontanus is here expressing, not only his own opinion, but that of the most learned and enlightened of his contemporaries, particularly among those humanists who took any interest at all in the sciences. A nearly contemporary humanist, Gregorio Tifernate (1419–1466), a poet and teacher of Greek and rhetoric, wrote a charming *Oratio de astrologia* in which he defends astrology against its critics, acclaims its distinguished practitioners—Egyptian priests, Moses and the Hebrews (the tabernacle was adorned with the signs of the zodiac), Greek poets, philosophers, and astronomers—and best of all, interprets classical myths astrologically, e.g., when Mars was discovered in adultery with Venus by Vulcan, it means that the planets Mars and Venus have come into conjunction near the star of Vulcan. One of the most striking lessons of Regiomontanus's oration is that astrology is not only one of the mathematical sciences, but is fully coequal with astronomy and absolutely supreme among them.

But what is perhaps of the highest interest to the history of science is that the oration shows with great clarity how the mathematical sciences and their value both pure and applied were seen by the one person who best understood them in the fifteenth century. It is as though one had a general appraisal of the mathematical sciences written in the second century by Ptolemy or the seventeenth century by Newton. Now this brings to mind an ingenious paper written by Professor Kuhn some years ago called "Mathematical versus Experimental Traditions in the Development of Physical Science," in which a distinction is made between the development of the "classical" or mathematical sciences—although astrology was understandably omitted—and the "Baconian" or experimental sciences in the scientific revolution of the seventeenth and eighteenth centuries. For in this period it was the mathematical sciences, cultivated since antiquity, that were transformed from what they had been in antiquity and the middle ages to something approaching their modern form—and astrology lost its respectability—while the experimental sciences, which barely existed at the beginning of the seventeenth century, were undergoing an entirely different process of what really amounted to gestation, and what we commonly call the Scientific Revolution, in the physical sciences at least, consists of these two for the most part separate traditions with very different histories. It occurred to me that Regiomontanus's oration, aside from its many other merits, might offer some insight into the state of the mathematical sciences and their position within the world of learning at what could be the earliest plausible date for the beginning of the Scientific Revolution.

Nevertheless, I will be the first to admit that the oration may really be a century and a half too early, which is itself a historiographical issue that should be addressed. Does the oration, for all its prescience, belong to the scientific revolution, and if not, then where? The question of when the scientific revolution began, and when its foundations were laid, which is not the same question, may be answered differently depending upon which sciences are examined. I believe we can all now agree that one should not pick out a single event, such as the heliocentric theory, and say there, that is it. But I hope we are also willing to agree that one should not hasten backwards through superficial similarities from Galileo to Buridan or Jordanus, and end up somewhere in the thirteenth century. The first suggestion is merely naive, the second is, I believe, pernicious. It is a late part of a historiographical tradition, at first having nothing to do with the history of science, originating in religious and political issues—Catholic, French, Italian—of the 1850s and 1860s in which the glorification of the middle ages and of scholasticism became a weapon on the side of illiberal and reactionary forces, politically, intellectually, and theologically, directed against liberalization of the Church and the secular states of Europe. The extraordinary demands of what was then called Ultramontanism, the doctrine of absolute papal authority in all jurisdictions both sacred and secular, the *Syllabus Containing the Principal Errors of Our Age* of Pius IX, with its eighty condemned propositions in effect anathematizing modern civilization and, not incidentally, maintaining the right of the Church to judge science and philosophy, and the establishment of Neo-scholasticism with its own condemnation of secular education and modern science are now mostly forgotten, but their historiographical legacy lives on, particularly in medieval studies. One consequence through the remainder of the century and beyond was a deluge of medieval scholarship, much of it of value, but another was the denigration of the Renaissance, identified as secular and liberal, as the crucial formative period in the history of modern Europe. At the same time northern European nationalism also encouraged study of the middle ages to show that the modern culture of the nations of northern Europe was independent of, and in some cases anterior to, the Renaissance. The fashion of medievalism in the late nineteenth century is too well known to require comment, and its motivation was not entirely aesthetic. Between the medievalism of the Church and the medievalism of German and French nationalism, the very idea of the Renaissance, especially Burckhardt's Italian Renaissance, was under attack, and it is not an exaggeration to say that by the 1920s and 1930s the principal defenders of Renaissance culture against the tide of medievalism were German Jews, concerned in particular with art history and humanism—one thinks above all of the Warburg—and Italians, some of whose motivation was less admirable.

A significant effect on the history of science came only in our own century through the influence of the pious and extremely conservative Pierre Duhem, who devoted his principal historical work once again to the glorification of the middle ages and scholasticism at the expense of the Renaissance and Leonardo, and of the early years of the seventeenth century and Galileo. His claims were extravagant and he did not mince words. In the condemnation by Bishop Etienne Tempier of Paris in 1277 of 219 erroneous propositions in philosophy and theology, a worthy ancestor in spirit of the *Syllabus,* Duhem saw nothing less than "the birthdate of modern science," which, he held, originated in the scholastic criticism of Aristotle and Averroes in the name of orthodoxy. "If therefore that science, of which we are so legitimately proud, could see birth," he wrote, "it was only because the Catholic Church was its midwife." And not only history, but also Duhem's formalist, anti-realist philosophy had the same motivation and the same consequence, a justification of the authority of the Church to judge science, something also shown historically in his polemical tract *To Save the Phenomena* with its startling claim that Cardinal Bellarmine, the patron saint of Ultramontanism, was a better philosopher than Galileo. As Arnaldo Momigliano frequently remarked, if we are going to understand history we must also understand historians. The influence of Duhem's work, even on those who have never read it, is still considerable, for the Renaissance is all but nonexistent as an independent period in the history of science, and endless controversy on much the same stubborn issues continues to rage around Galileo even in our own day. In the towering figure of Leonardo, however, more the province of art historians, he never succeeded in making a dent. It lies beyond my competence to comment on Duhem's influence on the philosophy of science—the philosophical issues are unresolvable in any case—but I would suggest that Momigliano's advice holds good there too, although it may be too much to expect that it will be followed in an age that canonizes Heidegger.

I believe the denigration of the Renaissance, or the absence of the Renaissance, in the history of science has been a grave mistake, as damaging to this particular field of study as the earlier belief that the middle ages were an undifferentiated dark age once was to the history of learning and art. Now is not the place to argue the point at length, and I do not wish to disparage the interest of the schoolmen within their own domain, provided that they are not read out of context and anachronistically as the progenitors of the scientific revolution, but to return to our principal subject, it seems to me as clear as day that Regiomontanus differs far more from Buridan and Oresme than from Kepler and Galileo. And I know that Kepler owed and acknowledged a debt to Regiomontanus while none was owed or acknowledged to the schoolmen, and the same can be said for Copernicus and Tycho. The description of the mathematical sciences in Regiomontanus's oration, although containing many traditional divisions and reputed origins familiar from ancient and medieval sources, is that of a humanist of the age of Nicholas V and Pius II, of that extraordinary renaissance of classical learning that marked the second half of the fifteenth century, and much the same spirit can be seen as late as Tycho and Kepler. Indeed, the oration, which is not itself science but about science, has more in common with Lord Bacon than with any philosophic or scientific writer of the middle ages, and the resemblance to Bacon in the age of Elizabeth and James—although not necessarily to what Bacon later came to represent—shows both a remarkable similarity in the reflections upon the sciences of two remarkable men and also how long was the reach of that very humanism and classical learning.

It may not be entirely impertinent to remark here that to my perception, the Renaissance must be taken as a distinct period in the history of science with its own characteristics, closely related to the humanism and scholarship of the age, and differing as much from the middle ages as from the more recent period conventionally, and correctly, called the Scientific Revolution. I believe this is evident in the oration, and is in fact the only way to make sense of the oration, that is, a characteristic humanist oration in praise of learning belonging as much to its age as an oration in praise of poetry or eloquence or the "good arts." And as such, it appears to have no predecessors, at least no surviving predecessors, in the history of science. Something like it could have been written a century later by Maurolico or Commandino, and perhaps under its influence Tycho wrote an *Oration on the Mathematical Sciences*, although devoted for the most part to the defense of astrology. But it could never have been written a century earlier by a schoolman, before the recovery of ancient science in its own language and before the recovery of ancient eloquence, both the accomplishment of humanists. The Renaissance of science, at least in the fields I know, extends from Regiomontanus to Kepler *and* Bacon, who repudiates scholasticism and praises science and learning and their applications much as Regiomontanus, although even more comprehensively and ambitiously. The similarities between Regiomontanus's oration and the preface to the *Rudolphine Tables,* even between the oration and the *Advancement of Learning,* are unmistakable. Together they mark the beginning and end of an alliance of science and humanism most notably characteristic of the Renaissance, after which each went its own way to the specific studies of the sciences and classical scholarship.

The background to the oration can be given briefly. In 1460 Cardinal Johannes Bessarion (1403–1472) was sent by Pius II as papal legate to Austria and Germany to enlist the aid of Friedrich III and the German princes for a crusade against the Turks to recapture Constantinople. Bessarion was then the outstanding Greek scholar and patron of Greek learning in Italy, and had assembled by far the largest collection of Greek manuscripts, many of the highest quality. In Vienna he met Regiomontanus and his former teacher and now colleague at the university Georg Peurbach (1423–1461). One of many projects that Bessarion wished to encourage was a new translation of the *Almagest*—a version he considered unsatisfactory had been made a few years earlier for Nicholas V by George of Trebizond—and also a textbook or abridgement based upon it. Peurbach, who according to Regiomontanus knew the *Almagest* almost by heart, although necessarily the Latin version of Gerard of Cremona, since he knew no Greek, was enlisted to carry out the latter, and he along with Regiomontanus were to go to Italy with Bessarion. While preparations for departure were under way, Peurbach died in April of 1461 after completing the first six books of the abridgement, and made Regiomontanus promise to complete the work. In the fall of that year Regiomontanus and Bessarion traveled to Rome, and there in the next year or so he wrote the following seven books, and I believe also revised Peurbach's original six books, of the *Epitome of the Almagest*, the completed work later being presented to Bessarion in a splendid manuscript (Venice, Mar. lat. 328). During this time he also learned Greek, and began his study of Greek mathematical texts in Bessarion's collection, among them Archimedes and Apollonius. Through Bessarion he made the acqaintance of humanists and learned men, some of them, such as Nicholas of Cusa, Leon Battista Alberti, and Paolo Toscanelli, of considerable scientific competence, even if nowhere near Regiomontanus's own level, and doubtless formed the progressive, humanist sensibility that pervades the inaugural oration. Likewise, the dismissal of scholastic philosophy *toto caelo* in the oration may also reflect the opinions of humanists around Bessarion, and perhaps of Bessarion himself who, as a Greek from Trebizond and a devotee of Plato, educated first in Constantinople and then under the worthy Gemistus Pletho, was always critical of Western scholasticism (although when advantageous he could make use of it for his own purposes).

In the summer of 1463 Bessarion was again appointed papal legate, this time to Venice to make further preparations for the crusade that Pius II intended to lead himself in the following year. Regiomontanus accompanied him, continuing his studies in Venice where he found a manuscript of Diophantus. At this time he carried on a correspondence on astronomy and mathematics with Giovanni Bianchini (ca. 1400–ca. 1470) of Ferrara, an industrious computer of astronomical tables and a veritable worshipper of Ptolemy for both his astronomy and astrology, which showed good taste. One of the letters, written from Venice in February or March of 1464, contains a devastating criticism of contemporary astronomy, represented principally by the then current *Alfonsine Tables*, all of which Regiomontanus wished to replace by an improved astronomy constructed on the foundations laid by Ptolemy and the few other writers he respected. Shortly afterwards Bessarion, who was to meet Pius II and the assembled armies of the crusade in Ancona, arranged for Regiomontanus to go to Padua, evidently to lecture at the university. This was obviously something that lay well within Bessarion's influence, since the preceding year he had arranged for Demetrius Chalcondyles to receive the first professorship of Greek studies at Padua, a position he held for nine years. There on 21 April 1464 Regiomontanus observed a total lunar eclipse, and perhaps even before that began the series of lectures on al-Farghānī for he remarks that he is about to speak unexpectedly (*repente*). The inaugural oration appears to have been delivered to the faculty and students of arts and medicine; physicians are directly addressed, but not jurists or theologians. At this time the number of the faculty of arts and medicine was as much as forty and of the students as much as eight hundred including law. How many attended the oration seems impossible to know. The lectures must have been finished by June at the latest, for from 26 June to 9 July in Venice Regiomontanus wrote several tracts on Nicholas of Cusa's faulty quadrature of the circle. On 15 August Pius II died in Ancona, which had become infested with plague, the crusade collapsed never to be revived, and Bessarion hastened to Rome to join the other cardinals in electing a new Pope. Regiomontanus accompanied his patron, for the scene of his amusing dialogue with Martin Ilkusch criticizing the old *Theorica planetarum* attributed to Gerard of Cremona is laid in Rome while awaiting the papal election.

Before considering the oration directly, there is one further point I wish to make, which is this. The oration contains a fair amount about the background of each of the sciences: where, by whom, and under what circumstances each was discovered, who were the foremost writers in each age and language, and who are the notable practitioners in Regiomontanus's own age. It looks like an early history of science, and I must confess that I once took it for that myself. Yet it is not, certainly not by intention, and if it appears so to us, it is only because we are unfamiliar with the differences between the genres of praise and history in the Renaissance. If Regiomontanus wished to write a history of the mathematical sciences, and I doubt whether it ever crossed his mind to do so, he would have written something altogether different, something characteristic of his age, almost certainly in the form of biographies of famous astronomers and mathematicians, however fanciful the information obtainable on Euclid or Archimedes or Ptolemy may have been. His exemplar would have been Diogenes Laertius's *Lives of the Philosophers*, translated about 1430 and already the prototype for a life of Aristotle by Leonardo Bruni and of Plato by Guarino of Verona, also written about 1430, and in the sixteenth century for Vasari's *Lives of the Artists* and, however quaint it may seem to us today, Bernardino Baldi's *Lives of the Mathematicians,* probably the first true history of science.

What Regiomontanus wrote was an oration in praise of the mathematical sciences, a species of *epideictic,* that is, *demonstrative,* oratory which is devoted to the praise or blame (*laus vel vituperatio*) of someone or something and showing why the subject of the oration is worthy of praise or blame. In a demonstration of excellence, it is proper to look back to the ancestry of the subject, and if it was illustrious to show that he is worthy of it, or if it was humble that he has flourished on his own, and likewise to look to his education to show that this too was honorable and worthy. The contrary for blame. Thus, when Regiomontanus tells us that Hipparchus was the founder of astronomy, Ptolemy the ruler, and Jābir ibn Aflah the corrector of Ptolemy, or when he says that according to some Jubal or Hermes discovered music, while he believes that Pythagoras, who examined the sounds of blacksmiths' hammers, has a better claim, he is not writing a history, but is proclaiming the worthiness of the ancestry of these sciences. And when he recounts the many contributions of Greeks, Indians, Arabs, and Latins, he is not so much investigating the transmission of knowledge as showing the genealogy, in a way the education, of the sciences practiced in his own time. This is not history, it is epideictic oratory, and the general characteristics of Regiomontanus's oration as oratory can easily be understood by reference to standard treatises like the *Rhetorica ad Herennium* and *De inventione.* But we shall leave such analysis to those who like to write about rhetoric (or is it now "discourse"?).

In what follows, I shall go through the oration, sometimes translating literally, sometimes paraphrasing or summarizing, without the distinction of quotation marks, along with whatever remarks seem appropriate by way of explanation or comment, although a detailed commentary is in no way intended. The text of the oration was published by Johann Schoener from Regiomontanus's papers in Nuremberg in 1537, presumably from a copy of the autograph which does not survive, along with Latin versions of al-Farghānī and al-Battānī, the latter from a manuscript annotated, although sparsely, by Regiomontanus. The lectures on al-Farghānī, if they were ever written down, have vanished without a trace.

An Inaugural Oration by Johannes Regiomontanus on all the Mathematical Sciences, Delivered in Padua When He Publicly Lectured on al-Farghānī

Regiomontanus's exordium is devoted to commending the good will of his learned audience for enduring his hastily prepared lecture. The style is grand, indeed, the most ornate of the entire oration, for example: "For my part, I who am unexpectedly about to speak appear to undertake no ordinary new trial—in this most notable city, in these highly esteemed and very justly celebrated seats of learning from which arises health for nearly all Christendom, before such learned gentlemen, before men accustomed to listen to profound sciences— inasmuch as I have disregarded the duty of such lecturing, not only for many days, but for more than two years." He is speaking boldly, he says, not out of desire for fame or ostentatious display or hope of gain, but at the request of his listeners and, of the highest importance, because the most illustrious ornament of the mathematical sciences, divine Astrology herself, has appointed him to declaim their virtues. Neither the pleasantness of Plato nor the eloquence of Cicero, even if they came back to life, would be sufficient to the beauty of this subject, so his listeners will understand that it is entirely due to their own indulgence if his praises are sufficiently fine.

I could, he continues, recall the origin of our sciences and their first cultivators, how they were transmitted from foreign languages to Latin, and who among our predecessors and contemporaries have achieved distinction. But first it seems appropriate to define the mathematical genus and its species so that the origin and progress of each one may be set out and suitably praised. "Mathematical" is defined as the science concerned with quantity. The name itself is Greek, for $\mu\alpha\nu\theta\acute{\alpha}\nu\omega$ in Greek means *disco* (I learn) in Latin, from which come $\mu\acute{\alpha}\theta\eta\sigma\iota\varsigma$ *disciplina* (instruction) and $\mu\alpha\theta\eta\mu\alpha\tau\iota\kappa\acute{\eta}$ *disciplinalis* (instructional, mathematical). The term *disciplina* is used consistently in the sense of "mathematical science," and we shall translate the word by "science." The words *doctrina* and *ars* are used in the same sense, although *ars* is sometimes best translated as "art" when the reference is to a "liberal art." Quantity itself, according to Aristotle, is of two kinds, continuous and discrete, and thus the most general division of the mathematical sciences is into geometry, which considers continuous quantity, and arithmetic, which considers the relations of numbers (*rationes numerorum*).

Geometry is so named, not just because it teaches the measuring of the earth—for γῆ means "earth" and μέτρον "measure"—for it also shows how other bodies are measured and the other attributes of continuous quantities. Nevertheless, it is pleasant to consider the origin of the name. It is not altogether clear how seriously Regiomontanus takes the following familiar story, which originated with Herodotus (2.109) and was elaborated by Diodorus, Strabo, Proclus, and many later writers. When the river Nile violently rose in flood and obliterated the boundaries of the fields of the Egyptians, the farmers would argue and attempt to claim more than was theirs. So the ruler of the country used some kind of just principles and accurate measures to restore to each his boundaries. And thus when men were compelled to perform measurements, they began to propose questions to each other, and whatever seemed to be correctly discovered in such exercises, they endeavored to commit to writing. And little by little it was done in the same way in neighboring countries.

Many such writings eventually came to the hands of Euclid "of Megara" (an old confusion), to which he added not a little by virtue of the acuteness of his intellect. And lest the enduring study of these ancient men be lost, he began to write down conclusions gathered from far and wide into thirteen books which he justly called the *Elements* because all sciences are based upon them, and to call the science of the whole volume "geometry" because it originated in connection with the measurements of the earth. Hypsicles in turn added two books treating the inscriptions and superscriptions of the five regular solids, and Boethius translated the whole into Latin, although without the commentary contained in the Greek text. Therefore Adelard, Alfred—an error for one of the Adelard of Bath versions in a Venice manuscript Regiomontanus had studied—and Campanus revised the fifteen books as though published by Euclid alone, the first two correctly and very briefly, the last very clearly.

Regiomontanus next turns to Archimedes and Apollonius, writers known to his listeners as names at best, as he himself had only just studied Archimedes in a manuscript of the translation made for Nicholas V by Jacobus Cremonensis and both Archimedes and Apollonius in Bessarion's Greek manuscripts. Which of the two is to be preferred to the other, I shall not easily say. For although Apollonius, on account of the greatness of his intellect usually called "divine," wrote *Elements of Conics* in eight books, yet the variety of books he published seems to have conferred pre-eminence upon Archimedes. Here Regiomontanus lists Archimedes's works and praises them, with a few remarks on illustrious men still awaiting the quadrature of the circle, for there was little more he could say of the greatest of ancient mathematicians in such a lecture, but as is known from his annotations in Archimedes, his understanding was very acute and he later intended to publish a corrected version of the translation of Jacobus Cremonensis. He mentions that some claim Archimedes wrote a *Mechanics* in which he described ingenious devices for use with weights, aqueducts, and other things, although he has not seen it, but he knows how Archimedes's mirrors are made. If any of you are interested in hidden things, you will soon be able to see everything just mentioned, and when Apollonius is translated from Greek into Latin, no one of you will not come to admire him.

After acknowledging Eutocius, Theodosius, and Menelaus, he takes up the other branch of pure mathematics, arithmetic or the theory of numbers, from the Greek ἀριθμός "number," which supposedly Pythagoras learned from the Egyptians and Arabians, and the entirely Arabic *algebra,* with Diophantus serving as an intermediary. Here the works of the greatest importance are the arithmetic books of Euclid, Jordanus's *Elements of Numbers* and *On Given Numbers,* and most important of all Diophantus, whose six surviving books Regiomontanus had himself discovered a few months earlier, as he had announced in a letter to Giovanni Bianchini in Ferrara. He had asked Bianchini whether the entire thirteen books might be in Ferrara, where there were scholars expert in Greek, since he would wish to translate the work, for which the Greek he has learned with Bessarion is sufficient. Other arithmetic or algebraic works mentioned are the *Quadripartitum numerorum,* presumably the work by Johannes de Muris, the *Algorismus demonstratus,* now attributed to Gerard of Brussels, the *Arithmetica* of Boethius, and a work in six books by Barlaam the Greek, the teacher of Petrarch, not yet translated into Latin.

He next considers the secondary mathematical sciences, sometimes called the "intermediate sciences," that we would call applied mathematics. These include astronomy, music, perspective (optics), and others less familiar such as the science of weights (statics), of aqueducts, and of the proportion of velocities in motions. This last is doubtless the scholastic study of motion, and it is of interest that after this brief allusion Regiomontanus says not a word about it, indicative, perhaps, of his lack of interest in scholastic science in general. Among all these, astronomy, like a pearl, far surpasses not only her sister intermediate sciences, but even geometry and arithmetic, the mothers of all sciences. On account of its great antiquity, we have not been able to discover the origin of astronomy, so that you would think it eternal or born together with the world. Stories of its origin that he does not take seriously are its discovery by Abraham, Moses, Prometheus, whose theft of divine fire was interpreted to mean that he delivered the light of astronomy to mortals, and Hercules, who bore the heavens on his shoulders in place of Atlas, meaning, of course, that he studied astronomy under Atlas. Instead he singles out Hipparchus, the progenitor of this science, for his discovery of the slow motion of the sphere of the fixed stars (the precession of the equinoxes) and Ptolemy, the authority and ruler, for confirming the rate of this motion at one degree per century. The brevity of this treatment of Hipparchus and Ptolemy is surprising since Regiomontanus certainly knew much more about their work from writing the *Epitome of the Almagest,* but again we must remember that this is not a history. He mentions Theon's commentary on the *Almagest* with the remark that it has not yet been translated with sufficient learning, a sly allusion to George of Trebizond's commentary on the *Almagest,* which had been partially plagiarized from Theon, whom George also attacked. Regiomontanus and Bessarion considered George of Trebizond, to put it very mildly, incompetent and dishonest, and Regiomontanus later wrote, and intended to publish, a work running to nearly six hundred pages in manuscript, surely one of the longest bad reviews ever written, defending Theon against George.

Many worthy witnesses demonstrate how highly proficient the Arabs were in this science. The Arabic writers mentioned are al-Battānī, translated by Plato of Tivoli, Jābir ibn Aflaḥ, translated by Gerard of Cremona, who Albertus Magnus called the "Corrector of Ptolemy" because in his preface he promises to correct thirteen of Ptolemy's errors (this is true), and al-Farghānī, the subject of the lectures to follow the oration. The most ancient Indians are commended for discovering the maximum declination of the sun (the obliquity of the ecliptic) to be 24 degrees, and the Persians for their astronomical tables, evidently a reference to the so-called *Persian Tables* translated into Greek by Gregory Chioniades.

At last astronomy reaches the Latins, and here Regiomontanus first singles out distinguished Paduans of whom there are three. First is Pietro d'Abano (d. 1318), whose large medical treatise, the *Conciliator,* the *Lucidator astronomiae,* and brief work on the motion of the eighth sphere are mentioned, all of which are not so much astronomical as astrological. Next is Giovanni de Dondi (14th cent.), whose *astrarium* (astronomical clock) is now kept by the Duke of Milan in the Castle of Pavia, to which countless prelates and princes resort as though about to behold a kind of miracle. Rejoice, therefore, O noble Paduans, to whom the studies of illustrious men have always been honorable. Finally Antonio de Monte Ulmi (late 14th cent.), another writer on astrology (and astrological necromancy), is mentioned briefly. English and French writers are acknowledged, but Spain and King Alfonso are not mentioned at all, and this is significant. Most of Regiomontanus's criticism of astronomy addressed to Bianchini was directed at the *Alfonsine Tables,* which Bianchini had used as the basis of his own tables, and I suspect that anything famous that Regiomontanus knows but leaves unmentioned does not meet with his approval.

Touching briefly on Albertus Magnus, whose works are already too well known to require comment, he praises at length Georg Peurbach, the everlasting ornament of the Germans whose knowledge of mathematics far surpassed all his contemporaries, a judgment that may well be true although hardly borne out by his surviving works, competent though they may be. Addressing his teacher in the second person, he says that not only the men of our nation, but also the French and Italians, have considered your name renowned. Some time ago you publicly taught astronomy in this celebrated city, and you are known to many present here both by fame and sight, evidence that Peurbach had himself earlier been in Padua. After enumerating the honors given to Peurbach by Ladislaus, the former king of Hungary (and a thorough scoundrel), by the Emperor Friedrich III, and by Nicholas of Cusa, he tells the story of how, following Peurbach's death—alas, at the unhappy memory of that event!—he came to Italy with Bessarion and was at last sent to Padua, while Bessarion is about to depart for Greece for the sake of the Christian religion. I hope you will grant indulgence by virtue of your kindness if I have seemed to digress a little too freely and have inappropriately described my travels.

Returning to the division of the sciences, the next to be considered is music. Stories of its discovery by Jubal (Gen. 4.21) and Hermes, when a shell resounded against his foot while he was walking on the shore, are passed over in favor of Pythagoras's measuring the weights of hammers used by blacksmiths and discovering the same proportions between the weights and the intervals of the sounds made by the hammers, a story known in Latin from Boethius's *De institutione musica* (1.10), and ubiquitous in medieval musical writers (there are also nice pictures of it). Regiomontanus says he knows only a few musical writers, mentioning Boethius, Johannes de Muris, and Ptolemy, whose work Nicholas of Cusa the previous summer wished to have translated into Latin. Either he does not know, or perhaps does not care for, the vast corpus of writing on harmonics from the thirteenth through fifteenth centuries. Finally, optics (*perspectiva*) is taken up, and here he singles out Ibn al-Haytham, our Thuringian Witelo, who has set out the subject *more geometrarum*, that is, in propositions with proofs, the writings of Euclid recently translated, and Roger Bacon, who is commended for his compendious *Perspectiva* (from the *Opus maius*) and little book on the burning mirror. And then Archimedes is recalled for his "philosophic" mirrors, specimens of which Regiomontanus promises to manufacture before long in Padua. Probably he is referring to a treatise on burning mirrors falsely ascribed to Archimedes.

This completes the first half or so of the oration which has consisted mostly of a review of the various branches of the mathematical sciences, their origins, and their outstanding works, as it were, the worthy ancestry of the subject of praise. The second half is devoted to praise and blame, praise of the sciences and their utility, censure of scholastic philosophy and fraudulent practitioners of astrology. First he recounts how much mechanics and artisans could learn from geometry if only they would study it. New vaults of churches have frequently collapsed because the architect chose an unsuitable form, and a tower in Venice recently buried some monks due to the builder's ignorance. A certain king would have attempted to connect the Red Sea and the Mediterranean except that a geometer warned him that Egypt would be submerged by a deluge. Bankers greatly increase their wealth by knowledge of computation. Armor bearers and soldiers use geometrical contrivances for hurling missiles and aiming siege engines, and could do well to learn the law of solar rays, meaning I suppose optics. What finally shall I recall of the makers of musical instruments, to whom I have so frequently pointed out their error in dividing measuring devices? Clearly Regiomontanus, like his friends Alberti and Toscanelli, has ideas about how mathematics could benefit the practical arts, a subject already developed to some degree in Italian treatises on practical geometry. Alberti of course had written on both architecture and the application of geometry to linear perspective in painting, and one may presume that Regiomontanus had some acquaintance with both subjects.

All of these are passed over briefly in order to show how useful mathematics is to liberal studies and especially Aristotelian philosophy, in which the University of Padua excelled. Nearly the whole of Aristotle's writings are fragrant with mathematical learning so that no one who has neglected the quadrivium may be considered capable of understanding Aristotle. The third book of the *Meteorology,* on the rainbow, cannot be understood without geometry and optics; the second and third books of *De caelo,* on the form and motions of the heavens, the earth, and the elements, require astronomy; and no one can follow the seventh book of the *Physics,* on motion and the comparison of motions, without knowledge of proportions. Referring to the review of the spheres of Eudoxus in the twelfth book of the *Metaphysics,* he asks, did it not appear difficult to Aristotle to demonstrate the nature of the celestial intelligences because he had not adequately studied astronomy? And did not Aristotle himself place the mathematical sciences in the first degree of certainty?

And a certain Academic, Plotinus, even said, "Would that all things were mathematical!" so disgusted was he with the other arts, which seem nothing more than a mass of discordant opinions. This is the beginning of a most startling criticism of speculative philosophy in general and scholasticism in particular. When discussing the substance of the soul, who did not adhere to Anaxagoras, or Democritus, or others before the sect of the Peripatetics grew up? Do not the followers of Aristotle today tear at his writings, uncertain whether he intended to speak of names or of things? And how many branches has this sect sent forth? Some follow Duns Scotus, some St. Thomas Aquinas, some now one, now the other, and the more leaders philosophy has, the less it is understood in our time. Meanwhile the Prince of Philosophers is altogether abandoned, while each one who excels the others in sophisms usurps his name for himself, and it is believed that not even Aristotle himself, if he came back to life, would adequately understand his disciples and followers. This is something no one unless mad has dared to assert of our sciences since neither age nor the customs of men can take anything away from them. The theorems of Euclid have the same certainty today as a thousand years ago. The discoveries of Archimedes will instill no less admiration in men to come after a thousand centuries than the delight instilled by our own reading.

If the preceding paragraph is to be taken seriously, and I see no reason why it should not be, it shows a complete repudiation of scholastic philosophy, and of its pretensions to knowledge, in favor of the mathematical sciences. Indeed, it appears that all speculations into subjects such as the substance of the soul are dismissed as nothing but a mass of discordant opinions (*diversarum opinionum congeries*) compared with the certain knowledge of mathematics. Ptolemy said something like this, although more politely, in the preface to the *Almagest*, a profound examination of Aristotle's divisions of theoretical philosophy, in which he compared the uncertainty of physics, concerned with the unstable properties of matter, and theology or metaphysics, concerned with invisible abstractions, with the certainty of mathematical demonstration. Regiomontanus does not so much go farther than Ptolemy in his criticism of philosophy—Ptolemy is already quite devastating enough—as turn it specifically on the scholasticism that still dominated university instruction and, as academic philosophy tends to do, goes on and on in endless disputations about unanswerable questions. Such repudiations of scholasticism became common enough in the sixteenth century, and are probably best known in various writings of Bacon, but Regiomontanus, here doubtless showing his humanist inclinations, is the earliest example I have discovered, and what is most important for our purpose, he puts the mathematical sciences in their place as the highest and, even after a thousand centuries, the most imperishable study. Ptolemy would have agreed and so too would Galileo, but one wonders what Regiomontanus's listeners, some of them certainly scholastically educated, made of it. It is clear enough from his other writings that Regiomontanus knew how to step on toes, but Padua, in which theology took last place behind the arts, medicine, and law, was sufficiently different from Paris that his remarks may have been taken without offense and even with some amusement at the expense of the Thomists and Scotists of the theological faculty, who presumably were not attending his lecture. In any case, the superiority of mathematics to scholastic philosophy is surely one of the principal messages of the oration along with the excitement in the recovery of texts, the beauty and intellectual power of mathematics, and the wonderful applications of mathematics to the improvement of knowledge.

And that is the subject he takes up next with the greatest enthusiasm, apostrophizing the mathematical sciences personified as goddesses. O constant companions of mortals, he declaims, not about to rest until the world itself comes to an end! O divine goddesses of philosophers, striving after the highest honors! O most delightful instructors of students, who fear to undertake nothing in all the world! You measure the depth of the earth, you ascend the heights of the heavens, you show the sun to be one hundred sixty-six times greater than the earth and the moon one-fortieth as great, you compare the stars to the volume of the earth, you investigate the dimensions of the celestial spheres, you promise to measure the size and distance of the earthly vapor ignited in the highest region of the air, which is called a comet.

And mathematics is even the way to that most coveted of humanist virtues, fame. For how many great men of an earlier age did you win fame? Through love of you, Hippocrates composed the quadrature of the circle (more likely than not, the wrong Hippocrates is meant). Avicenna wrote on lines and numbers. Hoping to receive glory, Averroes strove nearly all his life to discover an astronomy of concentrics, but at last acknowledged himself to have given up in despair. Have not physicians received the influential qualities of the signs of the zodiac from Hippocrates, as though no one can become a good physician unless he first study astrology? The reference here is probably to the Pseudo-Hippocrates *Libellus de medicorum astrologia* translated by Pietro d'Abano rather than to any ancient work of the Hippocratic corpus. Then, eyeing the physicians among his listeners, he says, but I shall restrain myself from saying that you are fallen far below your predecessors! And addressing the admirable nymphs of mathematics, who have brought forth many professors of canon and civil law, he asks: Will you refuse immortality to Bessarion of Nicaca, Cardinal and Patriarch of Constantinople, who, as in his youth under his learned teacher Gemistus Pletho he most wisely brought all of you together (collected manuscripts of the mathematical sciences), so now he undertakes to devote the remaining days of his life to the welfare of the Christian religion? And of his other contemporaries he mentions Nicholas of Cusa, for his quadrature of the circle (which two or three months later he demolished), Alberti, Toscanelli, and Bianchini, and would mention many more except that he must now turn to the praises of astronomy.

This is the last part and, I believe, intended as the culmination of the oration. By astronomy, he says, philosophers mean both the theory of motions and the prognostications of effects, and we shall acclaim the excellence of both parts together. Therefore I call upon you, divine goddess of astrology, I should be pleased if you would give assistance to your praises, if you would come forth to demonstrate your immeasurable benefits to mortals. You are without doubt the most trustworthy messenger of immortal God. By interpreting their secrets, you disclose the law for the sake of which the Almighty ordained to create the heavens, upon which he everywhere impressed glittering stars as monuments of the future. And on and on the praise continues. If your mind be carried away by the most beautiful and orderly succession of so many and such splendid lights, nothing you have ever perceived will be more delightful. Through this angelic science we are brought near to immortal God no less than through the other arts we are set apart from wild beasts. With age we are wont to become more remiss towards other kinds of studies; the desire for this science increases together with the years themselves. On this subject, I introduce the worthy testimony of Giovanni Bianchini, who very recently said, "Ten years ago I would have lain helpless, deprived of my life, were it not that the sweetness of astronomy maintained my spirit." Perhaps you desire honor, not delight. What, I ask, will you deem more honorable than this divine science, since it is desired for itself whereas the other arts mostly appoint an end foreign to themselves. A certain Peter of Cambrai (Pierre d'Ailly), Cardinal of the Holy Roman Church, constantly preached before clergy the honor as well as the beauty of our science, in this way arousing their enthusiasm for it, and several other bishops and noblemen are also called upon as witnesses to its worthiness.

What this means is that astrology, the judicial part of astronomy, is the highest and most valuable of all the mathematical sciences, and its great benefit is recognized by the testimony of many worthy princes and prelates, the highest commendation. But it has, alas, fallen into disrepute owing to its practice by fools and charlatans. After the praise of astrology comes the censor if its incompetent and dishonest practitioners. If you wish to apply knowledge of the heavens to useful ends, you you will find nothing more advantageous if only you profess it worthily, for if you instead pursue nonsense or, so to speak, the follies of old women (*aniles quasdem delirationes*), you will have only yourself to blame if no one has faith in you. For you will be a laughing stock if you but skim over scarcely the first rudiments of John of Sacrobosco and the first parts of al-Qabīsī, not even reaching his treatment of projections and rays, and then straightaway proclaim yourself an astronomer. O ridiculous man! O impudent men who, when they exhibit themselves to others expose their laziness to public shame! Who will imagine that they have through a kind of daydream attained to this most precious art, which the most diligent can scarcely master after much time and many nocturnal vigils? But many who were never students now put themselves forth as teachers, and it is for this reason that the splendor of this venerable science has nearly perished, so completely have fraudulent teachers overwhelmed it. And if they continue in this way, I shall launch a savage war against them. You therefore have the reason why this inestimable gift is thought to be worthless, since nearly all eagerly pursue riches. Indeed, like beasts they would rather gratify their stomachs and pursue wealth than practice virtue and study the good arts, and even the lovers of this art cannot achieve fulfillment for lack of good teachers. I could enumerate many teachers of the judicial part of astronomy, Indians, Greeks, Arabs, and Latins, so that it would be evident by what illustrious men the philosophy of heavenly influences was discovered, enlarged, and confirmed. I could show how necessary its knowledge is to physicians, how useful to teachers of law, how suitable to theologians. I could relate how we can avoid the imminent dangers of enemies in good time if we have adequately pursued this noble study. Finally, it would be proper to mention the assurance of immortality that the famous rulers of Venice and the venerable citizens of Padua would receive if they restore the provision for more frequent lecturing on astronomy, which has nearly perished, in this distinguished university.

So astrology, the most useful of all sciences, is taught very badly, if at all, but if the Venetians and Paduans will pay for more lectures, it will be taught very well, for which all will receive undying fame and gratitude. That, of course, is not the least point of the oration, which is not altogether unexpected since, as we earlier noted, inaugural orations were extended advertisements for the following course of lectures, for the subject in general, and even for the lecturer's qualifications for a job. But lest you be detained longer, he continues, I shall put these things aside and hasten to the end of the oration. Well now, distinguished gentlemen, I see you wake up, and in a manner befitting your extraordinary patience, be indulgent that our oration may return to the same place from which it began. And with a brief summary of what he has just said, Regiomontanus concludes with the announcement that he will now pass on to greet al-Farghānī, so to speak, at the door of his house.

And what did the distinguished listeners do? Some of them, I suppose, attended the lectures on Farghānī, and all must at least have thought that Regiomontanus was a very learned young man. They had been told that the mathematical sciences were the worthiest of studies, that they should waste no more time with scholastic philosophy, but learn geometry and arithmetic, music and optics, astronomy and, most of all, astrology. But how was anyone to follow this good advice? University courses in these sciences above the elementary level were almost nonexistent, even at Padua. Finding and purchasing manuscripts of such large works was difficult and expensive, not even one had yet been printed, and as a still greater obstacle, some of the most important were either in Greek or only very poorly translated. Regiomontanus may well proclaim the excellence of the mathematical sciences, for he had learned Greek and had Bessarion's manuscripts at his disposal, but aside from promising that the works of Archimedes will be made available and Apollonius translated, and the wish that Padua would institute more lectures on astrology, he nowhere said how anyone was to study these fine things.

It took seven years for him to come up with an answer. He remained in Italy until 1467 when he went to Hungary at the invitation of the Chancellor, Archbishop Johannes Vitez, earlier a patron of Peurbach's and a great devotee of astrology, to take a position at King Matthias Corvinus's newly founded university in Pressburg (Bratislava). In the dedication of the *Tabulae directionum* to Vitez, he commends the good Archbishop for raiding universities for the best faculty, we may presume by paying very high salaries, although Regiomontanus remained for only four years. During the ten years in Italy and Hungary, he completed an enormous quantity of work including the *Epitome of the Almagest,* the comprehensive textbook of Ptolemaic astronomy mentioned earlier; *De triangulis omnimodis,* a large treatise on plane and spherical trigonometry in five books; two complete sine tables to every minute of arc with unit radii of 6,000,000 and 10,000,000 parts; the extensive *Tabulae directionum,* which became the standard tables for spherical astronomy and the spherical parts of astrology for the next century-and-a-half; the *Tabulae primi mobilis,* a set of double-entry tables for direct solutions of right spherical triangles (the last two with instructions for solving many problems); the *Theonis Alexandrini Defensio in sex voluminibus contra Georgium Trapezuntium,* the defense of Theon's commentary on the *Almagest* against George of Trebizond's criticism; and finally, to name something less weighty, the *Disputationes contra Cremonensia in planetarum theoricas deliramenta* (Disputations against the Cremonese nonsense on the theories of the planets), the dialogue with Martin Ilkush criticizing the old *Theorica planetarum* attributed to the twelfth-century translator Gerard of Cremona.

But this was only the beginning. In 1471 he moved to Nuremberg where in July he wrote to Christian Roder, the rector of the University of Erfurt and, so Regiomontanus says, the only competent astronomer in Germany. The letter is very interesting. Among other things, like asking for an inventory of the Erfurt library which contains many mathematical books, he tries to get Roder's cooperation for the great reformation of astronomy, which he describes as a war to be waged

with observational instruments, and he asks Roder to exchange observations with him, something he intends to ask of other astronomers and universities. Such a reform had been on his mind at least
since he wrote to Bianchini in 1464, and perhaps since he made
observations in December of 1462, shortly after arriving in Italy, that
showed the faults of the *Alfonsine Tables,* or even since his days in
Vienna with Peurbach. The project requires observations with new
and improved instruments of the sun, moon, and planets, which in
turn are to be the basis for new derivations of elements, new tables,
and finally new ephemerides. This is something that ultimately took
the combined labors of Tycho, Kepler, various assistants, and fifty
years to accomplish, and left astronomy not just reformed but transformed. And even finding new elements for Ptolemy's planetary
theory, as Copernicus later did, required about twenty-five years of
observation and computation. Reforming astronomy was not easy.
He also poses for Roder's edification no less than thirty-six problems
in astronomy, geometry, algebra, etc., something he also did in his
letters to Bianchini, who mostly could not solve them. But our concern here is with his intention to publish books, for he says, "I am
undertaking to compose with a letter press all worthwhile books of
mathematics so that henceforth faulty texts will not annoy readers,
no matter how discriminating." The primary reason for printing is
the wish to produce reliable texts, and later in the letter Regiomontanus says that he also intends to print the ephemerides that he will
compute for more than thirty years.

This turned out to be the beginning of the first printing firm
devoted to scientific texts—scarcely twenty years after the invention
of printing from movable type—and in fact the most ambitious undertaking in scientific publishing for hundreds of years. I suppose
that by sometime in the late nineteenth century Teubner actually
produced more major scientific titles than Regiomontanus intended
to print, but it must have taken that long. He set up the press in his
own house, and by 1472 it was reported that he had printed a
calendar with the true motion of the sun and moon in an edition of
1000 copies (!), although no one except the typesetter had seen it.
That year he also went to Italy for several months in order to acquire
new books, and it appears that on his return publishing began in
earnest. Eight years earlier he had told the learned Paduans about

the glories of the mathematical sciences. Now he was going to make those sciences available to all of Europe.

After two titles were completed and two nearly so, he issued a prospectus, a double column sheet with the heading, "These works will be produced in the city of Nuremberg under the supervision of Johannes of Königsberg." Many of the works mentioned in the Padua oration are on the list, and much else besides, mainly of his own composition, but the number of medieval texts is sharply reduced and no Arabic works are included. The reasons were perhaps that the fundamental writings of antiquity necessarily received the highest priority and that he could not vouch for the reliability of the translations from Arabic. The first completed edition was Peurbach's *Theoricae novae planetarum,* the new planetary theory that was to replace the old faulty *Theorica*—and did, it was printed more than forty times in the next 150 years—an elegant folio of forty pages with twenty-nine woodblock figures, several of them hand colored. The second was an interesting choice, the *editio princeps* of Manilius's *Astronomica,* an astrological poem written early in the first century, now most famous for its extraordinary textual problems and its extraordinary editors, Scaliger, Bentley, and Housman. Since this was hardly a text of use for practical astrology, the reason for its publication was necessarily the humanist interest in recovering authentic ancient astrology from a classical Latin poet, and of course the poetry itself. Manilius was the subject of university lectures and was also of the greatest importance in the iconography of Italian art in the late fifteenth century. Nearly completed was a new *Calendar,* issued in Latin and German editions, giving new and full moons, solar and lunar eclipses, and the ecclesiastical calendar for 57 years, from 1475 to 1531, in both the conventional and a corrected computation. Vastly more ambitious were the *Ephemerides* already mentioned in the letter to Roder, daily positions of the sun, moon, and planets, with aspects of the planets to the moon and to each other, and eclipses for 32 years, from 1475 to 1506—no small job of computation—which Zinner remarks runs to 896 pages and contains something like 300,000 numbers—no small job of typesetting.

Three works not in the prospectus were also printed. The first was the *Disputationes contra Cremonensia deliramenta,* to which Regiomontanus added a preface defending both his intention to issue correct texts and new translations, and also his outspoken criticisms of incompetent editors, translators, and commentators—he named names, which was considered especially rude—against the accusation of arrogance, "as I live in Germany, not to say in the midst of barbarism, lacking books and distant from the concourse of learned men, and I dare to attack so many distinguished men." He was in fact defending the prospectus, and his reply is quite amusing and not a little arrogant, although with good reason "for who does not know that the wondrous art of printing, recently devised by our countrymen, is as harmful to men if faulty volumes of books are distributed as it is helpful when exemplars are properly corrected." The two other publications were not scientific. One was the famous exhortation to youths by St. Basil of Caesarea (4th century) *De legendis libris gentilium* on how profitably to read pagan authors, translated about 1400 by Leonardo Bruni with a preface to Coluccio Salutati, as one can imagine a favorite humanist text. The other was a "Dialogue between *Veritas* (Truth) and *Philalethes* (Lover of Truth)" by Maffeo Vegio, a poet and humanist best known for writing a thirteenth book of the *Aeneid* and, after he became devout, a life of St. Anthony in verse. A hand colored woodcut in the edition may show Regiomontanus himself as Philalethes.

The prospectus begins with "Ptolemy's *Geography* (*Cosmographia*) in a new translation, for the old one by Jacopo d'Angelo of Florence (1406), which is in common use, is faulty since the translator himself (no offense intended) has sufficient knowledge of neither the Greek language nor mathematics. In this verdict it will be proper to trust the best judges, Theodore of Gaza, an illustrious gentleman very learned in both Greek and Latin, and Paolo [Toscanelli] of Florence, by no means unacquainted with the language of the Greeks and extremely distinguished in mathematics." This is obviously the sort of thing that got people upset. It is interesting that the *Geography*, by far the most important work on cartography, surely a mathematical science, was not mentioned in the oration. Perhaps it was just an oversight, but a more likely reason is that Regiomontanus did not wish to say anything about the bad translation. Also planned are "large commentaries on Ptolemy's *Geography* in which the manufacture and use of a *meteoroscopic* instrument is described, by which Ptolemy himself derived nearly all the numbers in his entire work.

For one would believe incorrectly that the numbers of so many longitudes and latitudes were discovered through observations of the heavens. Further, a description of the armillary sphere together with the entire habitable world made so clear in a plane that most people will be able to learn everything, which no one has hitherto comprehended in Latin, since he has been hindered through the fault of the translator (Jacopo d'Angelo)." The *meteoroscope*, which Regiomontanus had described in a work dedicated to Bessarion, is like the armillary used by Ptolemy, but with altitude-azimuth rather than ecliptic coordinates and a movable equator ring with a perpendicular circle. It may be used to convert distances and directions from itineraries to geographical longitude and latitude. Also mentioned is a special treatise against the translation of Jacopo d'Angelo that will be sent to judges, evidently Theodore of Gaza and Paolo Toscanelli. Part of this work was published by Willibald Pirckheimer in 1525, and in it Regiomontanus explains that he and Peurbach were already concerned with correcting the translation of the *Geography* in Vienna, and that this was his original reason for wishing to learn Greek.

The list continues: a new translation of the *Almagest;* Campanus's edition of the *Elements* of Euclid including Hypsicles, with most errors removed, as will be explained in a short treatise; later he mentions a short commentary showing clearly that the opinions of Campanus should be removed from his edition of the *Elements;* Theon of Alexandria's *Commentary on the Almagest,* a work, by the way longer than the *Almagest,* and surely here intended to replace George of Trebizond's commentary; Proclus's very fine elementary text on Ptolemy's astronomy, the *Hypotyposis astronomicarum postitionum.*

It is not specified whether these are to be printed in Greek or translated, but I very much think Regiomontanus intended to translate everything, although he only mentions *new* translations, that is, where they replace old ones. The wide dissemination of the mathematical sciences he desired could not be brought about by Greek texts alone, but required translation. Further, no one had yet printed a book entirely in Greek anywhere. Thus far it was used only for quotations, and even then often filled in by hand, as Regiomontanus later did for two words in the preface to the *Disputationes,* so unobtainable was even a small amount of Greek type. Hence, if he intended to do something as unheard of as issue the first book ever printed in Greek, let alone print gigantic texts like Theon's commentary on the *Almagest,* he surely would have mentioned it in the prospectus. In fact, the first book printed in Greek was the modest and practical Greek grammar of Constantine Lascaris in Milan in 1476, and prior to Aldus in the early 1490s the number of printed Greek books can be counted on not many fingers.

Next comes astrology: Ptolemy's *Tetrabiblos* and the *Centiloquium* attributed to Ptolemy in a new translation; as much of Firmicus Maternus (4th century) as is found, evidently the *Mathesis,* the largest classical Latin astrological treatise (but one hopes not his *De errore profanarum religionum,* written after his conversion to Christianity, in which he exhorts Constantius and Constans to show no mercy in eradicating paganism); Leopold of Austria (14th century), evidently his *Compilatio de scientia astrorum,* and any other astrological prognosticators who seem worthy, for instance, any fragments of Antonio de Monte Ulmi, who was mentioned in the oration.

Next come *all* the works of Archimedes then known with the commentaries of Eutocius in the translation of Jacobus Cremonensis, although occasionally corrected; Regiomontanus later mentions commentaries of his own on books of Archimedes for which Eutocius is lacking; the *Optics* of Witelo, an excellent and celebrated work; the *Optics* of Ptolemy, surviving only in Latin and missing the first book, apparently unknown to Regiomontanus when he gave the oration; the *Musica,* that is, the *Harmonics* of Ptolemy, with Porphyry's commentary, the work Nicholas of Cusa wished to have translated; a new edition of the *Spherics* of Menelaus, which survives only in Arabic and Latin; a new translation of Theodosius's *Spherics, On habitations, On days and nights,* elementary works of spherical astronomy; Apollonius of Perga's *Conics,* Syrenus's *Cylindrics,* evidently Regiomontanus intended to translate both (his colleague Bernhard Walther did translate Apollonius, but it is lost); Hero's *Pneumatica,* a mechanical work of extraordinary delight.

Then some works on number theory and algebra: Jordanus's *Elementa arithmetica* and *De numeris datis;* the *Quadripartitum numerorum,* a work abounding in diverse subtleties, presumably that of Johannes de Muris. Curiously Diophantus is absent; perhaps Regiomontanus was still awaiting the seven missing books. Then comes the *Mechanical Problems* of Aristotle, a really important work for the history of mechanics in the Renaissance, and last in this part of the list, the *Astronomia* of Hyginus (2nd cent.), an elementary work on, among other things, the mythology of the stars, with figures of the constellations.

The following works by Regiomontanus himself are described as "The Attempts of the Artisan. Although natural modesty and the republic of letters have long debated with each other whether they should be published or not, reason has held them worth the risk." *Defense of Theon of Alexandria* in six books against George of Trebizond, from which anyone will understand that his commentary on the *Almagest* is worthless (*frivola*) and that his translation of Ptolemy's work is not free of error. *On the five equilateral bodies,* which are commonly called regular, that is, which of them occupy a corporeal space, and which not, against Averroes, the Commentator on Aristotle. *On the quadrature of the circle,* against Nicholas of Cusa; Regiomontanus wrote a lot on this including a treatise sent to Toscanelli and a dialogue in which the learned Cardinal's quadrature is refuted. Then two works on problems in spherical astronomy pertinent to astrology: *On directions,* against the Archdeacon of Parma, Matthaeus Guarimbertus (14th century); *On the distinction of the houses of the heavens,* against Campanus and Johannes Gazul of Ragusa (Dubrovnik, in Padua in 1430), whose other doctrines concerning seasonal hours are also corrected. *On the motion of the eighth sphere,* against Thābit ibn Qurra and his followers; Regiomontanus had criticized this in his letter to Bianchini on the problems of astronomy. *On the restoration of the ecclesiastical calendar;* in the *Calendar,* which contains a section on problems of the ecclesiastical calendar, Regiomontanus says that some Jews objected to him and to Bessarion that the Christians celebrated Easter at the wrong time. *Breviarium almaiesti,* the *Epitome of the Almagest. Five books on triangles of every kind;* the treatise on plane and spherical trigonometry. *Astronomical problems pertaining to the entire Almagest;* an important work, later described as in three books, that appears to be lost. *On the size and distance of a comet from the earth, on its true position, etc;* already in the oration Regiomontanus claimed these could be found, although his elegant little treatise is probably later. Note that he believed comets were in the upper air, not the heavens, so their parallaxes should be a few degrees. *Panonian Game (Ludus Panoniensis), which at another time was called the Tabulae directionum,* the tables for spherical astronomy and its applications to astrology. *Great table of the first movable,* with various uses and certain computations; the double-entry tables for solving right spherical triangles. *Sighting rods of many kinds, with their uses;* a treatise on the cross staff, an instrument for measuring angles of up to a few degrees. *On weights and aqueducts,* with descriptions of instruments necessary to them; very interesting looking but lost. *On burning mirrors and others of many kinds and of astonishing use;* this is the sort of thing referred to in the oration where Regiomontanus said he was prepared to make Archimedes's "philosophic" mirrors.

Books are not the only things that Regiomontanus intended to print. A "tree of Ciceronian rhetoric" (*arbor rhetoricae tullianae*) in a beautiful likeness had already been made, although no copies are known to survive. A map of the entire known habitable world, commonly called a *Mappa mundi,* will be made, and individual maps of Germany, Italy, Spain, all of France, and Greece, along with brief descriptions (*historias*) drawn from many authors, that is, such as seem to pertain to mountains, seas, lakes and rivers, and other particular places. Had this been carried out, it would have been the earliest series of printed maps. The first editions of Ptolemy's *Geography* with maps were Bologna 1477 and Rome 1478. In the workshop an *astrarium* is under construction, a work that clearly should be considered a miracle. Some astronomical instruments for observations of the heavens are being made, and likewise others for common everyday use of which it is lengthy to recite the names. Finally, he will set up a factory for casting type, which he regards as more enduring than even all the books he will print: "Last of all, for the sake of enduring written memorials, it has been resolved to commission that wonderful art, the casting of type (*litterarum formatricem*), whereby when completed, God willing, even if the artisan soon falls asleep, death will not be bitter, since he will have left so great a gift of inheritance to posterity, by which they will forever be able to deliver themselves from want of books."

The prospectus of books to be printed is an extremely valuable document. Just as in the oration the foremost mathematician of the age set out, for the learned and studious, for professors and students of arts and medicine, what he considered the virtues and achievements of the mathematical sciences, so the prospectus shows in detail and, as it were, for his own colleagues who have the dedication to master these subjects, the essential contributions both of the past and of his own age. The foresight in undertaking such a venture, as noted before, scarcely twenty years after the invention of printing, is simply astounding. Unfortunately, nothing beyond the eight titles mentioned earlier and the prospectus itself appeared, for Regiomontanus's death in Rome in the summer of 1476 at the age of forty terminated the whole gigantic project. Had he lived another twenty years, a fair number of these books might well have been printed, perhaps making Archimedes and Apollonius available nearly three-quarters of a century, and Diophantus, if he got around to it, nearly a century, before they actually appeared. It was of course through these three authors and Pappus, who Regiomontanus does not mention, that Europeans became proficient in mathematics and began to make original contributions of their own in the late sixteenth century. Perhaps this could have been hastened by fifty or so years, and in fact the first editions of most of the works of Archimedes in Latin, of the Greek *Almagest* with Theon's commentary, and several other works were originally printed from his manuscripts. A number of his own works, among them the *Tabulae directionum, Tabulae primi mobilis, De triangulis omnimodis,* the sine tables, and above all the *Epitome of the Almagest,* were printed in the late fifteenth and early sixteenth centuries, and established the foundation of spherical astronomy and the understanding of Ptolemy upon which all the astronomers of the next century built, not the least being Copernicus, Tycho, and Kepler. Indeed, most of the fine things Regiomontanus wished to accomplish actually came about in the second half of the sixteenth and the early seventeenth centuries, although in some cases, Kepler's planetary theory in particular, they took forms that would have amazed him. And some of the texts have still not been published.

In any case, I believe it is clear, first from the oration and then from the prospectus, that Regiomontanus had a vision of the importance and the potential for improvement and application of the mathematical sciences far beyond any of his contemporaries, at least those whose opinions are known, a vision requiring a century and a half until it was brought to fulfillment. Yet, given his extraordinary abilities and extraordinary capacity for work, there is no telling how much he might have accomplished on his own. Throughout the sixteenth century he was regarded as second only to Ptolemy, a correct judgment in light of the importance of the *Epitome of the Almagest* to the recovery of Ptolemy's astronomy and the constant use of the *Tabulae directionum* for spherical astronomy. The *Almagest* had been available in Latin since the late twelfth century, but it is hard to think of a single European who used it constructively before the publication of the *Epitome* in 1496, while Copernicus's new planetary theory then followed within less than twenty years. And although Regiomontanus's writings ceased to be of practical use by the middle of the seventeenth century, still even a century or so ago, when not much was known of him beyond Gassendi's fanciful biography (1654) and the account in Doppelmayr's quaint old book on the Nuremberg mathematicians and artists (1730), he was quite a celebrated figure, even a legend a bit like Faust, and rather admired by Germans. He is said to have constructed a mechanical fly that would leave the hand, fly around, and then return, and a mechanical eagle that could fly away from a city and back. There are other stories less preposterous, but no less false.

More recently he fell victim to the denigration of the Renaissance in the name of medievalism mentioned earlier, with two disparaging chapters by L. Thorndike, filled with errors and based upon reading little more than incipits, title pages, and prefaces, and a condescending chapter by Duhem, only published in 1959, claiming that he subscribed to Duhem's philosophy but did nothing new. Although Zinner's fundamental study, which is not free of German nationalism, originally appeared in 1938, the significance of Regiomontanus for the history of science has still not been adequately recognized except, I believe, by P. L. Rose in his valuable book on humanism and mathematics in the Renaissance, which makes much the same point as this paper about their importance, and by M. Claggett in his study of Regiomontanus's work on Archimedes and conic sections. For this general lack of recognition there are a few reasons. First of all, he did not propose a momentous hypothesis like Copernicus—although Copernicus may have received the most important lead to his own theory from Regiomontanus—and in fact he intended not so much new science, which no one then wanted anyway, as better science, the value of which should not be underestimated. Second, he was a very fine mathematician, but original mathematics was still a century away, and here too he looked upon his work principally as improving the knowledge of mathematics, both by his own writings and by recovering and translating the writings of antiquity. And third, his works are not merely technical but, with a few exceptions, the oration being the outstanding example, positively unreadable—tables with precepts on how to solve problems, however ingenious their design and application, are meant to be used, not read—and except for some of us odd historians, even the *Rudolphine Tables* have not had many readers. Yet, much as he said of Apollonius, when the *Epitome of the Almagest* is translated, no one will not come to admire him.

But there is something else, and it takes us back to a subject discussed in our introduction, and that is the historiographical evaluation of the science of the Renaissance, the mathematical sciences in particular. What may we learn of them from the oration? There was a scientific revolution in the seventeenth century, and its character is defined, as Professor Kuhn has pointed out, by the transformation of the classical, mathematical sciences studied since antiquity and by the new experimental sciences just coming into existence. At the same time scholasticism, while by no means defunct in university education—where traditions, no matter how superannuated, die very hard—had long since ceased to contribute anything now recognized as significant to the history of science (except by advocates of medievalism). The description of the mathematical sciences in the oration belongs to neither, but is in itself definitive of a Renaissance science allied with humanism for which the oration and the prospectus of books taken together constitute a virtual manifesto. Here is Regiomontanus showing the clearest possible appreciation of the mathematical sciences, both as science itself, of the greatest interest technically and aesthetically, and also of the greatest promise for intellectual and material benefit. "The discoveries of Archimedes will instill no less admiration in men to come after a thousand centuries than the delight instilled by our own reading. . . . When Apollonius is translated from Greek into Latin, no one of you will not come to admire him." And the same mathematics will measure the dimensions of the heavenly spheres and show how to build vaults of churches that will not fall down. But he also sees these sciences as a part of the "good arts," of learning within humanistic studies, recovering achievements of antiquity that can be admired beside poets, philosophers, and historians, and that encourage and ennoble us, and even bring us the most desirable of all virtues, fame. It was, after all, the sweetness of astronomy that maintained Bianchini's spirit when he lay helpless, deprived of his life. "Through this angelic science, we are brought near to immortal God no less than through the other arts we are set apart from wild beasts." And speaking of the wonderful *astrarium* of Giovanni de Dondi, he commends the fame that accomplishment in the sciences can bring. "Rejoice, therefore, O noble Paduans, to whom the studies of illustrious men have always been honorable."

The oration is optimistic, full of hope for the future under the guidance of the divine goddesses of the mathematical sciences. And the venture in printing and all the great plans in Nuremberg were to be a beginning to carrying out the prospects set out in the oration. But this is not the scientific revolution. What is surely central to the early years of the scientific revolution is the replacement of Aristotle as the foundation of natural science, in fact by a science based, at least in principle, on mathematics, but Regiomontanus does nothing of the sort. On the contrary, good humanist that he is, not only he does not depreciate the fine ancient author Aristotle anywhere in the oration, except for remarking that he should have studied more astronomy, but he explains how necessary mathematics is to a full appreciation of Aristotle's works, certainly good advice to the Paduans. And one largely mathematical work then attributed to Aristotle, the *Mechanical Problems,* he even included in the prospectus. Now this I believe is characteristic of the century or so after Regiomontanus. Copernicus, for example, if his first ten chapters be read without anachronism, in no way repudiates Aristotle, rather, he makes the least possible changes in the Aristotelian physics of the natural motions of bodies that may accommodate the motion of the earth. Nor, on the other hand, does the oration belong to the world of the schoolmen, who Regiomontanus disparages with wit and sarcasm while actually defending Aristotle against their misrepresentations. The schoolmen tear at Aristotle's writings, uncertain whether he intended to speak of names or things. Each one who excels the others more in sophisms takes Aristotle's name for himself, and not even Aristotle, if he came back to life, would understand his followers. One thinks of Bacon's "cobwebs of learning, admirable for the fineness of thread and work, but of no substance or profit." Anticipating the wisest of mankind by a century and a half is no mean feat. But it is not the whole of the middle ages that is to be cast aside in favor of the recovery of antiquity, for good mathematics is valuable whenever it was written. In the oration Arabic and European astronomers and mathematicians are praised, and in the prospectus also there are medieval works on number theory and optics and even some on astrology, but nothing of scholastic treatises on motion or proportions or quantification of forms. Regiomontanus knew the difference. The omission was not due to ignorance, since he had a manuscript of works of this kind by Bradwardine and Oresme; it was simply not what he considered of value to mathematics.

What we have in the oration, in the prospectus, and indeed in Regiomontanus's very technical works, is something that belongs to its own time, the Renaissance, with values and virtues of its own that cannot be understood if we regard it only as an early part of the scientific revolution, to which it contributed greatly, or as a continuation of the middle ages, which it was at pains to leave behind. And if we are to appreciate the oration and its spirit, appreciate Regiomontanus's excitement in the discovery of the old and of the new, appreciate the sense of wonder and admiration at the power of mathematics to fathom the heavens, map the earth, construct a siege engine, appreciate finally a science that at its highest level, astrology, teaches us of the stars that determine our character and our fate, we shall have to reform our own sense of history to recognize the distinct character of the Renaissance of science.

Bibliographical Note

The paper by Professor Kuhn from which this study began is "Mathematical versus Experimental Traditions in the Development of Physical Science," *Journal of Interdisciplinary History* 7 (1976): 1–31; repr. in *The Essential Tension* (Chicago, 1977), 31–65. Some of the same points are discussed within a more general consideration of the historiography of science in "The History of Science," *International Encyclopedia of the Social Sciences* 14 (New York, 1968), 74–83; repr. in *The Essential Tension*, 107–126, at 117–118.

The text of Regiomontanus's (R.) *Oration on the Mathematical Sciences* was first printed as *Oratio Iohannis de Monteregio, habita Patavij in praelectione Alfragani*, in *Rudimenta astronomica Alfragrani* (sic) *item Albategnius peritissimus de motu stellarum* . . . , Joh. Petreius, Nuremberg, 1537. Aside from chaotic punctuation, the text seems fairly good. It was reprinted with some alterations by Erasmus Reinhold (Wittenberg, 1549), whose edition was the basis of later printings in collections of Melanchthon's orations, from which comes the very poor text in the *Corpus Reformatorum* 11, ed. C. B. Bretschneider

Science and Humanism in the Renaissance

(Halle, 1843), 531–544. The prospectus of books to be printed in Nuremberg has been reproduced several times, most usefully with notes by E. Zinner, "Die wissenschaftlichen Bestrebungen Regiomontans," *Beiträge zur Incunabelkunde*, N. F. II (Leipzig, 1938), 88–103. It is also included in *Johanni Regiomontani Opera Collectanea*, ed. F. Schmeidler (Osnabrück, 1972), a collection of facsimiles of original editions of several of his works, and in Zinner's biography. A new text and translation of the oration and the prospectus, along with other writings of R. on science and humanism, will be published before long by A. Grafton and me. I would like to thank Professor Grafton for his many discerning remarks on this paper. R.'s correspondence has been edited by C. T. von Murr, *Memorabilia Bibliothecarum Publicarum Norimbergensium* (Nuremberg, 1786), 74–205, and less accurately by M. Curtze, "Der Briefwechsel Regiomontan's mit Giovanni Bianchini, Jacob von Speier and Christian Roder," *Abhandlungen zur Geschichte der mathematischen Wissenschaften* 12 (1902): 187–336. Except for a few excerpts, the are no modern editions of any of his works.

The fundamental study of R.'s life and work is E. Zinner, *Leben und Wirken des Joh. Müller von Königsberg gennant Regiomontanus*, 2nd ed. (Osnabrück, 1968), in which the oration is treated at 110–118. This book has its problems, but is an indispensable source. A less-than-accurate translation by E. Brown has recently appeared (Amsterdam, 1990). Papers on various aspects of his work are in *Regiomontanus-Studien*, ed. G. Hamann, Sitzungsberichte der österreichische Akademie der Wissenschaften, Phil.-Hist. Kl. 364 (1980). There is a chapter on R., considering in particular the relation of his work to humanism, in P. L. Rose, *The Italian Renaissance of Mathematics* (Geneva, 1975), 90–117 with the oration at 95–98. Rose's important study of the mathematical sciences and humanism in the Renaissance is a mine of information, and I have found it very helpful. R.'s relation to German humanism in particular is treated in H. Grossing, *Humanistische Naturwissenschaft* (Baden-Baden, 1983). While the oration is frequently mentioned in passing, the only other extended treatment is in M. Cantor, *Vorlesungen über Geschichte der Mathematik* 2, 2nd ed. (Leipzig, 1900), 260–264. Part of the correspondence with Bianchini has been translated with detailed notes by A. Gerl, *Trigonometrisch-astronomisches Rechnen hurz vor Copernicus* (Stuttgart, 1989). The criticism of contemporary astronomy sent to Bianchini is translated with a commentary by N. M. Swerdlow, "Regiomontanus on the Critical Problems of Astronomy," *Nature, Experiment, and the Sciences*, ed. T. H. Levere, W. R. Shea (Dordrecht, 1990), 165–195.

In writing this paper and in preparing the forthcoming edition and translation of the oration, I have consulted such a large number of modern publications on Renaissance science and humanism and on the various people and writings referred to by R. that it would be very lengthy to list them all here. Just to mention a pertinent selection: M. Clagett on R. and Archimedes and Alfred, M. Cantor, P. Duhem, G. Federici Vescovini on Pietro d'Abano, E. B. Fryde on Guarino, E. Garin, D. J. Geanakopolos on Greek humanists, A. Grafton and L. Jardine on humanist education, T. Hankins on Bessarion, N. Jardine on early history of science, J. Jervis on R. on comets, P. Kibre on universities, P. O. Kristeller, B. Nardi, L. Olschki on technical writers, J. Monfasani on George of Trebizond, J. S. Morse on R. and Diophantus, J. E. Murdoch on the the distinction between mathematics and philosophy and on the medieval Euclid, J. D. North on Renaissance astrology, R. E. Ohl and J. H. Randall on the University of Padua, P. L. Rose and S. Drake on the Pseudo-Aristotle *Mechanical Problems*, C. B. Schmitt, N. Siraisi, L. Thorndike. A study of particular importance on inaugural orations is C. Trinkaus, "A Humanist's Image of Humanism: the Inaugural Orations of Bartolommeo della Fonte," *Studies in the Renaissance* 7 (1960): 90–147. Gregorio Tifernate's inaugural orations *de astrologia* and *de studiis litterarum* are edited in K. Müllner, *Reden und Briefe italianischer Humanisten* (Vienna, 1899), repr. with add. by B. Gerl (Munich, 1970), 174–191. Tycho Brahe's *De disiciplinis mathematicis oratio* is in *Tychonis Brahe Dani Opera Omnia*, ed. J. L. E. Dreyer, 1 (Copenhagen, 1913), 145–173.

To take up the historiographical issue of medievalism and what used to be called "the problem of the Renaissance," Duhem on R. and the school of Vienna is in *Le Système du Monde* 10 (Paris, 1959), 349–367; Thorndike in *Science and Thought in the Fifteenth Century* (New York), 1929, 142–50, and *A History of Magic and Experimental Science* 5 (New York, 1941), 332–377. Galileo is of course a far more serious battleground, but that is a subject in itself. We have found surprisingly useful S. L. Jaki, O. B., *Uneasy Genius: the Life and Work of Pierre Duhem* (The Hague, 1984)—our quotations are from pp. 394 and 399—an apologia that rebukes its hero only when he shows a spark of humanity. Jaki was somewhat more moderate in his introduction to the translation of Duhem's *To Save the Phenomena* (Chicago, 1969).

The literature on the background to Catholic medievalism is very large—it is the history of the Church in the nineteenth century—and a brief treatment may be found in J. B. Bury, *History of the Papacy in the 19th Century, Liberty and Authority in the Roman Catholic Church* (London, 1930), augmented ed. (New York, 1964), which includes a discussion of the *Syllabus praecipuos nostrae aetatis errores* in chap. 1. The articles in the *Encyclopedia Britannica*, 11th ed. (1910) on Pius IX, Syllabus, Ultramontanism, and Vatican Council are still worth consulting. The Ultramontane, Catholic side is best represented in English by the numerous writings of Cardinal Henry Manning.

The historiography of the criticism of the Renaissance is considered in W. K. Ferguson, *The Renaissance in Historical Thought* (Boston, 1948), chaps. 10 and 11. There is a symposium on the originality of fifteenth-century science in the *Journal for the History of Ideas* 4 (1943): 1–74, which is worth consulting if only as a relic of its time. Two notable defenses of the Renaissance by E. Panofsky are, "Artist, Scientist, Genius: Notes on the 'Renaissance-Dämmerung'," *The Renaissance* (New York, 1962), 121–182, an interesting collection in which even Sarton comes around to recognizing the Renaissance, and the justly famous *Renaissance and Renascences in Western Art* (New York, 1969). There are serious reflections on the issues in a general sense in two essays of M. I. Finley, "The Ancestral Constitution" and "The Historical Tradition: the *Contributi* of Arnaldo Momigliano," *The Use and Abuse of History* (New York, 1987), 34–59, 75–86.

Design for Experimenting

Jed Z. Buchwald

1 Axiomatics and the Autonomy of Experiment

Several years ago the late historian of science Derek de Solla Price remarked,

Any effect or phenomenon like the Edison effect, Cerenkov radiation, and the creep of liquid helium, might be just the thing to measure or reveal that which we did not know before. Such experimentation is a sort of fishing expedition because you never quite know what you will catch, and always hope for the unexpected. It cannot be planned with an eye towards any particular objective, though of course it is common to say as a necessary condition for getting funding. (Price 1983)[1]

Price argued strongly for the autonomy of experimental investigation, for its substantial independence from theory. Among other examples of this he offered "radiotelegraphy":

The simple truth is that if one wishes to do something to something, what one uses is a technique rather than an idea. A hackneyed example of this is the application of Maxwell's electromagnetic theory to the invention of radiotelegraphy and all the techniques of radio and TV broadcasting. Maxwell's theory was a tremendous unifying concept that explained the nature of light and suggested one could make similar wave radiation electrically. *The trick was not knowing it might be done, but finding out ways to generate and detect such waves.* The early history of radio is not so much a matter of physics, but the control of experimental techniques like spark gaps and of detectors of such devices as coherors, surface magnetism, etc. Quite often the detecting devices in particular were known to work, but the reason why was not ascertained till much later. *There is simply no way to apply a theory. The crucial point is to acquire and operate with a technique or a new effect, even if one has no idea why it works. That can come later.* (Price 1983; emphasis added)

Price had consigned "Maxwell's theory" to the role of a possibility generator, and he implied that the discovery of how to produce and manipulate electric waves had little to do with physics. It depended rather on "the control of experimental techniques" of certain kinds. Although Price's remarks, if taken to an extreme, exaggerate historical reality, they contain a powerful kernel of truth, one that strongly connects to concepts enunciated by Thomas Kuhn. My goal here will be to develop that connection eventually in the particular context of Heinrich Hertz's work but immediately in more general terms.

The *Oxford English Dictionary* gives seven meanings for the word "theory" whose overall thrust is to distinguish it as something systematic, abstract, and quite likely shaky, from practice, which is unsystematic, concrete, and unlikely to be questioned. Price's remarks divide theory from experiment in essentially that way, and they were meant to redirect the historian's attention from the theoretical high ground to the nitty-gritty of laboratory experience. His assertions, moreover, seem to run against a contemporary cliché commonly held by philosophers as well as historians of science, to wit, that experiment necessarily reflects theory. And since Kuhn's work has, in one way or another, been seen as a powerful support for what is nowadays a near platitude, there would seem to be little room for agreement between the two of them.

Yet I think that Price was reaching for an understanding of experiment that makes contact with what Kuhn has taught historians.[2] In order eventually to reach this common ground, I will begin with a bald assertion that connects directly to Kuhn's work and that, we will see, leads indirectly to Price's remarks concerning experiment: namely, that living sciences cannot be corralled with exact generalizations and definitions.[3] Attempting to capture a vibrant science in a precise, logical structure produces much the same kind of information about it that dissection of corpses does about animal behavior; many things that depend upon activity are lost. When such a thing eventually happens, as it often does, the science either is very dead, is of interest only to a few people at the margins of the active discipline, or else has been transformed into a technical artifact and relocated to an engineering department. One might say that axiomatics and definitions are the logical mausoleums of physics. This is hardly a novel perception, since it was discussed in many ways by Kuhn and since it also, albeit in a considerably different sense, underpins Michael Polanyi's "personal knowledge." But I do not think either that its strength for historical understanding has been as fully conveyed as it might be or that it has been closely linked to experiment. I want, therefore, to begin straightaway with a brief example drawn from electromagnetism in the last quarter of the nineteenth century that illustrates why it is difficult, and perhaps impossible, to cage living science. This will seem to take us far from experiment proper for a time, but it will in the end carry us back to Price's claim.

During those years field theory, based on Maxwell's *Treatise on Electricity and Magnetism,* was extensively developed by a closely knit group of British physicists and occasionally by outlying group members in North America. Members of the group shared an understanding of what fields were and how to use them in practice. Although they did not hold rigid, fixed conceptions about fields, in many exemplary instances one finds them agreeing fully about the proper way to solve a problem, what kinds of mathematics to use, what problems were solvable, and what might practically be detected. Suppose, however, that we, as Maxwellians, decided to require that a field must have such and such a property and no others, and that it can only be examined in such and such a way. Suppose, that is, that we generated rules and definitions precisely to circumscribe the phenomenal range of the field and meticulously to prescribe how to use it. We might say, e.g., that a field always acts by soaking up and giving back energy (as a sponge does water), extracting it from one material structure and transferring it by propagation to another, distant structure. Certainly much Maxwellian work did presume such a thing. If we did require fields to behave in this fashion, we would a fortiori have to forbid the existence of fields that direct processes without transferring energy. Indeed, such a possibility would most probably never even occur to us. Yet this was precisely how the American physicist Edwin Hall thought about a new effect that he discovered in the early 1880s.

Some Maxwellians, spurred by Hall's mentor at Johns Hopkins, Henry Rowland, found this conception of a directional field compelling; others rejected it entirely. Or I should say, Maxwellian rejectors of Hall's new kind of field wrote articles in which they proposed what they thought were simple, conservative explanations of the Hall effect. Although these explanations were certainly intended explicitly as alternatives to Hall's, they never did grapple directly with the distinction between transferring and directing fields. These authors knew that Hall's field could not absorb or give off energy, and so they simply spurned it *as a field*. It did not fit their understanding of what such a thing should be. Those who accepted Hall's and Rowland's understanding went right from it to a special energy structure from which (reversing the analysis) they could deduce the field, but even they never explicitly discussed its novel *directive* character. These authors were, however, indubitably aware of the distinction, because it is right there in their mathematics, and they make assertions that depend upon it. The most voluble Maxwellian of them all, the brilliant eccentric Oliver Heaviside, insisted that the essence of a field was determined solely in relation to energy transfers. He never had much to say about Hall's effect or others in which the fields didn't seem to behave as he wanted them to. Yet Hall, Heaviside, and indeed all Maxwellians deployed field terminology and ideas in other ways that reflect a powerful, common underlying pattern in their thought, for there is a vast spectrum of problems about whose solutions they agreed. We can perceive this pattern by seeing how it works in particular circumstances, but our perception will always be difficult to capture in words and rules. The tremendous vitality of British field theory in the 1880s was due in large part to this very flexibility.

One might nevertheless be tempted to assert that a difference like the one between Hall and Heaviside can be pinned down by differentiating between higher- and lower-order elements in a scheme, with each kind of element being quite precise and well defined in itself and with the links between different-order elements also being neatly specifiable. In this case, however, as well as in many others, no such division is likely to work, because the different levels would have to impinge upon one another in a galaxy of conflicting ways in order to capture historical practice.

Consider Heaviside's unique conception of a special form of energy conservation, which powerfully molded his reformulation of Maxwellian field theory. If we make this conception part of a well-defined *superstructure* of field physics, we will have a difficult time interpreting a great deal of Heaviside's own work, for at least three reasons. First of all, Heaviside's deepest understanding of field relations was so closely bound to certain kinds of energy considerations that it cannot be divorced from them, whereas a superstructure could be removed without damaging underlying foundations. Second, in many cases Heaviside's energetics is not overtly present in his discussions, whereas a superstructure would be (and might hide a substructure). Finally, Heaviside rarely specifies clear rules for deploying energy relations, whereas a superstructure's form would be reasonably well defined. On the other hand, if we treat his energetics as an essential part of a well-defined *substructure* for field physics, we will be hard pressed to understand how Heaviside's contemporaries understood him and hard pressed as well to understand Heaviside's fluid, creative use of energy transfers and transformations. There is certainly a substructure present, but it is, with equal certainty, not well defined. Heaviside's contemporaries took his work very seriously indeed, yet they extracted what they wanted from it, not what he intended by it. Heaviside himself would have insisted that his understanding of field structure was inextricably bound to his special view of field energetics, even though particular analyses might seem to be independent of it and despite the fact that he used it in extraordinarily different ways, depending upon the circumstances. His British contemporaries generally tried to separate Heaviside on field structure from Heaviside on field energetics, but not always with success. Occasionally elements similar to his energetics were mixed into their discussions, but these elements usually reflected requirements that had little or no force for Heaviside, while his own understanding remained very hard for his contemporaries to penetrate.[1] In this case the connections between Heaviside's perception of the field and that of other Maxwellians, as well as Heaviside's own understanding, are simply too fluid to separate into well-defined higher- and lower-order levels of their common field physics.

Return now to Price's assertions. He argued that experiment has its own autonomy, that it often, and perhaps usually, proceeds without the guide of theory. The image against which Price contended was, it appears, that of a theoretician developing purely on paper something that an experimentalist (or a technologist) builds a device to embody. That image makes the relation between the two similar to a cliché of technology history: the successful Edison designed; his technicians built. The theorist, by analogy, draws the blueprint, and the experimenter constructs the thing. It may be that something like this occasionally occurs in experiments that reveal novel effects, but it seems unlikely. For by the time an experiment can be done according to a blueprint, the aspects of the science it makes contact with are dead, and the center of the discipline has shifted elsewhere. Consider Hall's work. The experiment that he designed was *initially* based on the fruitful conflation of two thoroughly different conceptions of electric current, namely James Clerk Maxwell's with Wilhelm Weber's. In his case there was certainly no well-drawn theoretical blueprint at all to work from. He worked much in the fashion that Price asserted, following a half-developed hunch or pressing the apparatus into new forms. Hall did not set out to test a theory, in any reasonably precise sense, because he had no reasonably precise theory to test.[5]

This brings me back to the question addressed by the example of Hall and Heaviside, namely whether or not a theory can be thoroughly codified. If it is indeed the case that a living science cannot be constrained by definitions and axioms, then Price *must* be correct. Without this kind of structure, there is nothing with which to draw the sorts of plans that, Price asserted, experimenters never use. Seen this way, Price's point concerning the autonomy of experiment changes from a descriptive generalization about what experiments have been like into something closer to a necessary characteristic of scientific behavior. He is correct not because experimenters simply pay little attention to theory but because there is usually no theory *of the right kind* for them to use. By the time such a theory freezes into shape, experimental interest has shifted elsewhere. But if theories of the right kind are rarely present, then what is? Or put another way, how does an experimenter decide to do something, and what kinds of connections exist between experimental practice and the kind of theory that does usually exist while the science is still vibrant?

No general answer to these questions can be given, because laboratory work and theoretical analysis vary tremendously from place to place and from time to time, not only in their contents but even in the kinds of things that they are. For example, it hardly seems worthwhile, or perhaps even meaningful, to compare an experiment done on a tabletop by one or two people over a couple of months with one that involves acres of apparatus and dozens or even hundreds of participants. The negotiations among the participants and between them and the wider scientific community, the tactile manipulation of apparatus, and the manifold purposes of experiment all eventually transform beyond recognition as the scale inflates. Consequently, in what follows my remarks are for the most part confined to the kinds of experiments that were performed in times before gigantism invaded physics. Given this limitation, it is possible to understand what experimenters were about in a way that captures, at least for nineteenth-century physics, elements of Kuhn's insight into the nature of scientific behavior.

2 Overt Theory and Unarticulated Knowledge

One of the great changes in physics began to take firm shape during the 1880s, though its roots go back at least a half century, at first primarily in Germany. Experiments before then were usually done by people who were just as deeply involved in shaping conceptual apparatus as they were in making and using laboratory devices. This began to change rapidly after the 1880s, and Heinrich Hertz is among the last of this long line of experimenting theorists, or theorizing experimentalists—it hardly makes historical sense to distinguish people in this way. These people thought of themselves and of their careers in very different ways from later physicists, who differentiated into distinct genera (theorist or experimentalist, with many associated species) early in their professional lives. For people like them—like Hertz, or his master, Hermann von Helmholtz, or, early in the century, Augustin Fresnel, the French founder of the wave theory of light—experiment was closely bound to theory. For later physicists, experimentalists, laboratory work may proceed largely on its own, though in ways that can be fit into the visions of theorists.[6]

However, it does not follow from this that nineteenth-century experimental practice among such people was determined by theory in Price's sense. On the contrary, the close connection between abstraction and laboratory practice strengthened or even exacerbated the quality that helped to make theories fruitful: their resistance to encapsulation in closed, well-ordered schemes. Unfortunately, theories are not the only things that resist encapsulation; so does talk about theories and experiment. Consequently, I will not attempt an abstract, logical analysis but will again turn to an example, drawn from the history of optics, that illustrates how experiments *are* molded by something, but not what the *OED* or many historians and philosophers evidently mean by the word "theory."

In 1809 Etienne Louis Malus, a French Polytechnicien, discovered that a property previously evinced by light only in crystals could also be produced by reflection. He named the property polarization and constructed a device to measure it. To work Malus's "polarimeter," you turned a mirror about a ray of light that was first reflected from another mirror and watched for the light to vanish. Malus was, like Hertz seventy years later, an experimenting theorist (or theorizing experimentalist). He built his device with a certain understanding of what was going on, but an understanding that did not generally prescribe what would happen in given circumstances. It did, however, carry a world of implications for how a polarimeter should be built, for how it should be used, and for how to think about the images that it might produce. In particular, Malus was thoroughly convinced that, except under certain very special circumstances, previously polarized light passing into his device can never be completely extinguished by it. He saw that this was so in the laboratory, and the same understanding that had guided the construction and use of his polarimeter *required* that it should be so. For him, this was not even a point worth explicitly making, because it was tightly bound to the deepest reaches of his optical conceptions. A decade later Fresnel took a polarimeter exactly like Malus's and used it to examine the same situation that Malus had. But where Malus had seen a significant splotch of light, Fresnel saw nothing at all. Indeed, he was puzzled by Malus's claim to have seen something, because he could not understand how such an able experimenter could have been so badly mistaken.

The overt conflict between these two claims, which could hardly be more closely tied, it would seem, to unmediated observation, reflects a divergence so great in their optical conceptions that even in so starkly simple an experiment as this agreement could not be reached about what the mirror showed. But we are not, let me quickly add, in the realm of irrationality or blindness here. The fact is that both Malus and Fresnel were correct, because Fresnel undoubtedly did see a little light. But for Fresnel, such a pale light was the trivial and unavoidable result of imperfections in the apparatus; for Malus, the light may have been weak, but it had nothing to do with instrumental problems. It had to be there because its presence betrayed something very important. At that time, and indeed for decades afterward, no instrument could have distinguished between the two assertions.

What is even more interesting for our present purposes was that Fresnel plainly did not understand how Malus could have been so thoroughly mistaken. Yet I claim that Malus was not mistaken, that he had *good reasons* for his assertion. For, according to Malus's calculus for rays, the situation in which (according to Fresnel) no light at all should be visible is precisely the one where the light should be extremely small. This was what Fresnel missed.[7] It was not the case, in other words, that Fresnel, having read Malus's work, saw how the latter had applied it to reach a contrary conclusion, but that Fresnel rejected Malus's way of doing optics. Not at all, for according to Fresnel, neither Malus nor indeed anyone else before Fresnel himself could possibly have computed anything at all in such an experiment. What Fresnel accordingly missed, and what accounts for his profound surprise at Malus's claims, was the very existence in Malus's work of a way to deal with polarization experiments. Fresnel did not disagree with Malus about the latter's theory. It simply never occurred to him that Malus had a theory.

In one sense of the word "theory," in Price's sense, Fresnel was right about Malus. Malus did not have anything based on the well-articulated principles of the emission theory that could be deployed to calculate, as Fresnel knew that wave principles (though not ether structure) could be. Fresnel missed Malus's ability to compute because he did not distinguish the overt principles of particles and forces from the unarticulated core of discrete rays, a distinction that neither Malus nor any of the large number of optical scientists in the 1810s who thought like him ever made explicit. This is hardly surprising, because it is extraordinarily difficult to bring these notions to the surface, or better put, to freeze them into a rigid set. Fresnel's own optical conceptions have this same characteristic, only it betrays itself in very different environments from the situations that reveal it in Malus's case.

Neither Malus nor Fresnel were guided in the laboratory by anything like a set of codified, or even codifiable, assertions.[8] Price was correct about that. On the other hand, something did guide each of them in their several ways. Something permitted Malus to convince himself through calculation that the light was what it should be. Something permitted Fresnel, in a very different context, to insist on distinctions between kinds of polarization that for many years made little sense to many of his contemporaries. This kind of thing could rarely be used to provide a blueprint for experiment, but it could powerfully structure the kinds of experiments that seemed important to do; it also provided tools for understanding, and for calculating things about, experiments. It sat for the most part well below explicit discourse, subtly influencing its tone and powerfully guiding the construction of sentences and their material realization in the laboratory.[9]

Kuhn's work was the first not only thoroughly to argue that such things do occur but to insist they must occur, and to insist also that they occur in context. The power of the conception that guided Malus derived in substantial part from the fact that it was continually reinforced in manifold ways by his teachers, by his colleagues, by contemporary texts, indeed by the competitive, striving environment of the French *physicien* of the day, an environment that fostered certain kinds of research. It was not easy to find out how to get to the cutting edge of French optics circa 1810, but whoever succeeded in doing so well understood how to deploy an array of concepts—some overt, some covert—in the laboratory as well as on paper. By that time or shortly after, the integrative power of the system was well embedded in Parisian scientific institutions in the persons of people like Pierre Simon de Laplace, Siméon Denis Poisson, Jean-Baptiste Biot, and others. It permeated their language and wound through their careers. But it was not so easy to learn just what was expected, particularly since so much was left unstated and since the kind of training that became common with the foundation of research seminars in German universities decades later did not then exist. Failure thoroughly to assimilate the scheme could be quite devastating to one's career prospects, as François Arago discovered to his despair in the early 1810s.

Unlike Malus, Augustin Fresnel was not a fellow of this intense, tightly-knit scientific community, though he had been trained by its members, or by collateral members of it, at the École Polytechnique. He was, however, driven by a powerful urge for recognition by them, one that eventually coupled fruitfully with his distance (both literal and metaphoric) from active optical researchers and with the wounded pride of a disgruntled François Arago, who used Fresnel for his own career goals. The unarticulated views that Malus and other researchers shared and deployed were accordingly not so deeply embedded in Fresnel's outlook, though they were certainly present and caused him a great deal of trouble in creating new structures based on wave concepts.

Many of the difficulties with terms like "paradigm" or even "disciplinary matrix," pointed out over the years not only by his critics but by Kuhn himself, are at least in part oblique reflections of parallel difficulties in codifying the subjects of these kinds of discussions. If it is difficult to imprison fertile scientific schemes in axioms and definitions, then it seems reasonable that claims about such things will be equally difficult to codify. Commentators have written of Kuhn's "quasi-poetic" style, which some readers find confusing and even irritating. Given the nature of his subject, I do not see how it could easily be otherwise. From this point of view, the following passage from Kuhn's *Structure* is particularly significant:

As the student proceeds from his freshman course to and through his doctoral dissertation, the problems assigned to him become more complex and less completely precedented. But they continue to be closely modeled on previous achievements as are the problems that normally occupy him during his subsequent independent scientific career. One is at liberty to suppose that somewhere along the way the scientist has intuitively abstracted *rules of the game* for himself, but there is little reason to believe it. Though many scientists talk easily and well about the particular individual hypotheses that underlie a concrete piece of current research, they are little better than laymen at characterizing the established bases of their field, its legitimate problems and methods. If they have learned such abstractions at all, they show it mainly through their ability to do successful research. That ability can, however, be understood without recourse to hypothetical rules of the game. (Kuhn 1970, 47; emphasis added)

The hypothetical "rules of the game" that Kuhn refers to go greatly beyond the notion of a codified theory—one that is or can be laid out in explicit definitions and axioms—but they include it.[10] According to Price, codifications hardly ever guide experiment; according to Kuhn, they scarcely exist at all.

The unarticulated structures that I have been referring to are uncodified and extremely flexible. They do not constitute "rules of the game" in a strict or even in a somewhat looser sense. They are essential for productive research to take place, but they do not exist independently of a set of particular problems whose solutions are treated as canonical.[11] For these things to become explicit, they must, as it were, be detached from the particular and made general, which inevitably means that they must be transformed from guidelines into rules. One might assert, in opposition to the claim that such rules do not exist, that it is a matter of maturity. Perhaps unarticulated structures are hidden only at early stages in a discipline's development, but they are eventually cast into a rigid form and brought to light. It is undoubtedly the case that something like this does often occur. However, the historical reality is that while the science lives, while it stimulates research and attracts new practitioners to compete in developing it, the fundamental structures remain very flexible and are *therefore* hardly ever visible. In the transition to the wave theory, for example, nobody ever did articulate the critical points at issue, despite the fact that individual practitioners were able very nicely to fit the implicit scheme to the demands of the moment by following the solution patterns for earlier problems.

Underlying conceptions rarely come into the open even when two groups come into direct conflict with one another. This is a rather startling claim, because after all there is ample evidence of a great deal of highly contentious argument throughout the history of science. The rise of the wave theory is particularly famous in this respect, because its adherents certainly argued vociferously with partisans of the old optical way, and vice versa. But it is always important to pay close attention to what the arguments are actually about. They are not in general about the kinds of things that I have been discussing. Arguments may revolve about physical issues (was ether a better physical foundation than light particles?) or about analytical complexity (were Fresnel integrals less acceptable as primary apparatus than equations for particle motion?) or about deductive integrity (did the wave or the particle adherent have to monkey extensively with basic physical principles to get where he wanted to go?) or about appropriate goals for the subject in context (was it, for example, important to provide formulas for optical intensity in situations that held out no contemporary hope for measuring such things?).

Arguments of this kind did not at first seem problematic to the participants because the issues, though difficult and even upsetting, seemed to be straightforward. It was possible, for example, to examine in careful detail the various physical assumptions of particle optics and to criticize the structure for not adhering consistently to them, which was precisely what Fresnel did.[12] Or conversely, one could argue that the physical principles underlying ether structure could not bear the weight that Fresnel's mathematics put upon them, which was often pointed out by even by early wave scientists (who, however, used it to establish a research tradition in ether structure, not to challenge wave principles). In either case the terms of the argument appeared, initially at least, to be relatively clear-cut, in part because the physical principles involved on either side were quite similar to one another.[13] Yet despite this comparative clarity, Fresnel did not thereby convince many people to change their optical beliefs, nor were critiques of ether mechanics devastating to early wave partisans. And the reason for this was that neither critique could reach very far into its target's true core, a core that in both cases was powerfully embodied by physical models but that nevertheless could survive extreme mutations in those models. When arguments did (rarely) reach the core, they inevitably became highly problematic and resulted in anger, exasperation, or eventually, in attempts to mix the immiscible.

The demarcation between the unarticulated core and the overt arguments that derive much of their strength from it can hardly be rigidly specified, since the core itself cannot be. Nevertheless, historical evidence does indicate that the core remains sufficiently rigid that its stamp can be discerned even when the overt structures evolve markedly. This is so in every kind of scientific analysis, from the most abstract to the most concrete. It holds for paper arguments divorced from the particulars of experiment, and it holds for laboratory life as well. There is, however, a difference in the ways in which the unarticulated core functions at either extreme. In paper arguments the core molds an abstract language that seems to be divorced from material instrumentalities, whereas in laboratory work the core molds something that directly concerns instrumentalities. In shaping high theory, the core operates on something that, though vastly more rigid, is like itself. In the laboratory the core functions in this way as well, but it also functions in a different way, because it shapes objects, laboratory devices, that do not exist only on paper.

There are certainly innumerable instances in which explicit theory powerfully guides experimental work, work in which the devices themselves directly reflect precise theoretical structures. However, these kinds of cases occur primarily, perhaps only, within a well-established but unarticulated tradition. Because the tradition is so well established, it can be exceedingly difficult to disentangle the tradition from the overt structures that the experiment embodies. Far more significant to that end are experiments undertaken before the implicit tradition has stabilized and so before it has been encrusted with an overlay of explicit concepts and concrete devices that depend upon it but that also tend to conceal it. Hermann Helmholtz's Berlin laboratory in the 1870s and early 1880s provides a striking instance of such a powerful but largely unarticulated tradition.

In the late 1860s Helmholtz developed a new way to think about physical interactions, one that differed not only from contemporary belief among German physicists but that differed as well from the various strains in British physics. Many German physicists, particularly those who were trained by Wilhelm Weber or by his students and assistants, sharply divided theory from experiment. For them, the objects of theory necessarily transcended laboratory experience, and the relationships between those objects held between things that, it was thought, could not be directly manipulated. Hermann Helmholtz and the group of assistants, visitors, and students that formed in Berlin during the 1870s and early 1880s made no such division in their practice, though they rarely, if ever, brought their understanding to the surface. For them, objects of theory and objects of experiment were, or at least should be, one and the same. This difference between the Weberians and the Helmholtzians, as I will call the two groups, has roots that run deep in contemporary ideological, social, and organizational issues, but I want here to concentrate on the ways in which the difference betrayed itself in the laboratory.

Mathematics serves very nicely to embed unarticulated knowledge and to disguise its presence. Most competing groups, among them Weberians and Helmholtzians, hold a great deal of mathematical structure in common, although the ways in which new propositions can legitimately be linked to the existing structure may be (and in this case are) strikingly different. Nevertheless, the very existence of a mathematical form tends to impart a sort of concreteness that anchors argument. When Helmholtz introduced his new electrodynamics in the early 1870s, he embodied it in something called a potential function, and he claimed that this function could, by appropriate choice of constants, represent all contemporary forms of electrodynamics.[14]

The essence of Helmholtz's new and largely unarticulated way for physics, put rather crudely, consisted in its assumption that interactions between objects are determined by only two things: first, the *states* of the objects at a given instant and, second, the distance between them at that same moment.[15] These states are for the most part not considered to be reducible to anything else; they are qualities of objects that can be assigned numerical values (such as a state of charge or a state of conduction or even a state of strain). Then, according to Helmholtz and his group, a particular interaction is represented by a so-called *potential function* that embodies the energy stored in the bipartite system formed by the objects in particular states at a given distance from one another. This determines as well the mathematical apparatus that Helmholtzians used. In order to find out what kinds of forces occur the potential function must be perturbed a little—in analytical terms, it must be varied. Variational techniques consequently lay at the heart of Helmholtzianism, and this fact has important implications for laboratory life in that environment. Translated into laboratory technique, the variational apparatus suggested that its instrumental analogue had to have either characteristics that changed over time or parts that changed their mutual positions. In either case the device could not be static, although the changes involved could be extremely small. During the 1870s the Berlin laboratory concentrated on investigating novel effects that depended on changes in an object's state of charge. The apparatus used (in particular oscillating currents and electrostatic generators) were not put together in this way anywhere else. The novelty extended to other fresh areas as well, such as the investigations of gaseous discharge by Eugen Goldstein, who built devices to look for things that revealed Helmholtzian properties.

It is scarcely an exaggeration to say that the unarticulated core of Helmholtzianism suggested ways to tailor devices for discovery. According to this core, nothing prevented the existence of hitherto unknown object states and interaction energies whose effects could only be discovered by fermenting changes in instrumental states and configurations. Indeed, the goal of research was precisely to discover new states and interaction energies through laboratory manipulation coupled to variational technique. Such things could not possibly be known a priori. For that reason Helmholtzianism was, or seemed to be, the apotheosis of empiricism, of a view that made the laboratory the essential locus of scientific knowledge and creativity. Helmholtzianism's core was not a blueprint for calculating; it was a *design for experimenting*.[16]

The experiments that were done in the Berlin laboratory did not succeed in eliciting the potential's novel effects. Helmholtz had them performed with great care and several times. Repeated failure in the laboratory eventually convinced him that something was not right. What did he do? Did he throw out the potential and the entire new physics that it embodied? Not at all. He turned for the first time to theory, to overt, articulated knowledge. Until he was convinced that the experiments had failed (and this took some time, as he had the experiments refined and redone), Helmholtz had kept the laboratory uncontaminated by theory, since in his eyes the potential itself was not truly theoretical. It could accommodate different theories by assigning particular values to the arbitrary constant that it contained, but he was and remained convinced that the potential was beyond suspicion. For to abandon the potential meant abandoning as well the galaxy of unarticulated views about objects, states, energies, and indeed about many wider things, including teleology, idealism, and perhaps even politics, that were central to his way of thinking.[17] Accordingly, the laboratory failure had to be transformed into positive rather than negative evidence, and the only way to do this was to introduce theory, which amounted to adding something else. That something else was the highly articulated, overt electrodynamic ether, which Helmholtz (though not his assistants, such as Eugen Goldstein) had previously kept at arm's length precisely because it was not easily connected to Helmholtzianism's implicit core. Given the ether Helmholtz could plausibly, though hardly definitively, transform failure into success.[18] This occurred shortly before the ambitious but stubborn young Hertz arrived in Berlin in 1878.

Consider the implications of these events for the distinction between implicit and explicit knowledge and for their appearance in the laboratory. Once carried out, the Berlin experiments could be understood without direct recourse to the unarticulated structure of Helmholtzian physics. Concentrating solely on the potential function, one could first see how it leads to certain forces, and one could also understand how the particular experiments tested for them. This much was comparatively unproblematic. However, the special design of the experiments, their use of unfamiliar apparatus in strange new ways, was itself buried deep in the unarticulated core. Given only the potential, the route to appropriate experiments was hardly well lit. But given the core's understanding that interactions depend upon variations in object states or distances, experimental designs of a certain common type are readily envisioned. One might put it (coarsely) in the following way. Interesting experiments for Helmholtzians require things to move about, at least a bit, or to change their states. Look around, Helmholtzianism's core suggested, for devices that will do the trick and see if they interact with one another. If you have problems eliciting effects then you had better look to theory to see where you went wrong. Theory, then, was Helmholtzianism's inevitable recourse in the event of instrumental recalcitrance. It is hardly surprising that the focus of subsequent work among Helmholtzians tended to be the polar ether, the theory, rather than the potential or the unarticulated apparatus that lay behind it.

Helmholtz's laboratory during the 1870s was, then, the site of attempts to stabilize the unarticulated core of Helmholtzianism. During these years the Berlin experimenters were trying to find ways to elicit effects that were closely bound to general Helmholtzian precepts through the particular properties of the electrodynamic potential. It would not be correct to think about their activity as in any sense an attempt to test these precepts. Far from it. The precepts were themselves hardly ever voiced, much less questioned. It is most fruitful to think about the Berlin experiments as attempts to freeze Helmholtzianism into laboratory objects. To do so, new devices had to be built, and old ones had to be used in new ways. The investigators were well aware that the behavior they were seeking lay at the margins of their technical abilities, but over time they refined their apparatus in ways that, they were convinced, could produce meaningful signals. Had they succeeded in this quest then Helmholtzianism would have been directly realized in the concrete. It would thereby have been stabilized, in the sense that future investigations would have taken it entirely for granted. The experiments did not, however, succeed in producing stabilization, with the result that Helmholtz himself had to inject the overt but difficult concept of the polar ether into the structure's implicit foundations.

One might say (using Kuhn's original terminology) that Helmholtzianism's failure to achieve stability means that it never passed fully beyond the "preparadigm" stage. It was coherent, and it most certainly could motivate special kinds of research, but because no one succeeded in eliciting effects that seemed to depend uniquely upon it, Helmholtzianism did not evolve the kind of research activity that, for example, Maxwellianism did. A community of Helmholtzian physicists did exist, one whose members had a special way of approaching problems, but the community never generated a set of canonical problems that clearly imparted structure by example and that suggested new research paths. The necessity of incorporating the problematic polar ether into the pattern evidently vitiated its coherence. These developments are no doubt unique, but they do suggest that the difference between the unarticulated substructure held by a discipline and a voiced superstructure can be particularly important in understanding the discipline's ability to generate results. The examples of Maxwellian physics, optics early in the century, and Helmholtzianism suggest that the distinction between sub- and superstructure must be very robust for progress to occur.

French optics in the 1810s was an extremely progressive discipline—it generated new research topics that investigators succeeded in carrying through in the laboratory—in major part because the boundary between the implicit substructure (which treated light as a set of discrete rays) and the explicit superstructure (which treated light as particles governed by forces) was quite strongly maintained. When the boundary did begin to waver, as Biot occasionally injected particles into ray discussions, a path to critique was opened for Fresnel. Maxwellianism derived its considerable empirical strength from its elaborately orchestrated distinction between field behavior and ether properties. It began to disintegrate only when Joseph Larmor attempted to pull these two things together in novel ways.

In both of these cases, empirical arguments were very strong, indeed central, factors in the process of disintegration, but not because laboratory effects were at the time in clear conflict with anything. What happened was much more complicated than that. An outsider to the discipline (Fresnel or Larmor) took difficulties that active researchers had not considered to be important and made them explicit: these difficulties all revolved about forced mergers between two areas that had been kept quite separate. This would not necessarily have destabilized the research traditions, but it did so in these two cases in part because both Fresnel and Larmor were able to acquire powerful allies who were directly involved in ongoing research but who nevertheless had some social and intellectual distance from it (George Francis FitzGerald and Lodge in Larmor's case), or else who were in positions of influence but who had strong reasons for wishing to destabilize the tradition (Français Arago in the case of Fresnel). In other words, these two research traditions disintegrated as a result of external pressures brought to bear at internal weak points.

In Helmholtzianism we have the extraordinarily interesting case of a tradition that remained permanently unstable but that nevertheless succeeded in motivating a great deal of research, though perhaps not in so direct a manner as Maxwellianism or French optics. But because it remained unstable, Helmholtzianism was peculiarly subject to being transformed into something else, and this is what eventually happened. By the early to mid 1890s German texts commonly began their route to field theory with Helmholtz's polar ether and proceeded to generate from it (by a process of physically meaningless passage to limits) equations that looked like the ones that Hertz had developed for Maxwell's field. Few of these texts discuss Helmholtz's potential in any detail, though most of them mention it. Here we have Helmholtzianism, whose purest expression was the electrodynamic potential, transformed through its destabilizing superstructure (the polar ether) into the Hertzian field. Nothing similar happened to Maxwellianism, which seems to have been abandoned between circa 1895 and 1905, or to French optics, which was thoroughly replaced by wave optics by circa 1840.

3 Translation

An attempt to understand

The distinction between unarticulated and overt knowledge that I have pursued through these examples raises the long-debated issue of *translation* between theories. It is at the level of unarticulated structure that attempts at translation are liable to be made, and then to fail markedly, for concepts here may appear to be separable from their particular context precisely because that context is not overt. Two decades ago Kuhn made the following remarks in a new postscript to the *Structure*:

Briefly put, what the participants in a communication breakdown can do is recognize each other as members of different language communities and then become translators. Taking the differences between their own intra- and inter-group discourse as itself a subject for study, they can first attempt to discover the terms and locutions that, used unproblematically within each community, are nevertheless foci of trouble for inter-group discussion. . . . Having isolated such areas of difficulty in scientific communication, they can next resort to their shared everyday vocabularies in an effort further to elucidate their troubles. Each may, that is, *try* to discover what the other would see and say when presented with a stimulus to which his own verbal response would be different. If they can sufficiently refrain from explaining anomalous behavior as the consequence of mere error or madness, *they may in time become very good predictors of each other's behavior.* Each will have learned to translate the other's theory and its consequences into his own language and simultaneously to describe the world to which that theory applies. That is what the historian of science regularly does (or should) when dealing with out-of-date scientific theories. (Kuhn 1970, 202; emphasis added)

The emphasized word and clause in this passage are particularly important because, I think, this kind of effective penetration into an alien vocabulary and grammar rarely occurs among practising, creative scientists. It certainly did not occur in any of the cases considered above, despite, in one instance at least, an intense effort to do precisely what Kuhn describes.

Heinrich Hertz, who remained in several important respects Helmholtzian even after his creation of electric waves, grappled mightily with Maxwell's *Treatise* and also (though the evidence here is indirect) with Maxwellianism. Despite his best efforts, he simply could not make consistent sense of Maxwell's remarks concerning charge. Hertz's comments in the introduction to his *Electric Waves* are certainly a confession of failure to comprehend Maxwell, but they are also more than that. Taken in context, they reflect as well his failure to translate Maxwell's language into his own language, or, one might say, to map the unarticulated core of Maxwell's scheme onto the unarticulated core of Helmholtz's. Yet Hertz was convinced that Maxwell did make consistent sense. He knew that he was missing something, but he simply could not make out what it was. He wrote, "If we read Maxwell's explanations and always interpret the meaning of the word 'electricity' in a suitable way, nearly all the contradictions which at first are so surprising can be made to disappear. Nevertheless, I must admit that I have not succeeded in doing this completely, or to my entire satisfaction; otherwise, instead of hesitating, I would speak more definitely" (Hertz 1962, 27). Concentrating on the one thing that seemed to give the most trouble as he read Maxwell's work, namely the word "electricity," Hertz tried to read it in context in a way that gave meaning to the sentences in which it occurred. Frequently he was able to do this. But as he read on, a subsequent passage seemed to require a new reading for the word, one that was not at all consistent with the previous one. This happened frequently enough, he confessed, that he "hesitated" to assert just what it was that Maxwell held.

Other passages in Hertz's introduction make clear that he tried to fit Maxwell's "electricity" somewhere into a scheme in which charge must be thought of as the state of some thing, whether that thing is a dielectric or a conductor. This leads to an inevitable dualism in the concept that is completely lacking in Maxwell's understanding of it. Hertz's difficulties undoubtedly occurred as he tried to map this dualistic structure onto Maxwell's unitary conception. It cannot be done consistently, although in a particular passage it is occasionally possible, as Hertz discovered. These sentences are accordingly a confession of Hertz's literal failure to *translate* Maxwell's language into something that he could make sense of. Yet no comparable difficulties occurred for British Maxwellians. They frequently resorted to analogies and examples in order to convey Maxwell's conception of charge; they were implicitly convinced that it was consistent, and they were indeed correct. Maxwell's conception, though utterly incompatible with a reading of Hertz's kind, fits without conflict into a scheme based entirely on field properties.

An attempt to expropriate

Hertz's attempt literally to translate Maxwell's "electricity" into something that he could understand is perhaps not as common as another kind of activity that often occurs when a new scheme has (for whatever reasons) made inroads. Attempts are made to *expropriate*, not to adopt, useful novelties. Whole categories may be taken over from the new scheme and linked to an alternative, existing structure without at the same time carrying over the novelty's polluting effects. Precisely this took place in early nineteenth-century optics when David Brewster attempted to expropriate Fresnel's concept of phase without abandoning the unarticulated core of rays, without, that is, adopting Fresnel's new scheme, in which phase, together with amplitude, distinguished between the main categories of polarized light.

Fresnel's wave scheme had markedly changed the role of polarization in optics. There were, according to its (largely implicit) scheme, two major categories for light: the polarized and the unpolarized, corresponding respectively to light with a fixed phase difference between its orthogonal components and light with a random phase difference between them. Each category had subdivisions. Polarized light divided into linear and elliptical; the latter in turn divided into circular and noncircular. Unpolarized light divided into completely unpolarized and partially unpolarized.[19] None of this would have much impressed Brewster were it not for the wave optician's ability to produce a device, the Fresnel rhomb, that behaved in a way that seemed closely tied to the demands of Fresnel's phase. Ray optics, however, sorted polarized light into only three categories: the polarized, the partly polarized, and the unpolarized. The problem for Brewster, given the apparent instrumental usefulness of 'phase', was

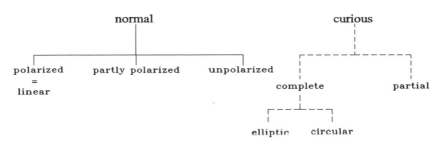

Figure 1
Brewster's novel structure for light

how to expropriate it without also adopting its original function in specifying the temporal relationships between the components of an oscillation, without, that is, adopting the wave scheme.

He did so in the following way (see figure 1).[20] Instead of modifying his existing, ray-based categories, Brewster decided to create a parallel form of light. He referred to it as "curious" light, and for it he built two subcategories, one of which in turn had two subdivisions: there was complete (curious) light, and partial (curious) light; complete light divided into elliptic (complete) and circular (complete). "Phase" in Brewster's scheme specified the degree of completeness: partial light had a "phase" less than 90°; complete light had a "phase" equal to 90°. This could be realized instrumentally by passing light that had been reflected at various angles from metal surfaces through a polarimeter. As for the difference between elliptic and circular (complete) light, Brewster had little to say, other than that the former is produced by metallic reflection, the latter by whatever happens in Fresnel's rhomb.

My point here is not to draw the reader into the rather complex mysteries of Brewster's scheme but rather to indicate the kind of activity he was engaged in.[21] He could not carry over the phase- and amplitude-based categories for polarization that wave optics used into his own scheme, because his categories could not subsume them. He could not, that is, make the novel wave categories of circular and elliptical, which he deemed to have meaningful laboratory life, subsets of his own categories, because e.g. according to the wave scheme circular light is polarized, whereas according to ray optics it is not polarized at all (since it is completely symmetric in a polarimeter). Accordingly, Brewster invented a new, parallel form of light, providing for it a mathematics designed to mimic the mathematics of phase in wave optics.

Brewster accordingly thought that he could rip "phase" loose from its moorings to wave concepts. The result was to permit certain kinds of manipulations that have no meaning at all in wave physics (but with which he was entirely comfortable). His use of the wave concept of elliptical polarization, for example, made no sense whatsoever in wave contexts; it did make a great deal of sense in ray contexts. This is why Brewster's work should be thought of as an *expropriation* rather than as a *translation* of wave vocabulary, for the sentences that can be constructed using Brewster's vocabulary cannot in their turn be translated into anything meaningful using wave vocabulary. Brewster took wave words and used them in his own way.

This sort of thing does, I think, regularly occur in science. Words or even sentences are pulled away from their context and attempts are made to translate them in near isolation into something meaningful. A new vocabulary may result from the attempt, even one (as in Brewster's, though not in Hertz's, case) that works; that is, consistently meaningful sentences can be constructed from it. But the process, if continued, will inevitably fail in one of at least two ways. Attempts to decode further sentences from the original using the new vocabulary will eventually produce nonsense (Hertz). Or, reversing the process, attempts to regenerate sentences in the original language by translating ones forged in the new vocabulary will eventually manufacture nonsense in the original (Brewster).

Practising physicists, particularly in the midst of debate, are not very much like historians. They are even less like philosophers. Nearly everything about their training, goals, career patterns, and beliefs presses them not to undertake the arduous effort of understanding a different point of view to such an extent that they can do different physics. During an individual's transition to something novel, it does happen, perhaps regularly, that a previous vocabulary and grammar are partially reworked, but this is not because the individual during this process successfully translates between two points of view. When, for example, John Herschel went through the formidable transition from rays to waves sometime between 1825 and

1827, he attempted for a time to adapt old language to the new environment. But in so doing, he was already operating almost entirely within the new structure, with the inevitable result that he found it impossible to produce a consistent dictionary. At some point he lost the ability to manipulate the old structure, or perhaps it would be better to say that he saw the old structure fracture as he tried to bend parts of it to the new framework.[22] He was well aware that something was wrong here, much as Hertz was aware of something wrong in a different context sixty years later.

4 Relativism

In a 1970 reply to his critics Kuhn had this to say about the charge that his views lead inevitably to relativism:

It must already be clear that my view of scientific development is fundamentally evolutionary. Imagine, therefore, an evolutionary tree representing the development of the scientific specialities from their common origin in, say, primitive natural philosophy. Imagine, in addition, a line drawn up that tree from the base of the trunk to the tip of some limb without doubling back on itself. Any two theories found along this line are related to each other by descent. Now consider two such theories, each chosen from a point not too near its origin. I believe it would be easy to design a set of criteria—including maximum accuracy of predictions, degree of specialization, number (but not scope) of concrete problem solutions—which would enable any observer involved with neither theory to tell which was the older, which the descendant. For me, therefore, scientific development is, like biological evolution, unidirectional and irreversible. One scientific theory is not as good as another for doing what scientists do normally. In that sense I am not a relativist. (Lakatos and Musgrave 1970, 264)

On the other hand, Kuhn continued,

There is another step, or kind of step, which many philosophers of science wish to take and which I refuse. They wish, that is, to compare theories as representations of nature, as statements about "what is really out there." Granting that neither theory of a historical pair is true, they nonetheless seek a sense in which the later is a better approximation to the truth. I believe nothing of that sort can be found. On the other hand, I no longer feel that anything is lost, least of all the ability to explain scientific progress, by taking this position. (Lakatos and Musgrave 1970, 265)

I know from one and a half decades of teaching graduate students that some of them find Kuhn's position to be disturbing, even profoundly confusing; others now take it almost as a matter beyond discussion. It is felt either that there must be some set of extrahistorical criteria for deciding between alternatives or that scientific change must be treated primarily as a matter of taste and social opportunism. This means that Kuhn's views are extraordinarily hard to grasp for many people, including those who trivialize them by running with arms spread wide into the embrace of full-blown relativism. Without claiming to offer anything at all new, I would like to close by briefly considering the problem.

Begin with the question posed by the first passage quoted above, namely whether it is possible to tell abstractly which of two schemes is the ancestor, supposing that one is in fact a descendant of the other. As a case in point, take geometric optics and the wave theory of light.[23] For doing what geometric optics did, to paraphrase Kuhn, the wave theory per se is useless. More than that, it avers that no such thing as geometric optics in the traditional sense can possibly exist. This seems to pose a difficulty, because it means that the *wave theory* "is not as good as [geometric optics] for doing what [geometric opticians] normally do."

The phrases in brackets respectively replace Kuhn's "another" and "scientists," and in so doing, they mislead, but, I think, in a fruitful way. The problem here is that what wave theorists do is simply not, in general, what geometric opticians did. Interesting wave problems, and the instruments and experiments that are bound inextricably to them, by and large do not intersect with interesting sets of problems in geometric optics. This is because the most fundamental notion in wave optics, which is the concept of phase, has no analog at all in geometric optics, and so the constellation of problems that embody it has no geometric analogs. But how, then, can one tell abstractly whether the wave scheme is a descendant of the geometric scheme or vice versa?

I think that a partial answer is fairly easy to give but extremely difficult to carry out. Moreover Kuhn in effect has already given it.[24] It is (differently formulated) a very old answer, namely, that the descendant will explicitly seek to accommodate certain results of the ancestor by attempting to force relations within (the descendant) scheme that can mimic relations in the ancestral scheme. This is exceedingly difficult to do, even in the apparently straightforward case of geometric and wave optics, in major part because subjects that were previously well understood become problematic, and as a result, their properties must be retrieved in rather complicated ways. Nevertheless, I know of no case in which such a thing (relation forcing) could be done in the ancestor to generate otherwise absent, unforced concepts in the descendant.[25] It should be clear from this that there are important properties that bear the unidirectional imprint of time's arrow.[26]

However, the kind of progressive incorporation represented by the absorption of geometric into wave optics is not, perhaps, what most philosophers and many historians have had in mind when grappling with the notion of progress in science. For the most part the intriguing issues do not revolve about how new schemes retrieve old relations but rather about how one scheme comes to replace another. Returning again to optics, interest has long centered on how *light as an ether wave* replaced *light as rapidly moving particles*, something that unquestionably began in both France and England sometime during the late 1820s. This change is not the same kind of thing as the one that we have been considering, because the core of the scheme that is replaced may not transfer in any way at all into its successor. It would, e.g., be thoroughly incorrect to assert that ether physics can in any meaningful sense reproduce physical relations between optical particles, though it is indubitably correct to say that periodic disturbances can be manipulated *nearly* to retrieve, in the proper circumstances, the geometric relations between rays that, according to particle optics, are always exactly correct.

In this case of optics, in fact, the wave scheme incorporated *nothing at all* that was particular to the system that prevailed in France during the early 1800s. Referring back to Kuhn's image of the branching tree, we might place the two forms of optics on parallel branches, both of which run back to geometric optics, despite the fact that French ray physics temporally precedes wave physics. For neither of these two branches incorporates anything present in the other that does not descend from their common ancestor. The historical events that culminated in the diffusion of wave concepts concern its conflict with ray physics, not with the ancestral geometric optics. Moreover, this diffusion was substantially complete *before* the production of phenomena that ray physics could not deal with at all, and indeed, it took place without elaborate experiments that depended unambiguously upon Fresnel's central novelty, his understanding of polarization.[27]

This pair of examples greatly complicate the issue of relativism, because we cannot in either of them point to something general and unambiguous (such as the retrieval of the results of one of them by the other) that permits us to distinguish between them on the basis of abstract criteria *near the time during which the one displaced the other.* Certainly the partisans of each had very good reasons for advancing their own schemes, and good reasons also for rejecting the other. In the case of wave optics the new phenomenal world that it eventually succeeded in generating makes its case retrospectively persuasive. But what eventually took place cannot possibly influence something that happened beforehand.

This, I think, means that relativism, of a kind, where the initial generation *and* diffusion of a novel scheme is concerned simply cannot be avoided in many cases.[28] Scientists do not know where they are going; they have no more privileged access to the future than do the necromancers who draw up national budgets. What they do however possess is a superbly organized system designed to birth and to control certain kinds of novel phenomena. The world with which they deal is independent; it strongly and frequently resists the molds into which they wish to force it. But it is also dependent upon scientists, because the processes with which they deal are for the most part accessible only through an elaborate, historically conditioned universe of instruments.

I do not by any means intend by this to deny that scientists often argue with deep conviction by citing such things as simplicity, fecundity, comprehensiveness, and so on, during periods when a novel scheme is being developed and diffused. It is however a *fact* that these kinds of arguments are rarely compelling to contemporaries who are intimately engaged in working with a competing scheme. Moreover, they can usually be countered with arguments of a similar kind, and they tend to miss much that is at issue because they generally concern only overt aspects. Consider, for example, what Humphrey Lloyd, an Irish partisan of wave optics, had to say in 1834:

I would observe that any well-imagined theory may be accommodated to phenomena, and seem to explain them, if only we increase the number of its *postulates*, so as still to embrace each new class of phenomena as it arises. In a certain sense, and to a certain extent, such a theory may be said to be true, so far as it is the mere expression of known laws. But it is no longer a *physical theory*, whose very essence is to connect these laws together, and to demonstrate their dependence on some higher principle:—it is an aggregate of separate principles, whose mutual relations are unknown. Thus the cycles and epicycles of the Ptolemaic system represented with fidelity the more obvious movements of the planetary bodies; but when the refinements of astronomical research laid bare new laws, new epicycles were added to the system, until at length its complication rendered it useless as a guide. Such appears to be the present state of the theory of emission. (Lloyd 1834, 349)

Here we see that, according to Lloyd, the "emission theory" cannot act as a "guide" to new physics because "it is an aggregate of separate principles." These principles all directly concern the behavior of optical particles.

Wave optics, Lloyd insists, is nothing like this aggregate, and this is why it can generate new physics. But Lloyd's vision was decidedly colored by his advocacy of the wave theory, as we can at once perceive from his discussion of partial reflection. He explains Biot's physics in great detail and castigates it for being too loose and disconnected, and most of all for not yielding formulae. Fresnel was much better. He could obtain formulae, and he could link them nicely to ether physics! In Lloyd's words, "In the development of his theory [of partial reflection] the character of Fresnel's genius is strongly marked. Our imperfect knowledge of the precise physical conditions of the question is supplied by bold, but highly probable assumptions: *the meaning of analysis is, as it were, intuitively discerned, where its language has failed to guide*; and the conclusions thus sagaciously reached are finally confirmed by experiments chosen in such a manner as to force Nature to bear testimony to the truth or falsehood of the theory" (Lloyd 1834, 363; emphasis added). In other words, Fresnel did not have anything more than Biot did in the way of perfect knowledge of physical conditions of ether structure, but he had something that, according to Lloyd, Biot lacked: intuition. Of course, Lloyd's unfavorable comparison of Biot with Fresnel depended upon Biot's not having any formulae at all, when in fact he (or rather Malus before him) did. Lloyd missed this because he did not perceive the implicit structure of Biot's optics, a structure that was substantially independent of the kinds of criticisms based on the adequacy of models that Lloyd directed at the emission theory proper.

In other examples Lloyd's critique depended upon an unfavorable comparison between the powers of two completely different kinds of things: between Biot's overt physics for optical particles on the one hand, with such things as the principle of interference, Huygens's principle, and phase properties for transverse waves, on the other. Lloyd avoided as best he could invoking ether mechanics, because he was well aware that it was not much superior in power to particle physics, though he also felt that it eventually would be. His arguments were simply not compelling in any absolute sense, or even (in 1834) in a comparative sense, because wave optics was only just then beginning to engender its own particular universe of phenomena. Fifteen years later Lloyd would have had no reason at all to contrast the two schemes, because by then the wave theory was thoroughly embedded in a complex web of optical instruments and practices from which it could scarcely be extricated. Difficulties with ether mechanics had become interesting research problems, not evidence that the wave theory lacked a sure guide to new physics.

I want to close by emphasizing two points. First, though I am strongly convinced that implicit structures are critical for understanding scientific argument and that proponents of different structures often fail in consequence to understand one another, nevertheless I do not argue that this *cannot* be done. It certainly can be done; historians do it. But scientists rarely do, primarily because creative science requires a tremendous effort of concentration and often a great deal of careful physical effort. There is scarcely time or energy left for the critical appraisal of alternative points of view. As Rachel Laudan recently noted in discussing the delayed acceptance of continental drift, "Because testing demands a commitment of limited resources of time, money and skill, the decision about whether to pursue a theory hinged on an assessment of whether such a commitment would produce the necessary results, and produce them more quickly or at less cost than the pursuit of competing theories" (Laudan 1987, 205–206).

To return again to optics, Biot before the early 1820s was not in any sense irrational in rejecting wave optics, for at the time it hardly provided clearly superior solutions to the problems he was most concerned with. He did not invest the time and effort necessary fully to understand the new scheme, except insofar as he could appropriate things from it that were useful to him. Conversely, Fresnel was convinced that he could either solve every problem dealt with by Biot or else that Biot's way of framing the problem was incorrect. Deeply engaged in creating an entire new physics for light, Fresnel had no energy left for subtly penetrating the inner core of Biot's work. His criticism of it was consequently not at all convincing to Biot. But neither was Fresnel irrational in so single-mindedly pursuing his own interests, for he made tremendous strides in developing a coherent scheme that could generate new processes and either appropriate existing ones or else relegate them to subsidiary status.

My second point depends upon the first. Implicit structures are extremely robust, in major part because they must be flexible.[29] Robustness entails that it is hardly ever clear for quite some time that one *ought* rationally to pursue one scheme rather than the other, because new kinds of instruments are at first isolated units, which, precisely because they are solitary, can often be adapted to. This changes over time as a new family of devices evolves, making it increasingly difficult to incorporate their behavior into a scheme that did not birth them. But only historical contingency can help us to understand why these particular people, in these particular circumstances, do what they do. Abstract evaluations, however liberal in admitting criteria, do not seem to me to be useful things, at least not for the early years of the events that I have examined in detail. In general, it seems to me that novel schemes regularly diffuse well before any normative system would grant their clear superiority to alternatives. This was certainly the case with wave optics; it was true for Helmholtzian electrodynamics, and for Maxwellian field theory as well. In each case it was entirely rational to take up and develop the new scheme; it was also quite reasonable not to do so.

However, over time it no doubt does become *irrational* to continue with past systems when a new one has been used to generate a thoroughly novel instrumental universe. The price to be paid is expulsion from the scientific community, or at least indifference from it. This happened to ray physicists like David Brewster and Richard Potter, who found themselves increasingly unable to adapt to the new experimental world that the wave scientists were generating. By the late 1840s and early 1850s their behavior can, I think, be called irrational. They regularly failed to produce new results, and they could neither participate in nor fully appreciate ongoing work in the field. Experiments may eventually emerge that are thoroughly inimical to the older scheme, such as Hippolyte Fizeau and Léon Foucault's measurement of the speed of light in water. But these latter do not generally appear until quite some time after a new scheme has already diffused. At least this was so for the cases I have mentioned, and I think it to be so in other cases as well.[30]

Notes

1. In writing this chapter, I have used examples drawn from my own works in field theory and optics, some of which are listed in the bibliography. Further details can be found there. In what follows, I have for the most part refrained from explicit reference either to them or to the primary material on which they are based.

2. It is I think essential, if difficult, to distinguish at once between two broad kinds of experiments, namely, those done to obtain numbers for parameters that already fit into a scheme and those done to elicit or probe novel effects. The former (measuring experiments) may and often do eventually mutate in directions that cannot be confined by the original scheme, but their history nevertheless begins with a model that permits an instrumentally significant computation. The latter (discovering experiments) may have no model at all behind them, or if they do, it may not yield any numerical estimates whatsoever. Price's remarks concern discovery, though I will below return to the difference between these two species of the experimental art

3. I am concerned here with the question of a science's logical structure, with, that is, the issue of whether it has one. One might say that attempts to set out a scheme for understanding how science develops also "corral" the subject, but not in the sense I intend here (although a scheme for development will nearly inevitably address the issue of logical makeup).

4. Heaviside attempted what no other field theorist thought reasonable, or perhaps even possible to do: namely, to link the Maxwell stress tensor to the Poynting energy flux through what Heaviside termed the field's "equation of activity." This, if successful, would have given the Poynting flux a precise dynamical connection to stress. Heaviside did succeed, after a fashion, but his contemporaries were for the most part confused by his analysis. Their confusion was in some measure due to Heaviside's willfully obscure and convoluted style, as well as to his novel vector and operational methods, but not only to these things. In other cases they were quite capable of grasping *and* *using* his results, despite the odd style and strange mathematics. Here, then, something else troubled them, and that was Heaviside's forcing the complex of ideas in 'field' in perplexing ways, in particular, by trying to blend field energetics with material energetics. Maxwellians generally insisted on keeping the two well separated to avoid sticky issues concerning the relationship between ether and matter. See Buchwald 1985b for technical details.

5. For a different kind of example, consider Fizeau and Foucault's famous experiment in 1851, which established that the speed of light in water is less than that in air. Here one has something, it seems, drawn to a blueprint, one that delineates clearly the distinction between the emission theory of light, according to which light speeds up on refraction, and the wave theory, according to which it slows down. Their results certainly contradicted the former and supported the latter, but by the time it was done, the only remaining adherents of the emission theory were thought of as cranks by nearly every practising and publishing optical scientist. There were a few holdouts from the wave theory, but they were by then old and very far from the center of ongoing research, which was not at all interested in looking into questions concerning the tenability of the wave theory per se. To understand what Fizeau and Foucault were about, consider that Foucault also performed in 1854 a pendulum experiment that directly witnessed the earth's rotation. He did not perform it to test anything at all, nor, certainly, to probe difficult or contentious issues. It was an experiment designed to illustrate and perhaps measure. Here there was certainly a blueprint, but there was no experiment in the sense that Price intended, no attempt to test or to discover.

6. See, e.g., Pinch 1986 for a pertinent example. Pinch shows how an experiment to detect solar neutrinos was carried forward and altered over time by its developer in ways that had little to do with its multiple meanings for theoreticians. The experimenter wanted to detect neutrinos; theoreticians wanted to use neutrinos as probes for stellar structure, among other things. These goals intersected in manifold ways over time as each followed the dictates and pressures of the moment. It would I think be difficult to find anything entirely comparable to this during the nineteenth century, primarily because it depends upon a fairly clear division between experimenter and theoretician. Physics has so greatly changed during the last century as its scale has inflated that I am not certain the sorts of considerations brought forward here carry over particularly well, for the following reasons. The central distinction between what is articulated and what is not no doubt persists in physics to this day. It is hard to see how it could not. Nevertheless, it also seems as though the core is perhaps more deeply buried than it was in past times and that conflicts occur almost entirely over well-articulated issues. For example, since the early 1930s, when quantum mechanics stabilized, there has not been a division at all comparable to that between, say, British Maxwellians and German Weberians, or even to that between Weberians and Helmholtzians. This undoubtedly has much to do with the present social structure of the discipline, or rather of the many subdisciplines that physics embraces, within which overt conflict occurs, but whose members share a much wider base of agreement than was for the most part the case a century ago.

7. The fact that, in retrospect, Malus was mistaken and Fresnel correct has no historical importance. If sensitive photometers had existed in the 1810s, then the issues would have been transformed. But such things did not exist, and until they did, it remained reasonable to think that Malus's formulae worked as well as Fresnel's. Or better put, it was hardly *irrational* to think so. It did eventually become so circa 1850, when photometric technique had vastly improved, but by that time there was no issue to be decided: Fresnel's formulae had long ago spread throughout optics. If behavior must be judged then it is at least essential to do so in the light of contemporary skill.

8. In Fresnel's case this is nicely illustrated by his struggle to develop a new understanding of polarization, but is is also apparent in his use of Huygen's principle. The principle was hardly well integrated with mechanical considerations, and indeed, Poisson for one insisted, on mechanical grounds, that it is either a tautology or else that it cannot be used in Fresnel's fashion. Subtle issues concerning distinctions between

rays and beams were involved here, distinctions that could not be, or at least certainly *were* not, brought to the surface.

9. When divorced from particular examples, statements like these have inevitably sounded mystical. To move ahead requires a way to represent this kind of knowledge that makes clear *how it works in practice*. Kuhn's recent work on kinds begins to provide a method. Ian Hacking's chapter in this book explicates the concept. Unarticulated knowledge corresponds very well to the taxonomic tree of natural kinds because it is precisely the tree structure that remains for the most part unvoiced but that nevertheless governs the scientist's work. Articulated knowledge is perhaps more difficult to capture, but certain forms of it (in the nineteenth century at least) seem to correspond to situations in which scientists attempt to generate a novel and contentious taxonomy by embedding it within the kinds of another taxonomic tree that is considered to be much less contentious. For a brief example of a tree structure, see the discussion below of Brewster's attempt to expropriate phase from wave optics.

10. If a *codification* exists, then the "bases of [the] field, its legitimate problems" and perhaps (but not necessarily) also its methods would be nicely laid out. Such things might however occur even in the absence of a codification (as distinct elements that together form the "rules of the game"), though Kuhn asserts that even this looser kind of structure need not be supposed in order to understand the scientist's behavior.

11. One might reasonably think of unarticulated structures as guidelines rather than as rules, since, unlike rules, guidelines direct rather than determine.

12. Though he did not deploy it in the way that its greatest contemporary exponent, Biot, did, which in the end produced intense frustration on Biot's part.

13. Since Fresnel's ether lattice was composed of particles that interacted with one another in much the same way that the emission theory's light particles interacted with matter.

14. Commentators fastened on this claim, which provided a mathematical framework for argument. Intense confusion resulted, both in Germany and elsewhere, because the claim is incorrect. Helmholtz's function, and indeed the new kind of physics that it embodied, could capture only Helmholtz's understanding, despite the fact that there indubitably are many suggestive parallels between it and the analytical structures of competing alternatives.

15. The limitation to simultaneous moments is critical, because it is one of the major differences between Helmholtzianism and field physics, one that caused Hertz a great deal of trouble as he tried in the early 1880s to find points of contact between the two.

16. Experiments do, however, quite frequently operate from blueprints, or at least they have done so since microphysical modeling became common around the turn of the twentieth century. Measuring experiments have exactly this characteristic. To take just one example, consider the several experiments done early in the century to measure the gyromagnetic ratio. In most of these cases the experimenter or experimenters had a blueprint to work from (Galison 1987, chap. 2). Indeed, the issues revolved intially about one or another aspect of quite restricted *models* based on orbiting (or, later, rotating) electrons. By contrast, both Maxwell's attempt to look for a current's angular momentum, and Hertz's attempt to find the kinetic energy that a current might possess did not *test* anything at all precise, because neither of them had anything

like a restricted model. When such a thing exists, then paths for independent computation generally open up and the experimenter's reasoning also becomes more focused. Models, one might say, can crystallize (give stricter form to) apparatus and theory. (This does not imply that a lineage of experiments formed initially about a set of restricted models must then evolve along with the models, provided only that the latter continue to give meaning to the former.) The kind of experiments that Price was thinking about, and the kind that concerns us here, do not test something nicely laid out but rather probe for something that is intensely problematic, imprecisely formulated, or both. These kinds of experiments have perhaps become more rare since the advent of microphysical modeling, which ties interesting experiments closely to their implications for microstructure.

17. See Buchwald 1990 for details.

18. The ether had to be treated, on the one hand, like any other interacting object, but on the other hand, it affected the observable interactions between all objects. This provoked a galaxy of problems, both technical and conceptual (such as the embodiment in a ubiquitous entity of the, for Helmholtzianism, problematic conception of polarization).

19. Just what these latter two categories involve remained (necessarily) vague in wave optics.

20. Brewster, 1830.

21. For details, see my "Kinds and the Wave Theory of Light," forthcoming in *Studies in the History and Philosophy of Science*.

22. The historical process by which many old results do get incorporated into new structures is not simple and deserves some careful study. It is not enough to say that the old things become approximate where once they may have been exact, because their subjects may themselves not have survived the transition. Consider, e.g., what happened to geometrical optics when it was eventually incorporated into a wave structure. It is certainly approximate in respect to the new environment. However, its subject, the geometric ray, is not itself an approximation to anything in the new scheme. Perfectly definite things called "rays" occur in wave optics, and they are in no sense approximate. If geometric optics is approximate in relation to wave optics, it is not because the geometric ray is an approximation to the ray of wave optics: both are linear entities (though only the geometric ray is physically real), so that a geometric ray can hardly be an approximation to a wave ray. One must invent a rather elaborate route to geometric optics from wave principles, along the following lines: The old ray, which was not in itself a beam of light but formed beams in collections, must be treated as though it were a beam. Wave optics can then introduce an "approximation" in which this beam can be replaced under the right conditions with the wave theory's ray. This is not an entirely straightforward process, and Poisson for one never could follow it.

23. By "wave theory" I mean the kind of structure developed by Fresnel, not Christian Huygens's pulse theory or even Thomas Young's quasi-wave theory ("quasi" because it was based on rays coupled to the principle of interference). Huygens's pulses are thoroughly compatible with geometric rays. Young's rays have a property that geometric rays lack (namely, the ability to interfere with one another), but in every other respect they behave exactly like them. Only Fresnel's periodic waves, when bound to Huygens's principle, breach geometric optics.

24. Particularly by including "number (but not scope) of concrete problem solutions" among the possible deciding criteria. In this case of wave and geometric optics, both the criterion and the limitation are essential. Wave optics *under certain special and conceptually difficult conditions* can retrieve geometric optics, but the scope of the latter's applicability is thereby reduced in a radical fashion. On the other hand, the number of problems, or better, the number of kinds of problems, that wave optics can treat vastly exceeds the ones that geometric optics can deal with. There is simply nothing in the latter to deal with situations that wave optics handles using phase relations.

25. Try, e.g., to generate, as Brewster did, an analogue of phase for ray-based optics without smuggling in problematic elements that are not originally present in the scheme.

26. One of Kuhn's other suggested criteria, maximum accuracy of predictions, seems to me problematic, at least in this case. If we say that the wave theory is more accurate than geometric optics, then we must, it seems, be able to compare their predictions in *similar* contexts. But there are no similar contexts. Either geometric optics can be used reasonably well according to wave precepts, or else it cannot be—it depends on the experimental cirumstances. In other words, it is not quite the case that wave optics limits the accuracy of geometric optics. Rather, it limits the latter's domain of application.
   Consider, e.g., light passing through a slit from a point source (both schemes, for this situation, share the latter notion). According to wave optics, a fringe pattern *always* exists on the other side of the slit, however wide it may be. According to geometric optics, only a point of light occurs. These are, however, domains in which the visible width of the fringe pattern cannot be made out, and under these conditions geometric optics can be applied. One might say that accuracy is involved here, since it is a question of being able to observe the fringe pattern that wave optics avers must exist and that geometric optics denies. But it is, I think, more instructive to consider the practical conditions that enable one to retrieve a previous object (the geometric ray) from a scheme in which the object no longer exists, properly speaking. To see by contrast what I mean by this, suppose that wave optics did not generate fringe patterns. Suppose instead that it merely displaced the image without multiplying it. Then one could easily speak of increased accuracy (supposing our instruments capable of discriminating between the old and new loci) because it is solely a question of the locus of a thing whose nature remains otherwise unchanged. But if, as in fact usually does occur, the objects of the two schemes differ from one another in manifold ways, then the question of accuracy becomes rather more complicated.

27. The only two exceptions to my claim concerning experiments might be Fresnel's analysis of the "Fresnel rhomb," which involves internal reflection and his work on birefringence. The former, however, depended upon a novel and tentative interpretation of the physical meaning of an imaginary quantity, whereas the latter rested on the slenderest empirical thread. By the late 1840s a range of carefully contrived experiments combined with a more developed understanding of the theory's analytical structure to produce a novel world of phenomena that could be understood *only* with wave concepts. In other words, the phenomenal range and power of the wave theory emerged well after its widespread acceptance. This is to be expected, because a theory has to be widely used in order to generate its own empirical universe.

28. What persuasive reasons were there, e.g., for Biot (the most influential of ray physics) to give up the fight to Arago (his old enemy) and Fresnel (whom he hardly knew)? He certainly did have reasons to keep quiet after a bitter public quarrel with

them, but this had little to do with the abstract inferiority of his scheme to theirs (Buchwald 1989b).

29. Flexibility means two things. First, schemes must be flexible enough to admit new subcategories and even new categories at the highest level. Second, the kinds themselves are not rigidly defined, in the sense that the characteristics of a kind are impossible *exhaustively* to specify (though it is possible to specify that behavior x in device y qualifies object z for membership in category c). The criteria that determine category membership are, I believe, generally tied at any given historical period to quite specific instrumentalities, but they are also not taken to be *identical* with that instrumental behavior. So, for example, partly polarized light for Malus was revealed by a particular behavior of a particular device, namely his polarimeter. However, Malus thought the instrumental behavior to reflect an underlying ontology of discrete rays, so that as far as he was concerned, any device that revealed the ray structure of partly polarized light could count as a polarimeter. Such devices could have novel structures, ones entirely unknown to Malus, in which case one would certainly not wish to *define* the partly polarized category in terms of a known instrument. One consequence of this seems to be that the categories must be sufficiently flexible to admit their being distinguished by as yet undiscovered devices, which may generate difficulties concerning the relationships between different instruments, as occurred in optics when Brewster, for one, tried to understand how to fit devices that, according to wave understanding, manifest phase into his categories.

30. I particularly thank my Toronto colleagues Brian Baigrie, Craig Fraser, and Margaret Morrison for their very useful comments on drafts of this chapter.

References

Brewster, David. 1830. "On the Phenomena and Laws of Elliptic Polarization, as Exhibited in the Actions of Metals on Light." *Phil. Trans.* 120:287–312.

Brewster, David. 1850. *A Treatise on Optics.* Philadelphia: Lea and Blanchard. Originally published in 1833.

Buchwald, Jed Z. 1985a. *From Maxwell to Microphysics.* Chicago: University of Chicago Press. Paper, 1988.

Buchwald, Jed Z. 1985b. "Oliver Heaviside, Maxwell's Apostle and Maxwellian Apostate." *Centaurus* 28:288–330.

Buchwald, Jed Z. 1989a. *The Rise of the Wave Theory of Light.* Chicago: University of Chicago Press.

Buchwald, Jed Z. 1989b. "The Battle between Biot and Arago over Fresnel." *J. Optics* (Paris), 20:109–117.

Buchwald, Jed Z. 1990. "The Background to Heinrich Hertz's Experiments in Electrodynamics." In T. H. Levere and W. Shea, eds., *Nature, Experiment, and the Sciences.* Netherlands: Kluwer Academic Publishers, pp. 275–306.

Buchwald, Jed Z. 1992. "Helmholtzianism in Context: Object States, Laboratory Practice, and Anti-idealism. Forthcoming.

Jed Z. Buchwald

Galison, Peter. 1987. *How Experiments End.* Chicago: University of Chicago Press.

Kuhn, T. S. 1970. *The Structure of Scientific Revolutions.* Chicago: University of Chicago Press.

Kuhn, T. S. 1982. "Commensurability, Comparability, Communicability." *Philosophy of Science Association,* vol. 2.

Kuhn, T. S. 1984. "Revisiting Planck." *Historical Studies in the Physical Sciences* 14:231–252.

Lakatos, I., and Musgrave, A. *Criticism and the Growth of Knowledge.* Cambridge: Cambridge University Press, 1970.

Laudan, Rachel. 1987. "The Rationality of Entertainment and Pursuit." In J. C. Pitt and M. Pera, eds., *Rational Changes in Science* Dordrecht: Reidel, pp. 203–220.

Lloyd, Humphrey. 1834. "Report on the Progress and Present State of Physical Optics." *British Association Reports* 4:295–413.

Pinch, T. J. 1986. *Confronting Nature: the Sociology of Solar-Neutrino Detection.* Boston: Reidel.

Price, D. de Solla. 1983. "Sealing Wax and String: A Philosophy of the Experimenter's Craft and Its Role in the Genesis of High Technology." AAAS Sarton lecture. Unpublished typescript.

Mediations: Enlightenment Balancing Acts, or the Technologies of Rationalism[1]

M. Norton Wise

The industry in local history that has occupied historians of science in the last decade has given rise to a new industry, namely attempts to tie local affairs together into larger cultural affairs. How does local knowledge get extended to, or incorporated into, systems and networks of knowledge? The question requires answers with different sizes, because the cultures of science come in different sizes, from a particular research group or laboratory to groupings identified by technology, discipline, ideology, language, nation, geography, and time. Using "culture" in a general anthropological sense, I here take up the familiar size of rationalist scientific culture in Enlightenment France from about 1770 to 1810. In contrast to characterizations of the rationalist community in terms of its ideas, I will describe its integration in terms of a diverse set of technologies called "balances." This account focuses on the way in which the technologies functioned as active *mediators*, constructing agreements of two different kinds, which, for analytic purposes, may be represented geometrically. In the first place, they mediated between different subcultures—chemistry, political economy, mathematics, etc.—regarded as on a single plane. Second, they connected ideas with realities, along an axis orthogonal to this plane (see figure 1).

This spatial representation will facilitate a description of "rationalism," which preserves its reality as a historical structure and yet denies that it can be subsumed under a single monolithic concept. The structure is constantly being produced as a set of interdependent actions. It maintains its stability so long as these actions are repro-

M. Norton Wise

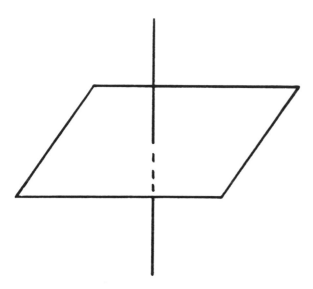

Figure 1
A geometry for horizontal and vertical mediation

duced, but the actions are always subject to contest. Consequently, the structure is always threatened by instability. I treat this structure as a cultural "space," somewhat after the idea of "fields" given by Pierre Bourdieu and also somewhat after the "networks" of Bruno Latour.[2] The most important aspect of the resulting heuristic model is that it is constructed from differential relations. Mediation, in this account, acts like a differential operator that generates the cultural space of rationalism. If excessively abstract, the model should nevertheless clarify some of the positive functions of technological mediation. Afterward its geometry can vanish, leaving the cultural network as a more vibrant and vulnerable object.

Consider, then, a very particular technological mediator, the calorimeter, which will begin to generate the space of rationalism.

The Calorimeter Was a Balance

When, in the early 1780s, Antoine-Laurent Lavoisier and Pierre-Simon de Laplace invented the device that they called a "machine" for measuring heat but that soon became the "calorimeter," they designed it as an analogue of that epitome of simple machines, the balance (figure 2). Our initial problem is to see these two instruments as similar.[3]

The measuring machine of Lavoisier and Laplace would balance quantities of heat against quantities of melted ice. The measurement actually involved weighing the water obtained on a standard mass balance. More profoundly, it assumed a balance in the water itself, an "equilibrium between the heat, which tends to separate the molecules of bodies, & their reciprocal affinities, which tend to reunite them." For a piece of ice to melt meant to change from one state of equilibrium of these two powers to another, which Lavoisier and Laplace likened to the different states of a simple rectangular block resting on its various faces under the action of gravity (figure 3). Similarly, they assumed that in the different states of water, its molecules occupied different positions with respect to one another. When the water was melted, the increased dissolving (and expansive) power of the heat taken up (caloric fluid by 1787) counterbalanced the rigidifying (and contractive) affinity of molecules of water more effectively than in the state of ice.[4] In this sense, the machine balanced expansive and contractive properties (and measured their amounts on a standard balance). Thus Lavoisier and Laplace in their joint work identified the calorimeter as a balance.

Mediation of the First Kind: Between Subcultures

Despite their collaboration, however, Lavoisier and Laplace recognized somewhat different balances in the calorimeter. With his primary interest in chemistry, Lavoisier saw a balance of chemical substances with their respective qualities and affinities, expansive and contractive substances to be sure, but nevertheless substances. Laplace saw a balance of forces, repulsive and attractive, by analogy with gravitational force acting at a distance between atoms of matter (even if caloric matter), as in rational mechanics and physical astronomy. The difference is between substances with qualities on the one hand and forces of atoms on the other.[5] Lavoisier too learned to employ the language of attraction and repulsion, but with extensive qualification: "We must allow that . . . the reason for caloric being elastic, remains still unexplained. Elasticity in the abstract is nothing more

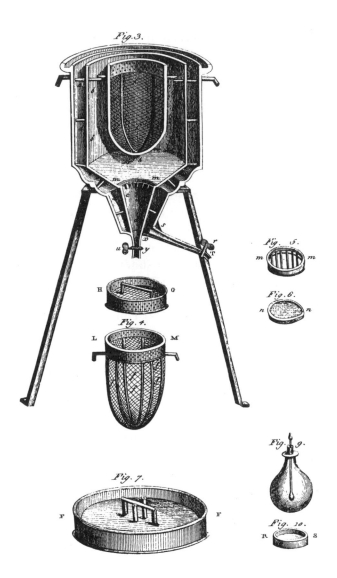

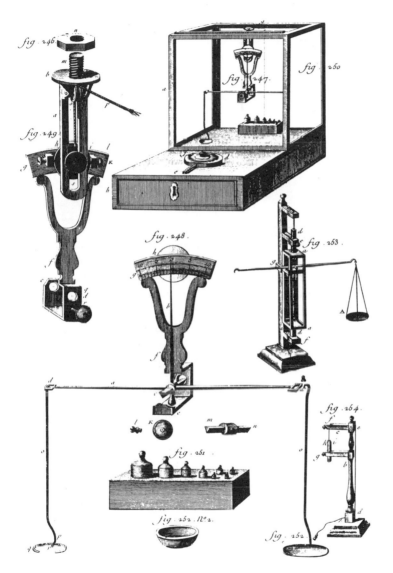

Figure 2
The calorimeter and the balance were similar.

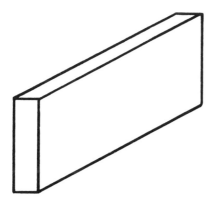

Figure 3
Ice and water might differ like the resting positions of a block.

than that quality of the particles of bodies by which they recede from each other when forced together." The distinction between Lavoisier's perspective and Laplace's is subtle, because the qualities of chemical substances and the actions of physical forces merged to a considerable degree in elective affinities. Indeed, the collaborators hoped to subject elective affinites in general to quantitative measurement with their calorimeter.[6] But outside of that common domain, the subculture of chemistry (a branch of physics) differed markedly from that of mechanics and physical astronomy (mathematics). Laplace tried at least once to extricate himself from his collaboration with Lavoisier in physics in order to pursue mathematics, assuming almost apologetic tones when explaining his continued participation to Lagrange.[7] The collaboration, therefore, was a tenuous one, one that depended on constructing a limited region of shared meanings, objects, and motivations, a region within which cooperation, competition, and exchange could occur (figure 4).

In this region of intersection the calorimeter mediated between the potentially divisive interests of Laplace and Lavoisier. It established a common ground where the discourse of chemistry partially overlapped with that of mechanics and physical astronomy and where force exchanged for substance through the medium of affinity. Prior to the existence of the calorimeter, the exchange had had only an idealized and qualitative existence. The new instrument gave it con-

Mediations

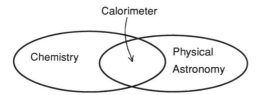

Figure 4
The calorimeter as horizontal mediator

crete location and quantitative measure. We can begin to develop this mediating role by placing the calorimeter on a different plane from the theoretical domains of chemistry and physical astronomy, thereby suggesting somewhat different roles for technology and for theory (or ideology) in the structure of rationalist culture (figure 5).

Mediation of the Second Kind: Between Ideology and Reality

If ideology and technology belong on different planes, then we require a third as well, a plane of reality, where the protagonists in the story imagined the furniture of the natural world to be positioned and moved about (figure 6). On the plane of reality Lavoisier placed chemical substances, such as caloric fluid, understood as a substance possessing the quality of expansivity and responsible for the melting of ice in the chemical compounding of the fluid state. Laplace placed there forces acting at a distance between particles of matter, in particular, the repulsive force of caloric fluid. In both cases the calorimeter established the measurability of the entities and thus their status as realities. "Reality" here has a phenomenological meaning, consonant with the epistemology of Étienne Bonnot de Condillac and friends, including Lavoisier and Laplace. If not ultimate realities, they were at least objectifiable and meaningful in "the present state of science," a favorite phrase among rational analysts who wished to leave room for future decompositions and to escape their own condemnation of metaphysics.

Once again, the calorimeter mediates horizontally between the two realities of fluid and force, as their common measure. But now more important, it also mediates vertically between theories and things. It

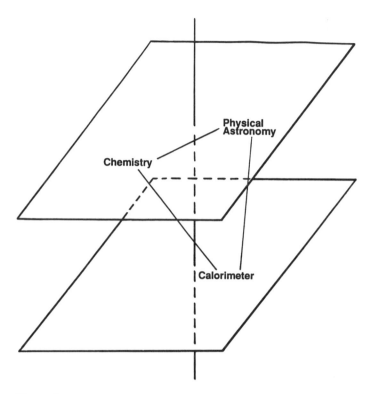

Figure 5
It is useful to represent theories and technologies on different planes.

exchanges theoretical entities for concrete realities. Laplace moves back and forth through the calorimeter from the ideas of physical astronomy to the tangible reality of repulsive force; Lavoisier similarly moves between chemical concepts and the real substance of heat. The instrument thus played the role assigned to "intervention" in Ian Hacking's formulation of the realism of theoretical entities.[8] I would stress that technological mediation works in both directions on the vertical axis. It reifies theoretical entities, but it also idealizes realities, in the sense that it often assigns to them the meanings they have in the theory. In everyday terms, an "electron," whose existence is evidenced by its hitting the flourescent screen of a television set, comes with much of the theoretical baggage that explains the operation of the television.

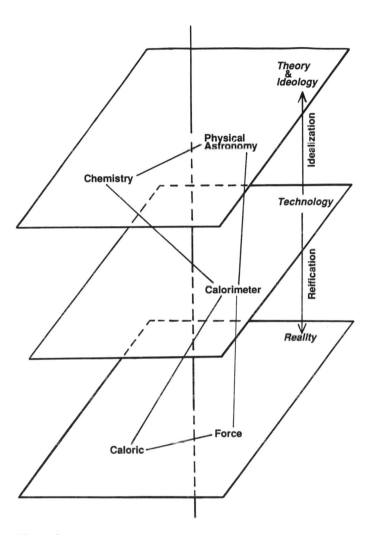

Figure 6
Technologies mediate vertically between theories and the entities whose reality the
theories assume.

M. Norton Wise

Extending the Technological Plane

We have already seen that the calorimeter did not stand as a balance on its own, but depended on its relation to the standard mass balance (figure 7). Use of the standard balance in turn depended on chemical and physical theories on the plane above (such as conservation of matter and universal gravitation, which do not appear on the diagram) and on their respective realities below (mass and gravitational force). Thus the meaning of the calorimeter, in its dependence on the balance, depended as well on these extended relations. To balance quantities of caloric and water would have made little sense without the prior balance of masses in relation to the conservation of matter above and the reification of mass below. Balancing attractive and repulsive forces would have made equally little sense without the prior balance of weights (as gravitational forces), the theory of forces, and their objectification.[9] This situation is typical. Instruments do not legitimize themselves. They borrow the legitimacy of other instruments, theories, and things. This reliance implies that we could speak of theories as mediators between technologies. The lines running from the mass balance to mechanics to the calorimeter would then show the theory of forces on the plane above mediating between the two instruments on the plane below. But stressing relations of this sort would lead to discussion of the power of theories to subsume a variety of applications, rather than to the action of technologies in integrating a cultural complex.

Sticking to the middle plane, we can follow Lavoisier and Laplace through a whole network of technologies that tie together their everyday activities with those of other prominent citizens of Paris. The network constitutes a veritable metropolis of balanced, and thus enlightened, architectures. Here are a number of them.

The balance sheet

Along with the standard balance and the calorimeter, Lavoisier adopted the balance sheet as his habitual accounting device for keeping track of and verifying the balance of substances in his experiments on chemical reactions, including the famous work relating combustion to respiration. Larry Holmes has given a rich account of the

Mediations

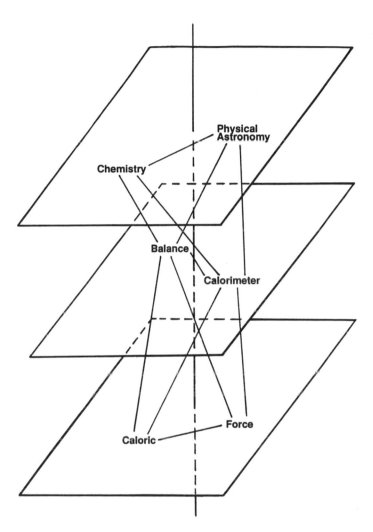

Figure 7
The calorimeter draws support from the mass balance, and vice versa.

Table 1
Lavoisier's balance sheet for fermentation

Materials of fermentation

		lbs.	oz.	gros	grains
Water		400	0	0	0
Sugar		100	0	0	0
Yeast in paste, 10 lbs., composed of	{ water	7	3	6	44
	{ dry yeast	2	12	1	28
Total		510			

Constituent elements of the materials of fermentation

		lbs.	oz.	gros	grains
407 lbs., 3 oz., 6 gros, 44 grs. of water, composed of	{ hydrogen	61	1	2	71.40
	{ oxygen	346	2	3	44.60
100 lbs. sugar, composed of	{ hydrogen	8	0	0	0
	{ oxygen	64	0	0	0
	{ charcoal	28	0	0	0
2 lbs., 12 oz., 1 gros, 28 grs. of dry yeast, composed of	{ hydrogen	0	4	5	9.30
	{ oxygen	1	10	2	28.76
	{ charcoal	0	12	4	59
	{ azote	0	0	5	2.94
Total weight		510	0	0	0

balance sheet in relation to the mass balance and balanced equations. Lavoisier's most famous balance sheet will stand as an icon for this account (see table 1).[10] The mass balance (and the calorimeter) quantified individual entries in such balance sheets. But the sheet as a whole balanced a complete set of input and output relations, presenting them in the standard format of an economy. In this way the balance sheet mediated between chemical equations and the "animal economy" in experiments on combustion and respiration. More generally, however, the balance sheets mediated between chemical equations and the political economies of farm, tax farm, and state, in which Lavoisier played a leading role. Along this road lie the houses of his friends among the physiocrats and ministers of state, such as P.-S. Du Pont de Nemours, A.-R.-J. Turgot, and M.-J.-A.-N. Condorcet.

Table 1
Continued

Products of fermentation

		lbs.	oz.	gros	grains
35 lbs., 5 oz., 4 gros, 19 grs. of carbonic acid, composed of	oxygen	25	7	1	34
	charcoal	9	14	2	57
408 lbs., 15 oz., 5 gros, 14 grs. of water, composed of	oxygen	347	10	0	59
	hydrogen	61	5	4	27
57 lbs., 11 oz., 1 gros, 58 grs. of dry alcohol, composed of	oxygen combined with hydrogen	31	6	1	64
	hydrogen combined with oxygen	5	8	5	3
	hydrogen combined with charcoal	4	0	5	0
	charcoal combined with hydrogen	16	11	5	63
2 lbs., 8 oz. of dry acetous acid, composed of	hydrogen	0	2	4	0
	oxygen	1	11	4	0
	charcoal	0	10	0	0
4 lbs., 1 oz., 4 gros, 3 grs. of residuum of sugar, composed of	hydrogen	0	5	1	67
	oxygen	2	9	7	27
	charcoal	1	2	2	53
1 lb., 6 oz., 0 gros, 5 grs. of dry yeast, composed of	hydrogen	0	2	2	41
	oxygen	0	13	1	14
	charcoal	0	6	2	30
	azote	0	0	2	37
510 lbs.		510	0	0	0

It would be difficult to overestimate the importance that Lavoisier assigned to the balance sheet as a tool of economic analysis and government policy, but also as the very embodiment of political economy. For example, in his report on a work entitled "The Territorial Riches of the Realm of France," published during the revolutionary days of 1791, when he served as temporary deputy to the National Assembly and as Commissioner of the Treasury, Lavoisier envisaged a grand balance sheet of the entire economy. The general balance sheet, he contended, would contain "the entire science of political economy, or rather, this science would cease to be one," because the perfected instrument would immediately resolve all questions in dispute.[11] To perfect knowledge was for Lavoisier to perfect a technology of knowledge, one that made philosophy agree with nature.

In practical terms, he sought a means of diagnosing the health of the political economy by analogy to the animal economy (and vice versa). The general balance sheet, he believed, "would form a true thermometer of the public prosperity," revealing at "the glance of an eye" the good or ill occasioned by previous legislatures, especially with respect to taxes, tariffs, and price levels. As a thermometer, the balance sheet would take the temperature of the political economy, revealing any fever that might be troubling its health, that is, its state of equilibrium from year to year.[12]

Such ideas on the similarity of the political economy to the animal economy, or of the "political machine" to the "animal machine," informed Lavoisier's career from beginning to end. "It is with the body politic approximately as with the human body," he wrote in an unfinished eulogy of Jean-Baptiste Colbert in 1771; "health depends on the perfect equilibrium of all the parts. The physician's aid must tend to restore to the weak part the degree of force proper to it."[13] And in 1786, when a shortage of forage for cattle "desolated" the country and the price of livestock soared, he supplied the following diagnosis as a member of the Committee of Agriculture: "One may compare the increase in price that accompanies the shortage to the fevers [crises] of illness; the fevers are nothing other than the effort that nature makes to reestablish in the animal economy the order that has been troubled. Woe to the physician, in administration as physically, who struggles against the law that nature has established and who adds to the sickness of the political body the sorry effects of his imperiousness."[14]

As components for his grand thermometer of the body politic, Lavoisier worked up balance sheets for several agricultural sectors. Table 2 is a summary one for the production and dispensation of wheat and rye (in pounds weight).[15] To use this accounting as a thermometer, the physician of the body politic would presumably compare balance sheets from year to year, watching their changing totals while keeping in mind the changing price of grain and changes in government policy. In a similar way he might have used the calorimeter to measure the changing amounts of caloric released per hour by a guinea pig, attempting thereby to register its relative tem-

Table 2
Lavoisier's balance sheet for the total production and dispensation of wheat and rye
in France

Comte des blés et seigles

Production		Emploi	
Produit total en blé des 233,333 charrues cultivées avec des chevaux	7,000,000,000	Portion réservée par les cultivateurs pour ensemencer les terres et préparer la récolte suivante	2,333,333,333
Produit des 466,669 charrues cultivées par des boeufs	7,000,000,000	Nourriture des cultivateurs	925,680,000
Total	14,000,000,000	Dépenses des moissons	1,068,340,000
		Frais de battage	420,030,000
		Dépenses diverses d'exploitation	1,971,620,000
		Dime du blé à la 20°	700,000,000
		Vingtiémes et sols pour livre	416,500,00
		Taille et accessoires, non compris les villes	1,120,000,000
		Droit représentatif de la corvée	186,666,667
		Gabelle, aides, tabac	463,700,000
		Part des propriétaires	4,395,160,000
		Total	14,000,000,000

perature. To put this a little too explicitly for Lavoisier (for whom
quantity of heat and degree of heat, or temperature, were not sharply
distinct), the calorimeter and the balance sheet both measure quantities—heat and produce, or its monetary value—but as "thermometers" they measure changes in these quantities. This is consistent with
Lavoisier's and Laplace's understanding of a mercury thermometer
as registering the (variable) state of equilibrium between the heat
tending to expand the mercury and the contractive affinity of its
particles.

For his entire scheme of economic measurement by balancing,
Lavoisier laid down as a first principle that "everything that is consumed every year is reproduced every year. . . . There is, therefore,
. . . an equation, an equality, between what is produced and what is
consumed. Thus, to know what is produced it suffices to know what

is consumed, and reciprocally."[16] Importantly, he specified that the equation holds only for a mean year and disregards imports and exports. That is, it holds in an idealized steady state of a healthy political economy, averaged over time. This is the same equation that Lavoisier postulated for continuous cycles of the animal economy (and is similar to that for chemical reactions in general): "A remarkable circumstance is that animals that are in a state of health and that have reached full growth return each day, at the end of digestion, to the same weight that they had the day before in similar circumstances, so that a sum of matter equal to what is received in the intestinal canal is consumed and expended, whether by transpiration, by respiration, or finally, by the different excretions."[17]

As for weight, so for heat. In the famous experiments on respiration and combustion, Lavoisier always confirmed an equation between heat produced and heat consumed, which led to the following conclusion: "Whenever an animal is in a permanent and tranquil state, . . . the conservation of animal heat is due, at least in large part, to the heat which the combination of the pure air respired by animals with the base of fixed air provided by the blood supplies to it."[18] In contrast to this healthy steady state, a fever registers the disequilibrium of disease, or better, it registers the processes that nature undertakes to restore its natural equilibrium state, whether in the political, animal, or chemical economy.

The balance sheet as fever thermometer thus informs, and is informed by, an extended analogy between the flow of heat and the flow of monetary value. An unhealthy economy is one with too much money flow (overheated) or an unusual flow, as produced, for example, by the shortage of livestock and attendant high prices. Just as heat flows from a hot body to a cold one to restore thermal equilibrium, so money flows to restore economic equilibrium.

And like heat flow, the movement of money is controlled by affinities or attractions, as between foodstuffs and money in foreign commerce: "Money attracts foodstuffs, and foodstuffs attract money, which is to say, for example, that if we consider two neighboring states, isolated like two particulars, of which one has all the money, the other all the foodstuffs, after a little time a necessary level will establish itself between them, and each will have approximately half of the money and half of the foodstuffs."[19] This image of economic "attractions" comes from Lavoisier's eulogy of Colbert in 1771, well before he developed his views on caloric or pursued his joint researches with Laplace. It reinforces the differences noted earlier between their disciplinary identities. It shows why we should understand Lavoisier's concept of attraction between caloric and normal matter as an affinity inherent in the economy of nature, "the order of things," to use one of his favorite phrases. This affinity is more nearly a material cause realizing a teleogical tendency toward equilibrium than it is a nonpurposive Laplacian attraction at a distance.

Thus the mediations between chemistry and political economy effected by the balance sheet involve exchanges of a somewhat different sort from that between forces and substances. In general, weights of chemical substances now exchange for weights of commodities, but more pointedly, caloric fluid exchanges for another fluid, money.[20] And now the balance sheet and the calorimeter—or, employed differently, the balance sheet and the fever thermometer—motivate and validate each other. Lavoisier could therefore rely on the strength of his chemical practice when conversing with Dupont de Nemours and the Committee of Agriculture and on his economic practice when discussing chemistry.

The economical table

In his remarks on the equilibrium level of money in the balance of commerce, Lavoisier went on to argue that the attempts of previous governments to increase national wealth by maintaining an imbalance of trade were pure folly: "This level, which the physical order of things does not permit to be troubled, exists very nearly in all the states of Europe. It would therefore be in vain that we form the project of attracting into France, by commerce, a [disproportionate] share of the hard currency that circulates in all of Europe." Putting his words in the mouth of Colbert speaking to Louis XIV, Lavoisier here ascribed to the bygone days of French wealth and power the fundamental policies of the new physiocratic economics, the vaunted economics of the natural order, whereby agriculture, not industry and commerce, provided the source of all real wealth in the nation, the motive force of the "political machine."[21]

Manufacturers and traders—the "sterile class," in a memorable phrase of the original physiocrats, François Quesnay and Victor Riqueti, Marquis de Mirabeau—produced only what they consumed. "The value of all manufactured merchandise consists," Lavoisier echoed, "(1) in the value of its raw material; (2) in the value of the foodstuffs of all kinds consumed by the different agents who have contributed to its fabrication." The "productive class" of farmers, by contrast, reproduced double the value of their expenditures, as Lavoisier confirmed in his balance sheet above for wheat and rye. The entire revenues of the state, therefore, consisting in government taxes, church tithes, and the rents of landlords, derived directly or indirectly from agriculture, which meant that the institutions and policies of the state should be reformed to promote a rich agriculture by promoting freedom of culture and export, nonexploitative taxation, and progressive farming techniques.

The primary means through which Quesnay and Mirabeau had advanced this program was another balancing instrument, the *tableau économique,* which went through a variety of versions from 1757 to 1766.[22] Figure 8 gives a late version with a few clarifying additions. This famous "zigzag" depicts a steady state of maximum prosperity, a state of continous circulation and distribution of monetary value. Farmers are the animating force that each year reproduce the net revenue of the state that each year is consumed and expended. But the action of this productive class is maintained by a balanced and reciprocal exchange (the zigzag) with the sterile class during the course of distribution of the revenue.[23]

The point of the table as an instrument, like the balance sheet, was both to reveal at a glance the operation of the "machine of circulation," the "economical machine," or "political machine" and to offer a means of analyzing it.[24] Beginning at the top, the state receives in taxes, tithes, and rent a net revenue of $2M$ paid by the farmers, who previously advanced $2M$ to yield $4M$ worth of produce. During the course of the year the state consumes its entire income in a balanced division, paying M to the productive class in return for bread, meat, and wine and M to the sterile class for clothing, furniture, and other manufactured goods. Following the same custom, the farmer and tradesman each spend the entirety of their own receipts in an equal division, half going to their own side and half to the other, which

Mediations

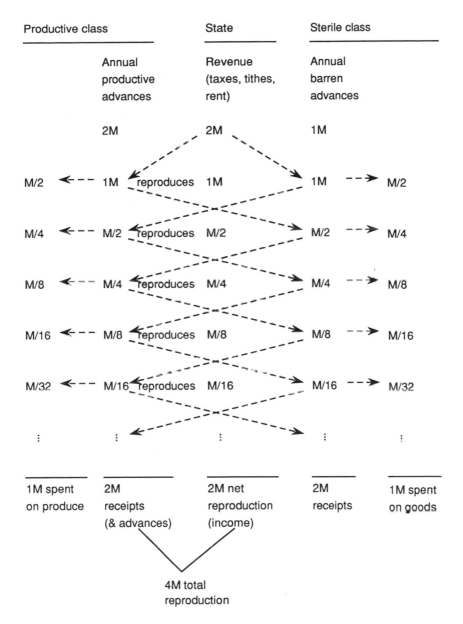

Figure 8
The economical table of Quesnay and Mirabeau

generates a flow of money back and forth between them that decreases asymptotically to zero. As a result, each side receives and expends a total of 2*M*. But the productive class always reproduces double its expenditures, giving birth (*faire naître*) to 4*M* worth of produce, while the sterile class manufactures only 2*M* worth of goods.

Quesnay and Mirabeau regarded this economical machine as a perpetual motion machine.[25] We might imagine a fall of water (revenue) that flows back and forth between two sides of a hydraulic balance, which drives a waterwheel, which continuously raises the water back to its original height (value). But the physiocratic machine is not self-perpetuating; it contains a vitalizing force, located in farmers working the fertile earth. Their restoration of value to the money circulating through the political body is not unlike the restorative action of respiration in Lavoisier's conception of conservation of animal heat via circulation of the blood, whereby vital air (oxygen) supplies the required caloric during combustion in the lungs. As a surgeon (and personal physician to Louis XV and his mistress Madame de Pompadour), Quesnay regularly mixed mechanical and physiological images in his prescriptions for the political body, so the principles of life and of a balancing machine seemed perfectly commensurate, if not quite identical. That relation reinforces the basic thrust of this section.

I am displaying an interconnected web of balancing technologies over a broad spectrum of rationalist pursuits, moving from mass balance to calorimeter to balance sheet to economical table. Through the balance sheet and the table, political economy and animal economy helped to construct chemistry as another such economy of nature, while the new chemistry and its balancing instruments lent the indubitable action of natural law to the representation of political and animal economies as balanced economies of nature. But these "balances" form an extremely diverse collection. The connections between them depend on their being susceptible of two or more interpretations, each legitimate in its own terms but only tenuously related to the other(s), just as in the case of the calorimeter in its mediating action between chemistry and rational mechanics. The balance sheet, standing between chemistry and economy, facilitates a subtle movement: from a static balance of two bodies placed simultaneously on a mass balance, to a static balance of the initial and final components of a chemical reaction, to a balanced exchange in commerce, to a balance of consumption and production in the cycles of an ongoing circulation. That is, the balance sheet relates the equality of two masses to the continuing self-identity of an economy. In this way it substitutes one sort of conservation for another, namely, conservation of substance for conservation (as preservation or restoration) of a living economy. The vital process becomes a material principle, pure air or oxygen, and in reverse, oxygen takes on the qualities of vitality.[26] The physiocrats' economical table displays this dual character in one picture, for the balanced zigzag in the distribution and exchange of value is simultaneouly the agent of restoration in the continuing circulation of value. In general, the balance sheet reduces temporal process to atemporal balance. Dynamics is reduced to statics. Change is eliminated from the natural order.

The appearance of the balance sheet and economical table on the technological plane requires a small digression, because an objection often arises to treating material instruments, like the calorimeter, as of a piece with analytic techniques, like the balance sheet. I offer two justifications, past and present. As a historical matter, the community of French Enlightenment rationalists regularly referred to their techniques of analysis as instruments. Lavoisier's reference to the balance sheet as a thermometer is typical. Mirabeau classed the economical table with print and money as one of the greatest inventions of humanity, thereby recognizing its dual status as both instrument and sign of the exchange economy he was analyzing.[27] More strikingly, Laplace and other mathematicians often referred to mathematical analysis itself (whether differential, integral, variational, or probabilistic calulus) as an instrument for revealing and embodying nature's own realities (see below). This usage corresponds well with our present term "software." Although we distinguish material from conceptual techniques, as hardware and software, we freely recognize languages and programs as technologies. I mean to reinforce that habit.[28]

Algebra

If we now follow Lavoisier along a different route, we come to *algebra* and an intense interaction with the philosophy of the Abbé de Con-

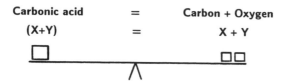

Figure 9
Three representations of a chemical equation

dillac. Lavoisier's balanced chemical reactions were to be balanced equations, with Condillac providing the required logic. It was the logic of analysis, put into operation as algebraic analysis. And, of course, the reformed language of chemistry was to be just as analytic as the algebra of weighing, with the meanings of the words for the constituents adding up to the meanings of the words for the products.[29] The relation of language, algebra, and balance may be represented as three parallel equations (figure 9).

The balanced lever in this picture is by no means too explicit, even for Condillac. Algebra to him meant not merely the abstract balance of symbolic equations but also a machine of discovery. "It is properly the lever of the mind." Indeed, in Condillac's version of Christian morality, before the fall of man into the darkness of rhetorical deceptions and the domination of despots and dogmas, one did not question whether "the word *thought* meant anything other than *to weigh, balance, compare.*"[30] Enlightened and reborn man had rediscovered in algebra the lost art of analytic thought, "the lever of the mind" that lifted unknowns by the weight of knowns. For example, "Having some tokens in my two hands, if I pass one from the right hand into the left, I will have as many in the one as in the other; and if I pass one from the left into the right, I will have twice as many in the latter. I ask you, what is the number of tokens [x and y] that I have in each hand."[31] The simultaneous equations (to which I have added levers) are shown in figure 10. Solving these equations, Condillac obtained $x = 7$ and $y = 5$.

Just as pure thought here employed the lever of the mind, so wise government rested on the practice of social statics, for "a people is an artificial body" and requires governance by a mechanic who can connect its parts "so as to make them move in concert and by one

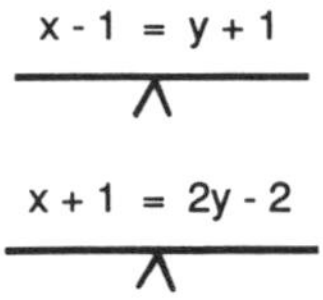

$$x - 1 = y + 1$$
$$\overline{\hspace{3em}\Lambda\hspace{3em}}$$

$$x + 1 = 2y - 2$$
$$\overline{\hspace{3em}\Lambda\hspace{3em}}$$

Figure 10
Simultaneous equations as simultaneous balances

and the same spring," which would maintain "perfect equilibrium among all the classes."[32] Such equilibrium required avoiding injustices and inequalities.

The probability calculus

An elaborate technology for the task of repairing inequalities came from Condorcet, who promoted a specialized algebra, the probability calculus, as the proper tool of analysis and decision for the "social art." In the recent book on reason and probability by Lorraine Daston it is apparent that enlightened planners of the public good thought of the probability calculus not only as the epitome of reason but also as the *balance* of reason.[33] In political economy Condorcet subscribed to the doctrines of his patron Anne-Robert-Jacques Turgot, who regarded the conditions of equal exchange in the market as an expression of a balance of psychological needs and desires. Condorcet took this view over into the probability calculus, where the concept of expectation values, or mean results, expressed those needs and desires. As Daston renders Condorcet, "Just as the 'relation of reciprocal needs' established equality among the free agents of the marketplace, so in probabilistic expectation, 'neither he who exchanges a certain value for an uncertain one, or reciprocally; nor he who accepts the exchange, find in this change any advantage independent of the particular motive of convenience which determined the preference.' Averaged over many such exchanges, the expectations, like the common price, would tend toward 'the greatest equality possible' between parties."[34]

Turgot and Condorcet thus expressed in expectation values a standard emphasis of the physiocrats on "reciprocal commerce," whereby

Figure 11
An expectation value X is a "balance point."

domestic and foreign trade maintained balance in the economy and thereby maximized monetary values, although on average they could yield no profit: "The trade, called reciprocal foreign commerce, consists in the sales of the merchant ballancing exactly the purchases made by him."[35] More especially, however, Condorcet's use of probability to calculate mean values now promised to make sense of the chaotic world of man, showing that apparently inexplicable variations in human behavior average out to a harmonious balance in the long run. This reification of rational man by the probability calculus shows the variations to be unnatural; they represent the accidents of irrationality, misfortune, and inequality. True human nature, if allowed to develop under its natural laws, would yield a harmonious social order. The problem was to discover the laws: "What are the laws of that equilibrium which tends without ceasing to establish itself between needs and resources . . . ? How, in that amazing variety of labour and productions, of needs and resources; in that frightening complexity of interests which ties the subsistence, the well being, of an isolated individual to the general system of societies, which makes him dependent on all the accidents of nature, on all political events . . . ; how in that apparent chaos does one nevertheless observe, by a general law of the moral world, that the efforts of each for himself serve the well-being of all?"[36]

The appropriate technology for unlocking these puzzles, according to Condorcet, was the probability calculus, and first of all the calculation of mean values and expectation values. Of course, a mean value or an expectation value X has precisely the same form as the balance point X of a lever (figure 11), or in general, of the center of gravity of a body:

$$X = (m_1x_1 + m_2x_2 + m_3x_3 + \cdots)/(m_1 + m_2 + m_3 + \cdots).$$

The variables are either lever arm x and mass m or a possible value x and its probability m. Large inequalities in social value x (wealth,

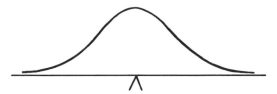

Figure 12
The normal distribution represents a balance of errors.

status, or education) were to Condorcet the primary source of imperfection in the social order. The great goal of the social art was to measure and analyze these inequalities and then to discover what policies conformed more closely to nature's own tendency to reduce them to their minimum, to consolidate the values of x so as "to bring together, to identify, the common interest of each person with the common interest of all." Without joining the probability calculus to the social sciences, those sciences "would remain forever gross and narrow, for want of instruments delicate enough to grasp the fleeting truth, of machines unfailing enough to reach the depths of the mine where part of their riches are hidden."[37]

In the hands of Laplace and others, the machine for mining the truth of mean values became the calculus of balanced errors, depicted as the bell-shaped error curve, or normal distribution (figure 12). Mathematical astronomers made powerful use of this depiction of the probability of errors in extracting mean values from astronomical observations. Its more popular fame derives from Adolphe Quetelet's use of it in social statistics to construct during the 1830s and 1840s the concept of the "mean man."[38] That strange homunculus inside the body politic hearkens directly back to the Enlightenment context of social analysis.

One interesting place to see this heritage is in a rather obscure piece of political arithmetic by the renowned mathematician Joseph-Louis Lagrange, because it displays both the ideological and technological content of the "mean." Closely following Lavoisier's "Richesse" of 1791, including its physiocratic assumptions, Lagrange attempted in 1796 to calculate the "first necessaries" of the new republic, completing only the food requirements or "mean consumption." For this purpose he constructed a "calculus of nourishment," based on an extension of mean values to "mean species." For example, "we will from the first reduce all vegetable nourishment to those grains which are cultivated on a large scale, and these we reduce to a single mean species which we will name simply *corn*, comprehending wheat, rye, and barley, which one eats in bread." Similarly, the sources of animal nourishment: beef, mutton, pork, etc., reduce to a mean species called "butcher's meat," and all drinks become the mean "wine."[39]

Now this may seem a strange notion of averaging indeed. It literally sums apples and oranges. It is not numerical averaging over units of the same thing but stockpot averaging, in which different sorts of meat are thrown into a common pot to make the mean broth. Lagrange, however, had in mind that different meats have a common measure, that one can substitute for the other in terms of nutritive matter, and that they enter into the calculus of nourishment only in the ratio of their nutritive value. He insists, therefore, that "this reduction is founded on the nature of things." But how does one locate nutritive value in the nature of things? Simple! By finding their mean monetary value or price, what Lavoisier called their "natural value" (with no reference to Adam Smith). "Relative to general and ordinary objects of nourishment," said Lagrange, "I believe . . . their nutritive value proportional to their price." A half-pound of dry cheese is equivalent to a pound of meat because their price is the same.[40]

Here is a fascinating instance of the physiocrat's economical table revealing the natural order and thereby supporting another instrument of the natural order, the calculus of nourishment. Because the table locates all value in agriculture, makes production equal to consumption in industry, and equalizes all trade, it ultimately bases all prices on the price of foodstuffs. Lagrange has identified this balance in the economic order with the nutritive order, so that in general people pay prices proportional to nutritive value. The calculus of nourishment borrows the force of the economical table and uses it to construct an object in nature, a mean species.

"These things supposed, the question is reduced to determining approximately the mean quantity of corn and of meat necessary for the subsistence of the Republic." That is, Lagrange wants to calculate the "mean consumption of each individual." To that end he constructs a kind of "mean man," who results from equating one man with a woman plus three children under ten. This average of humanity consumes the mean amounts of the mean species of "corn," "butcher's meat," and "wine." Lagrange's calculations, based on three different sorts of data, show considerable consistency, which "proves that men have need, in general, of a definite weight of food, like a kind of ballast that depends on the human constitution." Apparently, we must regard the mean consumption as being as definite a part of the order of nature as the mean species of corn. The ballast metaphor signals that calculating the mean consumption is like calculating the conditions of stability of a ship on the sea, or that the mean consumption is defined by the balance of another object in the order of nature, "the human constitution." The stability of these objects, revealed by the mathematical construction of the mean, guarantees their reality. Here again is the reification of theoretical entities by a technology. Quetelet's fully dressed mean man stands not far off. So too does Quetelet's analogy to the orbital motion of the center of mass of a planet.[41]

The variational calculus

As the probability calculus diagnosed inequalities in the social system, so another newly developed algebra, the variational calculus, probed inequalities in the system of the world, the solar system. In an accomplishment that epitomizes Enlightenment strategies, Lagrange and Laplace made the variational calculus reveal that the secular inequalities in the motions of the planets, which had threatened by accumulation to disrupt their stable order, were actually only the periodic oscillations of an eternal balance. "Their secular inequalities," Laplace announced, "will be periodic, and contained within narrow limits, so that the planetary system will only oscillate about a mean state, from which it will deviate but by a very small quantity; the planetary ellipses therefore always have been, and always will be nearly circular."[42] He recognized that it was the new tools of analysis that had made this stability into a reality for man. "It is chiefly in the application of *analysis* to the system of the world, that we perceive the power of this wonderful instrument; without which it would have been impossible to have discovered a mechanism which is so complicated in its effects, while it is so simple in its cause."[43]

In the celestial mechanism the perfections of equality and uniformity are guaranteed to time itself, as measured out by the world's clock. This clockwork exchanges the infinite line of time for timeless cycles in space (figure 13).[44] The self-regulating balance of nature's laws, furthermore, the inevitable consequence of gravity and inertia working in opposition, maintains the equilibrium.

Comparing the systems analyzed by the probability calculus and the variational calculus, we see that they both involve stability underlying variability. But while the variations in the planetary motions result from "constant" "natural" forces, those in human society result from "accidental" "unnatural" ones. The two technologies, consequently, are very different. I will take up some of those differences below. It is worth noting, however, that in physical astronomy the two techniques produced complimentary averaging effects. Probability calculations extracted the mean (true) value of planetary locations from the variable readings of observers subject to accidental errors of circumstance and psychology. The variational calculus similarly revealed the stable mean motions of the planets beneath apparently lawless inequalities.

This relation was, of course, no accident. From early to late in his career Laplace connected probability with astronomy in his mathematical papers. He envisaged using the probability calculus to fill in for human ignorance about complicated systems, thereby allowing predictions of future events from uncertain knowledge of data and laws in the way in which one predicts future planetary motions from certain knowledge of data and laws. The famous lines in the *Essai philosophique sur la probabilité* that make this comparison, and which define "Laplacian determinism" in terms of an intelligence who could have infinite knowledge, appeared already in a paper of 1776.[45] Condorcet too had long looked toward this predictive capacity as the perfection of probability. It provided the conceptual justification for his predictions of the "Future Progress of Mankind." "If man can predict, almost with certainty, those appearances of which he understands the laws; if, even when the laws are unkown to him, experience of the past enables him to foresee, with considerable probability,

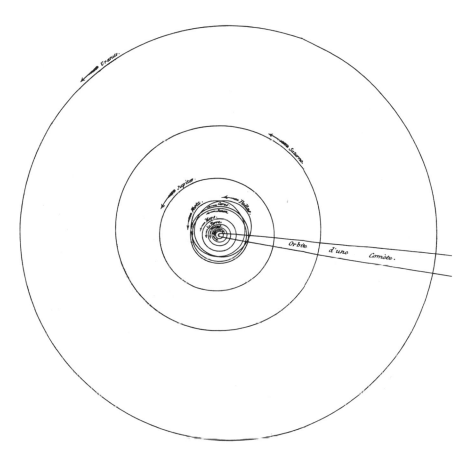

Figure 13
The system of the world

future appearances, why should we suppose it a chimerical undertaking to delineate, with some degree of truth, the picture of the
future destiny of mankind from the results of its history?" And by
implicit analogy with the removal of inequalities in the motions of
the heavenly bodies, Condorcet confidently predicted that once the
operations of natural law had been restored, the future of society
would see the reduction of inequalities in the social body.[46] As usual,
the variational calculus supported the probabilistic one, which mediated between the physical and social worlds.

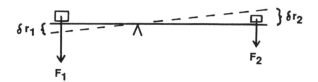

Figure 14
Virtual velocities for a balance

Virtual velocities

The variational calculus served another, even more fundamental role than removing secular inequalities. It provided the instrument that Lagrange used to reduce all of mechanics to the principle of the balance, the "principle of virtual velocities." This ancient principle of statics says that a system is in equilibrium if and only if, for every possible motion the system could be imagined to take, a certain quantity vanishes. The quantity is the sum of the forces acting (F), each multiplied by the imagined displacement (virtual velocity) in the direction of the force (δr). For the balance in figure 14, this gives, $F_1 \delta r_1 = F_2 \delta r_2$.

Lagrange, developing a principle of Jean Le Rond d'Alembert, applied the idea to dynamics.[47] He argued that if resistance to acceleration (reversed acceleration multiplied by mass) were treated as a reaction force, then any mechanical system whatsoever would be a balance, a balance of the applied forces with the reaction forces. All accelerated systems whatsoever then became balanced systems. For the simple case of a block on a frictionless table pulled by a string attached to a falling object of weight $W_2 = m_2 g$ (figure 15), the principle of virtual velocities is $m_1 a_1 \delta r_1 + m_2 a_2 \delta r_2 = W_2 \delta r_2$, or, since $a_1 = a_2 = a$ and $\delta r_1 = \delta r_2$, more simply, $a = m_2 g/(m_1 + m_2)$. The balance condition thus gives the acceleration directly.

This clean example does not perhaps capture the full flavor of the new program for equilibrium mechanics. Even an unfortunate skier, tumbling down a slope with legs, arms, skis, and poles flying, would now have to be seen as a system in equilibrium. The world, in fact, so far as mechanics extends, would be nothing but a complex balance.

The full meaning of these remarks depends on explicitly bringing the calculus of variations into the analytic apparatus, with the under-

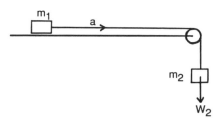

Figure 15
Every accelerated system is in balance, in Lagrange's mechanics.

standing that virtual velocities include all possible variations of the motion of a system under given constraints and that the virtual motions do not involve time. In summary, the calculus of variations was the technology that made the principle of virtual velocities work as a foundation for all of mechanics. In terms of horizontal mediation, it translated developing motion into continuing equilibrium and thereby exchanged dynamics for statics. Vertically, it established mechanical balance as the truth of nature's systems in the most general manner yet attained.

Border Crossings

The preceding discussion shows how a network of new technologies, each regarded as a balancing instrument by a community of Enlighteners, acted in important ways to tie together the system of ideologies and realities that constituted rationalism. It would be easy to extend the connections further to include, for example, electricity and magnetism by showing how Charles-Augustin de Coulomb's material balancing instruments, the torsion balances, articulated with Siméon-Denis Poisson's mathematical instruments to reify electric and magnetic fluids while idealizing new theories of action at a distance. This would illustrate once again how complex was the web of interdependent technologies that supported rationalism, as suggested in figure 16. The very complexity of connections in this picture argues for stability. I would note, however, that in each case, to represent a new instrument as a balance and to use it to mediate horizontally between different realms of nature was not unproblematic; it required an act

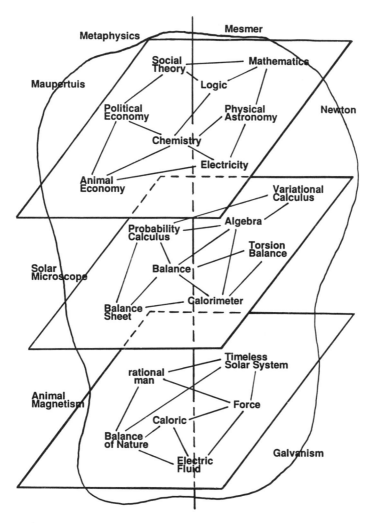

Figure 16
The theories, technologies, and realities of rationalist science formed a (temporarily) stable network, with each element drawing support from the stability of the whole. (Vertical connections are omitted.)

of construction and/or reduction. This seems to be typical of mediations, whether in settling disputes or, as here, in bringing two material
and conceptual structures into coherent relation. To make one side
agree with another requires attenuation of some qualities, to reduce
potential conflict, and enhancement of others, to produce accord.
Two such moves in the mediations above merit special attention.
They are the reduction of time to space and the representation of
fluctuations as oscillations.

To take up time first, all of the balances I have discussed eliminate
temporality from natural systems, at least in the sense of essential
development in time. They do so by collapsing relations generated
in time into simultaneous relations or into relations in space. The
effect is most obvious in the reduction of planetary inequalities developing in time to periodic oscillations, or closed cycles in space,
such that the timekeeper of the world acquires the timelessness of
eternity. Most esoteric, on the other hand, is the reduction of dynamics to statics through virtual velocities, where the virtual motions are
completely atemporal. But the balance sheet too converts a balance
developed in an economy over time into one visible all at once, as
though in space. Change is made *invisible* in order to make constancy
visible. Conservation and equilibrium appear, while the process of
reaching them disappears. This is the standard stuff of Lavoisier's
input-output chemistry, as opposed to the preceding chemistry of
Georg Ernst Stahl, but its relation to other eliminations of temporal
dynamics has not been sufficiently emphasized.

To see how deeply the strategy cut, it may be well to quote Condillac, who made it a matter of mental logic: "We decompose only in
order to recompose, and when knowledge is acquired, the things,
instead of being successive, have in the mind the same simultaneous
order that they have outside. It is in this simultaneous order that the
knowledge we have of them consists, because, if we could not recall
them all together, we could never judge the relations between them,
and we would know them badly."[48] This passage refers to the mental
process of decomposing a system in order to analyze it, but it applies
as well to events in time. Knowledge, for Condillac, meant knowledge
of God's truths, and God existed not in time but all at once, in
eternity. Thus algebra, supposed to be an atemporal language of
pure relation, best represented the condition of knowing well. Utter-

ing similar beliefs, but with less theology, Lagrange attempted to provide algebraic foundations for the differential calculus. He aimed thereby to eliminate Newton's calculus of fluctions, with its intuitive dependence on curves in space developed as motions through time.

In general, these various eliminations of temporality played a key role in identifying quite diverse systems as systems of the same kind, that is, as balances. They strengthened the network by diluting the conceptual boundaries between different subject areas, statics and dynamics, mechanics and chemistry, chemistry and economies, etc. This was an active function of the balancing technologies as mediators. In performing their function, however, the technologies themselves faded into the background, projecting their action into the subjects they related.

A second such reinforcement of the network occured through the treatment of *irregular* variations or fluctuations as though they were *regular* oscillations. Oscillations, like the swings of a pendulum, are typically periodic motions that range equally on both sides of an equilibrium position. They average out at the equilibrium position, which they thereby serve to locate. Balances balance by oscillating. Fluctuations, on the other hand, are typically aperiodic deviations that may or may not average out. Enlighteners regularly distinguished the oscillations produced by natural causes from the fluctuations produced by accidental causes. And yet they also treated these averaging effects as of the same kind, so that the social world of human variability and chance took on the same systematic cast as the oscillations of the planets.

A good place to see this is in Laplace's *Essai philosophique sur les probabilités* (1814), where the laws of chance produce harmony and stability in everything from astronomical observations to gambling, political economy, and the weather. Laplace always assumes that "the variable causes of . . . irregularity produce effects alternatively favorable and contrary to the regular march of events, which thus destroy each other mutually."[49] This assumption that accidental causes act equally in all directions, which is the assumption of balance over time, depends on the principle of insufficient reason: there is no reason to think that chance deviations are not symmetric. The assumption seems to depend also on believing that all natural systems are self-balancing, so that each accidental kick by itself will produce

oscillations. Accidents of weather, for example, cause wild disturbances of the sea, but these disturbances generate waves, which average out because they oscillate equally above and below the level of
perfect calm. Nevertheless, explicit justification is not the burden of
the *Essai*, and the technical role of the balance remains largely hidden.
The averaging effect of fluctuations follows almost imperceptibly by
a rhetorical slide from fluctuations to oscillations, which dilutes the
boundary between them. This rhetorical slide carries Laplace and
company freely back and forth.

Or rather, somewhat freely. For the policy of constructing analogies
does not make things the same. It does not subsume them under the
same set of concepts in any rigorous manner. The "balance," as the
technology of rationalism, remained a bagful of technologies called
by one name, resembling one another only in limited respects, but
presented so as to make those respects stand out and the differences
disappear. The result resembles a system of mutually reinforcing
alliances that remains stable so long as the interests of the various
parties remain relatively constant and not too many defections occur.

Border Guards

If coherence in the rationalist network depended on diluting internal
boundaries, the opposite was true at external boundaries, where
coherence depended on excluding foreign agitators and naturalizing
desirable immigrants. The border required protection. Most problematic was the indispensible but inconvenient Isaac Newton, whose
natural philosophy of forces grounded the whole system of the world
but who also required the forces to be active powers, the agencies of
a governing God, who restored the steady loss of motion in the world
that accompanied collisions. The only available solution was to cut
Newton in half, as Lagrange and Laplace did, leaving the rational
part inside and the unbalanced part outside, as in figure 16. They
accepted his mathematics, while completely ignoring his vitalist natural theology.

Similarly disruptive were those agents of Berlin metaphysics,
Pierre-Louis-Moreau Maupertuis and Leonhard Euler, whose principle of least action would subordinate mechanical forces to divine
wisdom, spreading teleology throughout the world.[50] Lagrange and

Laplace conquered this disease for a time by arguing that the principle of least action was no principle at all, that it was merely a derivative result of the direct action of forces and of the assumption of universal conservation of vis viva.[51] Their show of French muscle, however, suffered from the fact that the great mechanicians themselves required conservation of vis viva to repulse Newton's active powers, and given conservation, the force laws followed equally well from the Berlin principle of least action. One suspects that the temporary French victory may have depended as much on Napoleon's army as on his academicians. In any case, their dominance died with them. In the new centers of scientific power, in Britain and the German realms, the principle of least action would play an ever more prominent role.

Some aliens were French, of course. Fraud and irrationality at home had also to be exposed. An outstanding example is Jean-Paul Marat's rejection by the Academy after several committees narrowly judged his claims to be exaggerated. In 1779 and 1780 he had reported discovery of a universal igneous fluid, which would explain fire, light, and electricity and which he manipulated in a wide variety of experiments using instruments of his own design. But they were not balances and produced qualitative images rather than quantitative measurements. Unfortunately for his scientific career, his "solar microscope" would have to compete with the calorimeter.[52]

More dramatic were official attempts to undermine the popular craze for Mesmeric salons, where disciples of the master cured various illnesses by restoring the free flow and harmonious balance of magnetic fluids in the body. Lavoisier participated in these public exposures in 1784 as one of five members of the Academy serving on a royal commission, which included a master balancer of nature and society from abroad, Benjamin Franklin. They debunked mesmerized trees in Franklin's garden, while at Lavoisier's house they discredited similarly activated water.[53]

Such debunking was by no means unproblematic. It would not do to simply exclude anyone who relied on active or vital forces in natural explanation. Too many major propagandists of reason entertained these ideas. Quesnay and Mirabeau are good examples. Even better is Lavoisier himself, who once spoke of the academies of science and of art that Colbert had established as "so many little republics, of which the active force perpetuates itself from age to age." He compared their self-preserving power to that of the planets:

These bodies, endowed with an active force, conserve from age to age not only the original impulse that a great minister has given to them, but their active force will destroy the resistance which ignorance, superstition, and barbery may oppose to them.

I would gladly compare these bodies to those immense masses that revolve over our heads and on which the creator of all things primitively impressed a force of projection which they have conserved since the origin of the world.[54]

Does this formulation imply that a republic of letters has, in its active force, the same conserving power as gravity has, by a passive law of nature alone, or does it imply that Lavoisier regarded gravity as an active force in Newton's sense, with the capacity to overcome resistance and restore losses? Probably some of both and with considerable ambiguity, in the same way as Lavoisier's remarks about chemical affinity maintained an ambiguity between attractive force and a tendency in the order of things.

More effective negotiations were required to distinguish reason from unreason. Often they involved a decision about whether active or vital powers could be shown to have a probable material existence. Mesmeric fluid failed the test in the eyes of the dominant rationalists, along with Marat's fire and water witching. So did phlogiston. But vital air (oxygen) passed, as did caloric and electric and magnetic fluids. Key negotiators were the mediating technologies that stabilized the reality of these agencies. But even material instruments were not unproblematic. Marat and his friends in the Academy certainly regarded his apparatus as displaying reality. Laplace engaged in a polemic with Jacques Pierre Brissot over the issue. At the same time his and Lavoisier's calorimeter produced an uncertain measure of "fire" in some eyes.[55]

Thus the weight of prestige and measurement together had limited power to enforce consensus at the boundaries of reason. But this problem is not a matter of borders alone. It amplifies interior conditions. Put succinctly, the *tenuous* nature of the mediated agreements that established the network was reflected in the *contested* nature of exclusion at the boundary.[56]

M. Norton Wise

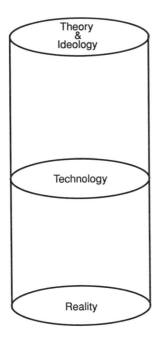

Figure 17
The cylindrical representation of theories, technologies, and realities

Maintaining control of the balancing network therefore proved a difficult business. It required a continuing campaign to define the shifting boundaries of rationality by inclusion and exclusion of new candidates for admission. For this reason, in figure 16, I have drawn a hazy boundary around the entire network and have indicated the divided and excluded positions of some of its challengers. This boundary corresponds to what Charles Gillispie has presented from a different viewpoint as *The Edge of Objectivity* in his well-known book of that title.

Closing the Circle

The geometry of the network so far generated still seems to be incomplete. It might be depicted as a cylinder, with two gaping holes at either end (figure 17). What seems to be missing is any direct connection between ideology and reality. In fact, most of what the

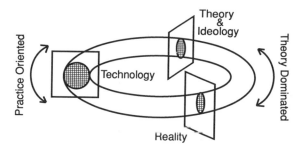

Figure 18
Symmetrical closure of the balancing network

history and philosophy of science has been about in the twentieth century is excluded. An obvious way to fill that hole is to wrap the cylinder on itself, forming a donut and a direct route between theories and things (figure 18).

The short path between theories and things has been well trodden by representatives of idealism, the hypothetico-deductive method, rational reconstruction, and other forms of theory-dominated science. Constructivists of all kinds travel the longer trail through technology, thereby emphasizing practices and instruments. We know well how to damn the shortcut by showing that it makes the means by which realities are actually constructed disappear and creates the illusion of an autonomous science in which knowledge is a thing of the mind and nature stands on its own, independent of culture. Very likely we have taken this critique too far and have begun to lose sight of the power of theories to create realities, much as we have taken local knowledge too far and now have to reconstruct the power of larger cultural units to shape individual experience.

The point I wish to make here, however, is that the cultural network I have been describing made extensive use of the short path. Without it the rationalist label would not apply. If more practical and engineering-oriented workers like Lavoisier and Coulomb emphasized the instrumental construction of knowledge, they consorted openly with friends whose final appeal rested on idealist criteria like simplicity and mathematical form. Laplace's argument for the inverse square law of gravity, for example, ultimately invoked "simplicity" and "analogy." Acutely conscious of the problem of measurement errors, he acknowledged that no amount of inductive evidence could ever establish with certainty that the law was exactly the inverse square.[57] Nevertheless, he relied on his faith that the underlying structure of nature was both mathematical and simple in insisting that the power of the distance was a pure 2 rather than a messy 2.0001. This view contrasts sharply with Newton's analogy of nature, which also invoked simplicity. Newton made atoms extended, hard, impenetrable, massive, and mobile by transfering the observeable properties of macroscopic bodies to the microscopic realm, which is precisely what Laplace did not do. Giving priority to the underlying mathematical forms, Laplace, like Ruggiero Giuseppe Boscovich, made atoms into mathematical points, to which the inverse square law would apply throughout its mathematical domain, down to an infinitesimal distance. These points, at least implicitly, could never collide, so that vis viva would always be conserved (assuming that forces are dependent only on distance). And their "mass" acquired a meaning independent of extended matter, being merely the ratio of applied force to acceleration produced. Laplace's atoms, then, were theoretical entities in a strong sense. Their properties derived in large part from theoretical considerations, rather than from any instruments or experiments.

Yet Laplace's explicit use of the direct route from theory to reality, despite its exemplary character, is not typical in my story of rationalist culture. More important were after-the-fact reconstructions, whereby the technologies of rational analysis were seen to work just because they mirrored an underlying rational reality. Condillac, for example, although he regularly described algebraic analysis as an instrument, also regarded it as the true machinery of the rational mind. It not only probed rationality; it was rationality. And herein lay its ultimate validity. Instrumentality disappeared in reality. Thereby the short path took over from the long one. Similar remarks apply to Condorcet's and Laplace's use of the probability calculus. It would work to analyze human society just because it reproduced the natural operations of rational man (which is a central lesson of Daston's account of probability in the Enlightenment).

This recalls a key point about mediation, now in relation to the donut (figure 18). In establishing an agreement between two parties, the work of mediation is successful only if it makes the mediator disappear, so that the agreement connects the two parties directly. Similarly, mediating technologies, to the degree that they function well as mediators between theory and reality, make themselves transparent, thereby working to replace the long path with the short one. The balancing acts become disappearing acts. To see this is to see immediately how the story of technological rationalism, which I have been telling, connects with the more traditional story of rationalism as a history of ideas. I conclude with a few methodological remarks on this connection.

Conclusion

The historiography of how the Enlightenment transformed the enterprise of science has had two poles: the history of ideas of Cassirer, Lovejoy, and Gillispie's *Edge* (though not his later books, which stress technologies) and the history of measurement suggested in seminal articles by Tom Kuhn and executed in an importantly revised form by John Heilbron.[58] Very recently, those two genres, ideas and measurements, have been revitalized by Raine Daston and Larry Holmes, respectively. Following more constructivist lines, I have been attempting to sketch out how the two perspectives might be wrapped up together in a cultural history that focuses on the interplay of ideological and technological aspects. This requires renovation of both stories, beginning with the measurement story.

Measurements are not self-justifying. They employ particular sorts of instruments constructed for the purpose of attaching quantitative values to valued things. Heilbron has extensively documented, with wit and penetration, how the measurement revolution of the late eighteenth century, Kuhn's "second scientific revolution," united the mathematical subjects of mechanics and optics with experimental electricity, magnetism, and heat to create the subject we call physics. The difficulty with his account (until the most recent version) has been that it takes the superior virtues of measurement to be manifest and has made the measurement revolution, so far as the positive content of science goes, a purely instrumental affair independent of metaphysics and ideology. I would stress that many of the measuring instruments Heilbron describes were designed as balances, intended to reveal simultaneously the balance of nature and the balance of

Figure 19
La France Republicaine, by A. Clément (after Boizot)

analytic rationality. Not measurements per se but measurements carried out on balances united mathematics with physics. The legitimacy of any one of these measurements was mediated by a whole network of balances in both vertical and horizontal dimensions, extending from mathematics and physical astronomy to psychology and social theory. The ideological cast of this network appears straightforwardly in a Jacobin representation of Republican France by A. Clément (figure 19).[59] Her pendant is another instrument of technical balancing, a carpenter's level. This icon of the new order suggests that in a different cultural network celebrating different values, different things would be measured using differently designed instruments.[60] In short, measurements are cultural expressions. Their significance must be understood in cultural terms.

To turn to the other historiographical pole, the history that empowers ideas has problems similar to the history that empowers measurements. The history of ideas is not wrong; it simply leaves out much of what makes cultures work. The three-level scheme suggests that a network of ideas extends little farther than the network of technologies for realizing them. Ideas and technologies empower each other reciprocally, so that neither can move far without the other. For that reason, crossing cultural boundaries is usually difficult. Tom Hughes has presented this problem for electrical power systems in his *Networks of Power*. At the mathematical end, Andrew Warwick discusses the nonreception of Einstein's ideas on special relativity in Britain, using the notion of "theoretical technologies" that structure the conceptual world of local communities of practitioners.[61]

The network approach suggests something else about traditional historiography of ideas. It indicates why particular individuals have played the role of "great men" with great thoughts. Their status could be redescribed in terms of the nodal positions they occupy in the network. That is, their power as individuals resides not so much in their creative genius, which is genuine enough, but in their having constructed themselves, their ideas, their technologies, and their realities so as to play a crucial role in stabilizing a cultural network. Once stabilized, the network acquires an elastic resistance to deformation palpable to anyone who would enter it from outside or who would threaten it. The resistance typically does not reside in any particular individual, however central, but in the dissemination of power through the network.

This resistance, furthermore, is perhaps the best indicator of why phrases like "rationalist culture" and "technology of rationalism" refer to actual structures, even though they encompass a wide variety of locally varying and competing elements and even though they consist in nothing other than the interrelations of individuals through ideas, instruments, and social practices. The interrelations maintain the stability and resilience of the network so that "rationalist" may be said to be one of the emergent properties of its structure. This structure as a whole puts severe constraints on what it means to know or to explain, or indeed, for a thing to exist, in the culture whose experience it organizes. It may therefore be said to specify a cultural epistemology, even though, once again, the range of acceptable meanings is diverse, fluid, and cannot be subsumed under a single set of criteria.

M. Norton Wise

I would therefore like to end with a more ambitious title: "Depicting Cultural Epistemology: Mediations." Reading the paper backward under this title, one would find daunting the task of an outsider to the network in designing a new instrument to measure heat, like the calorimeter, because it would have to fit into Enlightenment rationalist culture without violating its constraints on knowledge. Success would require the construction of a new node with numerous mediating links. On the other hand, from the position of an insider like Lavoisier, who already moved widely within that culture, the range of balancing strategies available as resources and the motivations for employing them were equally compelling. The constraints and the resources worked together in the creation of a new technology of knowledge.

Notes

1. This discussion expands a model developed in M. N. Wise, "Mediating Machines," *Science in Context* 2 (1988): 77–113, and uses it to restructure the balancing strategies in M. N. Wise, with the collaboration of Crosbie Smith, "Work and Waste: Political Economy and Natural Philosophy in Nineteenth-Century Britain. Part I," *History of Science* 27 (1989): 263–301. For intensive discussion I would like to thank Mario Biagioli, Raine Daston, Bruno Latour, Joseph O'Connell, Simon Schaffer, Tom Kuhn, and Elaine Wise, as well as colloquium participants at UCLA, Princeton, and Johns Hopkins. Readers of Otto Mayr's *Authority, Liberty, and Automatic Machinery in Early Modern Europe* (Baltimore, 1986) will recognize significant parallels in this paper (and I thank Paul Forman for drawing my attention to them). Mayr argues that self-balancing mechanisms provided important metaphors for liberal conceptions of political and economic order in eighteenth-century Britain. I make related claims about Enlightenment France. But I am less concerned with metaphors than with a specific network of balancing instruments and people that directly connected technical work in physics and mathematics with technical work in the human sciences.

2. Pierre Bourdieu analyzes the academic "field" of present-day France in *Homo Academicus,* trans. Peter Collier (Stanford, 1988). Bruno Latour and his coworkers have long stressed the generative nature of "action" in producing networks of knowledge, e.g., Latour, *Science in Action* (Cambridge, Mass., 1987), part 3. While his network-generating actions are very attractive, I have as yet been unable to see how they establish structural stability in a network without referring the actions to the network they generate. Thus I find it necessary to assume that actions presuppose and express the network they are reproducing or transforming. In current phrasing, the network is always already there. It supplies the resources, constraints, and motivations that structure propensities to act under contingent circumstances; it constitutes cultural entities. This scholastic disagreement aside, I have also learned much from Latour's insistence on treating human and nonhuman actors in a network as in many respects on an equal footing.

3. The calorimeter is from A.-L. Lavoisier, *Elements of Chemistry* (1789), trans. R. Kerr (New York, 1965), endplate. The main analytical balance is one of Rouelle's, from

Denis Diderot, *Encyclopédie, ou Dictionaire raisonné des science, des arts et de metier* . . . , vol. 24 (vol. 3 of Plates), (Stuttgart, 1967), facsimile of 1st ed. (1763), "Chemie," plate xv. For a rehtorical analysis of Lavoisier's presentation and use of the calorimeter, see Lissa Roberts, "A Word and the World: The Significance of Naming the Calorimeter," *Isis* 82 (1991): 199–222. Among the several recent proposals for reinterpreting Lavoisier's role in the "Chemical revolution," none challenges the central role of the mass balance and other balancing instruments, which I assume here without comment. See the papers collected in *The Chemical Revolution: Essays in Reinterpretation*, ed. Arthur Donovan, *Osiris*, 2nd series, 4 (1988).

4. Antoine-Laurent Lavoisier and Pierre-Simon, marquis de Laplace, *Mémoire sur la chaleur* (1783), reprinted with a translation and notes by Henry Guerlac in *Memoir on Heat* (New York, 1982), pp. 37, 39 (quotation). See also the similar statements in Lavoisier, *Elements*, pp. 3, 16. Lavoisier and Laplace apparently did not regard the fact that water expands on freezing as critically relevant to the question of how tightly the molecules are joined by their mutual affinity. They speak of the "force of affinity" being "exerted with most advantage" in the disposition of the molecules in ice where they would be farther apart (p. 39). Presumably they supposed that, on melting, the chemical combination of caloric with water (latent heat) could produce a contraction.

5. Mi Gyung Kim has analyzed the difference in chemical usage between "substances" possessed of specific chemical qualities and undifferentiated "matter," in "The Layers of Chemical Language. I: Constitution of Bodies versus Structure of Matter," *History of Science* 30 (1992): 69–96, esp. 71–77. Charles Gillispie, in C. C. Gillispie, Robert Fox, and I. Grattin-Guinness, "Laplace, Pierre-Simon, Marquis de," *Dictionary of Scientific Biography*, (New York, 1978), vol. 15, 273–403, p. 286, points out that Laplace entertained the idea of propagating forces in a seminal work of 1774. By the 1880s he seems to have abandoned such ideas and to have worked entirely within the action-at-a-distance formulation.

6 Lavoisier, *Elements*, p. 22. For Laplace as the source of the attempt to mathematize affinities by analogy with gravity and celestial mechanics, see Henry Guerlac, "Chemistry as a Branch of Physics: Laplace's Collaboration with Lavoisier," *Historical Studies in the Physical Sciences* 7 (1976): 193–276, esp. pp. 267–276, and note the quotation (p. 274) from the first edition of Laplace's *Exposition du système du monde*, 2 vols. (Paris, 1796), 2:197: "The simplest way [to determine the laws of affinity] is to compare these forces with the repulsive force of heat, which one may itself compare with weight." This describes the balancing operation of the calorimeter from Laplace's perspective.

7. Guerlac, "Chemistry as a Branch of Physics," p. 240. Denis I. Duveen and Roger Hahn, "Deux lettres de Laplace à Lavoisier," *Revue d'histoire des sciences* 11 (1958): 338–340. On physics (including chemistry) versus mathematics, see John L. Heilbron, "A Mathematicians' Mutiny, with Morals," this volume. On the further distinction between chemistry and physics, in the narrower sense of experimental physics, and on Lavoisier's program for reforming chemistry by applying the methods of experimental physics, see, in addition to Guerlac, Arthur Donovan, "Lavoisier and the Origins of Modern Chemistry," in Donovan, ed., *Chemical Revolution*, pp. 214–231, esp. pp. 219–228, and the exchange between C. E. Perrin, Donovan, and Evan Melhado in *Isis* 81 (1990): 259–276.

8. Ian Hacking, *Representing and Intervening*, (Cambridge, 1983), pp. 27–29 and chap. 11.

9. Tom Kuhn argues convincingly that the conception and measure of forces employed in mechanics after Newton depended on reference to the spring balance, in contrast

to the mass balance, as well as to Newton's third Law (action-reaction) and to Hooke's law of linear stretching in response to force. If so, the spring balance should sit here as an implicit middle term between the mass balance and the calorimeter, in order that the calorimeter can measure both quantity of substance and strength of force. See T. S. Kuhn, "Possible Worlds in History of Science," in Sture Allén, ed., *Possible Worlds in Humanities, Arts, and Sciences* (Proceedings of Nobel Symposium 65, Berlin, 1989), pp. 9–32, esp. pp. 16–18.

10. Frederic Lawrence Holmes, *Lavoisier and the Chemistry of Life: An Exploration of Scientific Creativity* (Madison, 1985), e.g., pp. 270–283, 388–402. Balance sheet from Lavoisier, *Elements*, pp. 133–137. Bernadette Bensaude-Vincent, "The Balance: From Chemistry to Politics," forthcoming in a special issue of *The Eighteenth Century: Theory and Interpretation*, gives an account of Lavoisier's several balancing strategies that agrees rather fully with the horizontal mediations discussed below, especially those for political economy, and adds important context.

11. Lavoisier, *Résultats extraits d'un ouvrage intitulé de la richesse territoriale du royaume de France*, *Œuvres de Lavoisier* (Paris, 1893), 6:403–463, pp. 415–416.

12. Ibid., p. 416. For Lavoisier's regular emphasis on the constancy of temperature in the healthy animal economy (measured to fractions of a degree), see Holmes, *Lavoisier*, pp. 193–196, 445, 466–467. Lavoisier and Laplace describe their Réaumeur thermometer (80° between freezing and boiling) in *Mémoire sur la chaleur*, p. 8. See also Guerlac, "Chemistry as a Branch of Physics," p. 254 n.; W. E. K. Middleton, *A History of the Thermometer* (Baltimore, 1966), p. 119, for Lavoisier's enthusiasm for thermometers; and Maurice Daumas, *Lavoisier, théoricien et expérimentateur* (Paris, 1955), pp. 124–126, for the numerous thermometers he had constructed. Although fever thermometers were not yet in common use for medical diagnosis, they had been studied and advocated since the early eighteenth century by "progressive" physicians, including Hermann Boerhaave and his associates (S. J. Reiser, *Medicine and the Reign of Technology* [Cambridge, 1978], pp. 113–114).

13. Lavoisier, "Éloge de M. de Colbert" (1771), *Œuvres*, 6:109–124, p. 117; "political machine," p. 112. For "animal machine," see Holmes, *Lavoisier*, e.g., pp. 458, 473. Lavoisier's eulogy, which portrayed Colbert as a protophysiocrat, would have responded to a prize competition set by the Academy of Sciences in the midst of a conflict over free trade, which threatened physiocrats in general and his wife's uncle in particular, J. M. Terray, who had become Controller General in 1769 (Bensaude-Vincent, "The Balance").

14. Lavoisier, "Mémoire sur la disette des bestiaux" (1786), *Œuvres*, 6:191–194, p. 192.

15. Lavoisier, "Richesse," pp. 445–446.

16. Lavoisier, "Richesse," p. 408.

17. Lavoisier, "Prix proposé par l'Academie des Sciences pour l'année 1794" (1792), *Œuvres*, 6:33–38, p. 34.

18. Lavoisier and Laplace, *Mémoire sur la chaleur*, p. 55.

19. Lavoisier, "Colbert," p. 118.

20. Lavoisier, "Colbert," p. 116: "Hard currency is a fluid, not, it is true, so mobile as water, but which with time necessarily takes its level."

21. Lavoisier, "Colbert," pp. 116, 118.

22. Elizabeth Fox-Genovese, *The Origins of Physiocracy: Economic Revolution and Social Order in Eighteenth-Century France* (Cornell, 1976), provides an invaluable historical account, with this qualification. Her insistence that the tableau represents Quesnay's vision of a dynamic rather than a static economy (e.g., pp. 97f., 121, 290–298) seems to misrepresent the role of balancing models in the physiocratic doctrine of natural order, particularly when considered in the context of neighboring disciplines.

23. This table derives from those published in the compilation of Quesnay's and his own ideas by Victor Riqueti, Marquis de Mirabeau, "Le Tableau œconomique avec ses explications," in his *L'ami des hommes,* part 4 (1759), translated as *The Œconomical Table, an Attempt towards Ascertaining and Exhibiting the Source, Progress, and Employment of Riches, with Explanations, by the Friend of Mankind, the Celebrated Marquis de Mirabeau* (London, 1766), endplates.

24. See Mirabeau, *The Œconomical Table,* pp. 15, 16, 22, 83, 99, 117, 196, for various uses of "machine."

25. Ibid., p. 196.

26. Holmes, *Lavoisier,* has also stressed the move from static balance to steady-state flow. On the vitality of dephlogisticated air and oxygen as measured by yet another balancing device, the eudiometer, see Simon Schaffer, "Measuring Virtue: Eudiometry, Enlightenment, and Pneumatic Medicine," in *The Medical Enlightenment of the Eighteenth Century* (Cambridge, 1990), pp. 281–318.

27. Fox-Genovese, *Physiocracy,* pp. 246, 249–251, interpreting Mirabeau, *La philosophie rurale* (Amsterdam, 1763), p. 4. Lavoisier expounds on the dual status of money in "Colbert," p. 112.

28. I intend also to reinforce the usage of "literary" and "social" technologies by Steven Shapin and Simon Schaffer, *Leviathan and the Air Pump: Hobbes, Boyle, and the Experimental Life* (Princeton, 1985), e.g., pp. 60–79. Accordingly, I use "technology" in general to refer to any means applied in a regular manner to attain ends; instruments are specialized bits of technology. Theories can be technologies when they are employed as means of investigation rather than as objects of investigation, although the distinction is often problematic, as in Lavoisier's view that the general balance sheet would contain the entire science of political economy.

29. W. R. Albury, *The Logic of Condillac and the Structure of French Chemical and Biological Theory, 1780–1801* (diss., Johns Hopkins University, 1972; University Microfilms 76–8469), pp. 110–192, gives an extended discussion of Lavoisier's algebraic method in relation to Condillac's, albeit without the balance. On Condillac's algebraic ideal, see Robin E. Ryder, "Measure of Ideas, Rule of Language: Mathematics and Language in the Eighteenth Century," in Tore Frängsmyr, J. L. Heilbron, and Robin Ryder, eds., *The Quantifying Spirit in the Eighteenth Century* (Berkeley, 1990), pp. 113–140, esp. pp. 115–120.

30. Étienne Bonnot de Condillac, *La logique, ou les premiers développemens de l'art de penser* (1780), in *Œuvres complètes de Condillac* (Paris, 1798), 22:127. The sentiment was common. See Lavoisier, "Colbert," 114, describing Colbert as one "balanced in the silence of his room the advantages and the interests of commerce, while he calculated

the effect of the different counterweights which maintain the equilibrium of the political machine."

31. Condillac, *Logique,* p. 157. On the lever, see pp. 2–3, 95, 138.

32. Étienne Bonnot de Condillac, *Traité des système,* in *Œuvres,* 2:375–377.

33. Lorraine Daston, *Classical Probability in the Enlightenment* (Princeton, 1988), esp. chaps. 2 and 6.

34. Ibid., p. 101; see also pp. 345, 349, 357.

35 Mirabeau, *Œconomical Table,* p. 37; see pp. 30 f. and 36 on maxima.

36. Marie-Jean-Antoine-Nicholas Caritat de Condorcet, *Esquisse d'un tableau historique des progrès de l'esprit humain* (1793), presented by O. H. Prior and Y. Belaval (Paris, 1970), p. 152.

37. Ibid., pp. 225, 227. A succinct sketch of Condorcet's "social mathematics," his physiocratic assumptions, and the prominent role of mean values is his unfinished "General View of the Science of Social Mathematics," in Keith Baker, ed. and trans., *Condorcet: Selected Writings* (Indianapolis, 1976), pp. 183–206.

38. Theodore M. Porter, *The Rise of Statistical Thinking, 1820–1900* (Princeton, 1986), pp. 100–109.

39. Joseph-Louis Lagrange, "Essai d'arithmetique politique sur les premiers besoins de l'intérieur de la république," *Œuvres de Lagrange,* ed. J.-A. Serret (Paris, 1877), 7:571–579, p. 572; first published in *Collection de divers ouvrages d'arithmétique politique, par Lavoisier, de Lagrange, et autres,* ed. Roederer (Paris: C.-C. Corancez & Roederer, 1796). As Simon Schaffer has noticed, Lagrange could have learned to relate groceries to algebra in Germany, from Lichtenberg: "The grocer who weighs something is as much engaged in putting the unknown quantity on one side and the known on the other as is the algebraist" (Georg Cristoph Lichtenberg, *Aphorisms,* edited by R. J. Hollingdale [Harmondsworth, 1990], p. 26, aphorism 28 [Notebook A, dated 1765–1770]).

40. Lagrange, "Arithmétique politique," pp. 572–573. Lavoisier, *Richesse,* p. 422.

41. Lagrange, "Arithmétique politique," pp. 573, 578.

42. Pierre-Simon de Laplace, *The System of the World,* trans. J. Pond, 2 vols. (London, 1809), 2:45. Words to this effect first appeared in one of Laplace's papers of 1788 (see Gillispie, *Laplace,* p. 333) and reappear in his *Mécanique céleste* vol. 3 (1802) Eng. trans. Nathaniel Bowditch, 4 vols. (Boston, 1829–1839), 3:x. These versions both add that nature has arranged the eternal duration of the heavens in the same way as she has arranged on earth for the "preservation of individuals and for the perpetuity of the species." Gillispie (pp. 286, 323) points out that Laplace seems not to have been worried about stability in his early papers but rather took it for granted. He only began to advertise his accomplishment as a grand proof of stability in the 1780s.

43. Laplace, *Mécanique céleste,* 3:xiii.

44. Figure 14 is from J.-B. Biot, *Traité élementaire d'astronomie physique,* 2nd ed., 3 vols. (Paris and St. Petersburg, 1810–1811), 3: endplate. The work closely follows Laplace,

to whom Biot dedicated it. On the celestial clock, see Biot's chapter "De l'égalité des révolutions du ciel, et de leur usage pour la mesure du temps," 1:57–72.

45. Gillispie, "Laplace," pp. 280, 285.

46. Condorcet, *Esquisse,* p. 203. The most specific analogy here may be to Laplace's nebular hypothesis for the origin of the solar system and its progressive attainment of an eternal state through the action of gravity and the laws of motion, although Laplace apparently did not publish this view until 1796 at the end of his *System.*

47. Joseph-Louis Lagrange, *Mécanique analytique* (1788), 4th ed., 2 vols. (Paris, 1888–1889), 1:21–22, 267. Laplace also grounds mechanics on the principle in *Mécanique céleste,* 1:78–82, 96–136. He gives a nonmathematical account in *System,* 1:337–346, 369–379.

48. Condillac, *Logique.*

49. Pierre-Simon de Laplace, *Essai philosophique sur les probabilités,* 4th ed. (1819), reprinted as the Introduction to *Théorie analytique des probabilités,* 3rd ed. (Paris, 1820), p. xli.

50. Mary Terrall, "Maupertuis and Eighteenth-Century Scientific Culture" (Ph.D. diss., University of California, Los Angeles, 1987), chap. 2.

51. Lagrange, *Mécanique analytique,* 1:262; Laplace, *System,* 1:326–328, 360, 377; Laplace, *Mécanique céleste,* 1:123. Lagrange seems to have reversed his view of least action, since he originally followed Euler's methods. See Craig Fraser, "J. L. Lagrange's Early Contributions to the Principles and Methods of Mechanics, "*Archive for History of Exact Science* 28 (1983): 197–241.

52. C. C. Gillispie, *Science and Polity in France at the End of the Old Regime* (Princeton, 1980) pp. 290–330, esp.pp. 308–312.

53. Gillispie, *Science and Polity,* pp. 314–315. Robert Darnton, *Mesmerism and the End of the Enlightenment in France* (Cambridge, Mass., 1968), pp. 62–64.

54. Lavoisier, "Colbert," p. 124.

55. Gillispie, *Science and Polity,* pp. 261–289. T. H. Lodwig and W. A. Smeaton, "The Ice Calorimeter of Lavoisier and Laplace and Some of Its Critics," *Annals of Science* 31 (1974): 1–18.

56. I thank Mario Biagioli for insisting on the continuity between negotiations inside and at the boundary.

57. Laplace, *System,* 2:12–26.

58. Thomas S. Kuhn, "The Function of Measurement in Modern Physical Science," *Isis,* 52 (1961): 161–190, and "Mathematical versus Experimental Traditions in the Development of Physical Science," *Journal of Interdisciplinary History* 7 (1976): 1–36; both reprinted in *The Essential Tension* (Chicago, 1977), pp. 178–224 and 31–65. John Heilbron, *Electricity in the Seventeenth and Eighteenth Centuries: A Study of Early Modern Physics* (Berkeley, 1979), pp. 73–97, 449–489. Heilbron's most far-reaching revisions appear in this volume, "A Mathematicians' Mutiny." Some excellent examples of the

compulsion to count and measure are contained in Frängsmyr, Heilbron, and Ryder, *The Quantifying Spirit.*

59. From Simon Schama, *Citizens: A Chronicle of the French Revolution* (New York, 1989), p. 769.

60. Consider the construction of instruments in mid-nineteenth century Britain as "engines" rather than "balances," as discussed in M. Norton Wise and Crosbie Smith, "Measurement, Work, and Industry in Lord Kelvin's Britain," *Historical Studies in the Physical and Biological Sciences,* 17 (1986): 147–173.

61. Thomas Park Hughes, *Networks of Power: Electrification in Western Society, 1880–1930* (Baltimore, 1983). Andrew Warwick, "Cambridge Mathematics and Cavendish Physics: The British Reception of Relativity, 1905–1911," to appear in *Studies in History and Philosophy of Science.*

The Philosophers Look Ahead

How We Relate Theory to Observation

Nancy Cartwright

1 Symbolic Generalizations and Practice Problems

There is no theory/observation distinction. This is commonly taken to be one of the principle lessons we have learned from Thomas Kuhn, both from his historical account of the structure of scientific knowledge and its changes and from his more philosophical reflections on the meaning and reference of scientific terms. What is this slogan supposed to deny? Usually it turns out to be a rejection of a foundational picture of knowledge: observation cannot provide epistemologically secure claims on which we can build an edifice of increasingly more general and more widely applicable laws and methods. We can be mistaken about even the most mundane claims about sensible properties, and once these are called into question, their defense will rest on a complicated and sophisticated network of general claims about how sensations are caused, what kinds of things can go wrong in the process, and what kinds of things can and cannot be legitimately adduced as interferences. This is all the more so in the case of scientific observations, where we need a clear idea of what our instruments are doing before we can describe the outcomes in any way that would bear on a scientific generalization. It is these kinds of worries that we summarize in the dictum, All observation is theory-laden.

Yet to admit that observation is theory-laden is a long way from denying that there is a theory/observation distinction. How far "down" do we have to go to reach a level of pure observation independent of theory: instrument readings, everyday observations, sense data, protocol sentences? We may admit that the question does not even make sense once we give up a foundational picture and accept instead that no claim about an individual matter of fact can be validated without already presupposing the truth of general laws, and vice versa. But that should not obscure the conspicuous fact that the terms of modern science, particularly of modern mathematical physics, are peculiarly recondite in comparison with those we are more used to in our day-to-day lives. C. G. Hempel marked the difference by distinguishing *theoretical* terms from those that are *antecendently understood*, remaining neutral about whether these latter described sense data or physical things or were merely terms from a more familiar and well-entrenched scientific discipline that may itself be highly abstract and mathematical. We know that a major part of Hempel's efforts have been devoted to trying to understand the relationship between these two levels of description.

Pierre Duhem drew much the same distinction as Hempel between the quantitative and abstract concepts of mathematical physics, which he described as symbolic, and those designated by the common terms of everyday life. He illustrates the difference for even such familiar scientific terms as "temperature" and "pressure."

Let us put ourselves in front of a real, concrete gas to which we wish to apply Mariotte's (Boyle's) law; we shall not be dealing with a certain concrete temperature embodying the general idea of temperature, but with some more or less warm gas; we shall not be facing a certain particular pressure embodying the general idea of pressure, but a certain pump on which a weight is brought to bear in a certain manner. No doubt, a certain temperature corresponds to this effort exerted on the pump, but this correspondence is that of a sign to the thing signified and replaced by it, or of a reality to the symbol representing it. (1974, 166)

The point of this essay is not to discuss Pierre Duhem but to discuss Thomas Kuhn. Duhem enters because he serves as a venerable predecessor who makes the same distinction, with the very same language, that Kuhn makes. For Kuhn too, the laws of physics, its theoretical claims, are *symbolic generalizations* to be contrasted with the different and more concrete descriptions needed to treat a real physical situation. Kuhn is far from denying the distinction between the theoretical or symbolic and the more observational or concrete, which so worried Hempel and Duhem. Indeed, considerations about how the two levels relate are at the core of his thinking about paradigms and thereby of *The Structure of Scientific Revolutions*. As Kuhn himself tells us in section 3 of the Postscript, "The paradigm as shared example is the central element of what I now take to be the most novel and least understood aspect of this book." This remark stands as the opening sentence to Kuhn's discussion of symbolic generalizations. I here reproduce the entire discussion, so that we can see exactly what Kuhn had to say.

Scientific knowledge (on the traditional view) is embedded in theory and rules; problems are supplied to gain facility in their application. I have tried to argue, however, that this localization of the cognitive content of science is wrong. At the start and for some time after, doing problems is learning consequential things about nature. In the absence of such exemplars, the laws and theories the student has previously learned would have little empirical content.

To indicate what I have in mind I revert briefly to symbolic generalizations. One widely shared example is Newton's Second Law of Motion, generally written as $f = ma$. The sociologist, say, or the linguist who discovers that the corresponding expression is unproblematically uttered and received by the members of a given community will not, without much additional investigation, have learned a great deal about what either the expression or the terms in it mean, about how the scientists of the community attach the expression to nature. Indeed, the fact that they accept without questions and use it as a point at which to introduce logical and mathematical manipulation does not of itself imply that they agree at all about such matters as meaning and application. Of course they do agree to a considerable extent, or the fact would rapidly emerge from their subsequent conversation. But one may well ask at what point and by what means they have come to do so. How have they learned, faced with a given experimental situation, to pick out the relevant forces, masses, and accelerations?

In practice, though this aspect of the situation is seldom or never noted, what students have to learn is even more complex than that. It is not quite the case that local and mathematical manipulation are applied directly to $f = ma$. That expression proves on examination to be a law-sketch or a law-schema. As the student or practicing scientist moves from one problem situation to the next, the symbolic generalization to which such manipulations apply changes. For the case of free fall, $f = ma$ becomes $mg = md^2s/dt^2$; for the simple pendulum it is transformed to $mg \sin\theta = -mld^2\theta/dt^2$; for a pair of interacting harmonic oscillators it becomes two equations, the first of which may be written $md^2s_1/dt^2 + k_1s_1 = k_2(s_2 - s_1 + d)$; and for more complex situations, such as the gyroscope, it take still other forms, the family resemblance of which to $f = ma$ is still harder to discover. Yet, while learning to identify forces, masses, and accelerations in a variety of physical situations not previously encountered, the student has also learned to design the appropriate version of $f = ma$ through which to interrelate them, often a version for which he has encountered no literal equivalent before. How has he learned to do this?

Later, Kuhn proposes a solution based on *similarity*; this is a feature that I will take up in this discussion. Here is what he says:

Science students regularly report that they have read through a chapter of their text, understood perfectly, but nonetheless had difficulty solving a number of problems at the chapter's end. Ordinarily, also, those difficulties dissolve in the same way. The student discovers a way to see his problem as *like* a problem he has already encountered. Having seen the resemblance, grasped the analogy between two or more distinct problems, he can interrelate symbols and attach them to nature in the ways that have proved effective before. The law-sketch, say $f = ma$, has functioned as a tool, informing the student what similarities to look for, signaling the gestalt in which the situation is to be seen. The resultant ability to see a variety of situations as like each other, as subject for $f = ma$ or some other symbolic genralization, is, I think, the main thing a student aquires by doing exemplary problems, whether with pencil and paper or in a well-designed laboratory. (Kuhn 1970, 187–189)

What I will do here is to provide the beginnings of an account of the relationship to which Kuhn points between symbolic generalizations, like $f = ma$, and the more specific versions of them used to describe concrete objects in concrete situations, such as the simple pendulum, a pair of interacting harmonic oscillators, or two masses separated by a distance. My language is already a clue to the view I will defend. The relationship between them, I will argue, is that of the abstract to the more concrete. Seeing the relationship in this way can account, I will argue, for those aspects of the relationship that Kuhn marks out as especially important: first, Kuhn's claim that the symbolic generalization is not a law but rather "a law-sketch or a law-schema"; second, Kuhn's interest in how symbolic generalizations are understood and his claim that our understanding of them derives from, or better, consists in, our understanding of the concrete versions of these generalizations that we use to treat individual cases; third, Kuhn's claim that the symbolic generalization serves to collect together a number of distinct cases, all of which are analogous one to another.

2 Duhem on the Symbols of Physics

The thesis I want to maintain is much like Duhem's. So I begin by explaining some doctrines of Duhem that I will utilize. Duhem, like Kuhn, is concerned with the significance of the symbols that appear in the equations and formulas of modern mathematical physics. He uses the term "translation" to describe how the symbols are connected with the concrete reality of our laboratory or everyday experience.

Thus as both its starting and terminal points, the mathematical derivation of a physical theory cannot be wedded to observable facts except by a translation. In order to introduce the circumstances of an experiment into the calculations, we must make a version which replaces the language of concrete observation by the language of numbers; in order to verify the results that a theory predicts for that experiment, a translation exercise must transform a numerical value into a reading formulated in experimental language. But translation is treacherous. Between the concrete facts, as the physicist observes them, and the numerical symbols by which these facts are represented in the calculations of its theorists, there is an extremely great difference. (Duhem 1974, 133)

On a cursory reading, this passage may seem to offer the beginnings of a verifiability criterion of meaning, or a demand for operationalization. In a sense this is right, but the crucial question is, Why do the concepts of mathematical physics need to be operationalized? In the writings of the Vienna Circle, verifiability or operationalizability was required to connect us and our language with the world, and it was a requirement that in principle every one of our concepts must meet. For Duhem, the need for operationalization is generated not by our epistemological problems but rather by an ontological problem. All language, he admits, is abstract. General terms like "man" pick out "the abstract idea of man in general, rather than the concrete idea of this or that man in particular" (Duhem 1974, 165). The symbols of physics work differently because of the ontological problem just referred to. Crudely put, for most of the symbols of mathematical physics, there are no quantities in nature for them to name. Unlike the general terms of our more everyday language, the concepts of physics do not refer to aspects that individual things have in common. (Notice that Duhem differs here from Kuhn and from the account I present below. That is because Duhem does not take symbols to represent "real" properties of the world, whereas, it seems to me, these properties may be as real or unreal as any others. All systems subject to a force of a given size do have something in common: that they are subject to a force of that size.) The concepts of physics are merely a *symbolic representation* for the disparate happenings that "wed" the terms of mathematical physics to "the language of concrete observation." This includes not only the ragbag of experimental procedures that we use to assign numerical values to the concepts but often the "qualitative" experience that we associate with the concepts as well. (I use scare quotes around "qualitative" to note that I do not think that the central point requires either going into or endorsing Duhem's way of distinguishing qualities from quantities.) Recall from the remarks of Duhem quoted in section 1 that in using Boyle's law we are not "dealing with a certain concrete temperature embodying the general idea of temperature, but with some more or less warm gas" (1974, 133).

The contrast is Descartes, who declared, "I admit no principles in physics which are not also accepted in mathematics" (quoted by Duhem, 1974, 114). Physics was to be entirely the study of quantities. But that is not the point of difference from Duhem, who can agree that "we wish our theoretical physics to be a mathematical physics starting with symbols that are algebraic symbols or numerical combinations" (1974, 115). The disagreement between Duhem and Descartes comes over the interpretation of the symbols. Descartes must find a real property in nature corresponding to each fundamental sign of his theory. But for Duhem, "Thoretical physics, as we conceive it, does not have the power to grasp the real properties of bodies underneath the observable appearances; it cannot, therefore, without going beyond the legitimate scope of its methods, decide whether these properties are qualitative or quantitative" (1974, 115). That is why Duhem calls the terms of mathematical physics "symbolic representations," as opposed to the more literally descriptive and more qualitative language of everyday life. "Theoretical physics," he insists, "does not grasp the reality of things; it is limited to representing observable appearances by signs and symbols" (1974, 115). In this way Duhem, like Descartes, can insist on a purely quantitative physics, but go on to provide a much different interpreation for it: "Without asserting that everything at the very bottom of material things is merely quantity, we should admit nothing but what is quantitative in the picture we make of the totality of physical law" (1974, 115).

The interpretation I want to defend of the symbols of physics in this discussion is much like Duhem's, but without the antirealism that I have introduced in trying to explain Duhem's ideas. "Crudely put," I said above, "for most of the symbols of mathematical physics, there are no quantities in nature for them to name." It is this aspect of what I have so far said that I want to let go of, while holding on to Duhem's insight, shared by Kuhn, that symbolic formulas, like $f = ma$, need the more concrete concepts of the laboratory or of everyday life—Kuhn's examples are of the simple pendulum, the pair of interacting harmonic oscillators, and the gyroscope—to connect them with the physical world.

In discussing a different attempt of mine to explain this view, (Cartwright 1991), Robin Le Poidevin highlights two features involved in realism: "On the common interpretation of realism, [to be a realist over a certain kind of item] is to say that (i) such items are mind-independent; and that (ii) statements taken to be about such items are not reducible to statements about other items" (1991). With respect to both these theses, I want here to defend a robust kind of realism about symbols and concepts like those of force (f), inertial mass (m), acceleration (a), or, in different domains, electromagnetic field strength (E) and the quantum Hamiltonian (H). (Though for different reasons involving problems of approximation that do not disappear even in ideal laboratory experiments, ultimately I am suspicious of this kind of realism.) Forces, I maintain here, are just as mind-independent as are the angle and length in a simple pendulum or the height above the earth of a freely falling object. Nor is the concept of force reducible to that of separations, masses, and motions: the meaning of "force" is largely determined by the roles forces are assigned in abstract theory, beginning with Newton's three laws. Nevertheless, I maintain, no force can exist except through some appropriate configuration of more mechanical qualities like angles, distances, and motions, qualities that characterize objects as pendula, harmonic oscillators, gyroscopes, and the like.

Le Poidevin distinguishes between *meaning reduction* and *fact reduction*. This distinction may be of help. "In meaning reduction, statements taken to be about *F*'s are held to be equivalent in meaning to statements which are in fact about *G*'s" (Le Poidevin 1991). It is meaning reduction that I have just dismissed. "In fact reduction," Le Poidevin explains, "statements taken to be about *F*'s are made true by facts describable by statements about *G*'s. There need be no meaning equivalence here. Token physicalism is an example of fact reduction" (1991). I think that in my intended sense, the view I want to defend rejects fact reduction as well. In part I think this because of the implications that follow from associating fact reduction with token physicalism. My thesis is not about facts but about their descriptions: a description ascribing a force is never true of a situation unless some different description involving more concrete concepts is true of it as well. We might say that force descriptions "piggy back" on more concrete descriptions. Originally, the concepts that could be used in these more concrete descriptions included only mechanical ones, and, of course, we inherit the legacy of the well-known debates about exactly what mechanical qualities comprises. In the late nineteenth century the admissible concepts were expanded to include electric and magnetic charge, and now in the twentieth century, characteristics of the nucleus have been added as well.

The relation that I am pointing to between forces on the one hand and mechanical, electric, magnetic, and nuclear concepts on the other is characteristic of the way abstract descriptions relate to their more concrete counterparts. So here I will use the ideas of the abstract and the concrete to describe the relation between what Duhem and Kuhn have characterized as symbolic on the one hand and what they count as more everyday and readily accessible on the other. In doing so, I depart from Duhem's own usage, since, recall, he described ordinary general terms as "merely abstract," in contrast to the terms of physics, which are "purely symbolic." Where Duhem uses the label "abstract terms," I talk instead about "general terms," and I reserve the concept of abstractness to explicate his idea of the symbolic. That is because I have in mind a particular account of the relation of the abstract and the concrete, from which, I believe, the characteristics that Kuhn notes follow quite naturally. This account plays a central role in the fable theory of the German Enlightenment playwright Gotthold

Ephraim Lessing. The key to the account is Lessing's claim that "the general exists only in the more specific, and is only made visualizable by it" (1967).

Before turning to a discussion of fables, I should first apologize for talking about the reality of properties. I did so because it seems natural to use this vocabulary in describing the views of Duhem, and also because Le Poidevin has introduced these distinctions in discussing the relation I am trying to characterize between symbolic claims and more everyday ones. But I should not have. Questions about the reality of properties, and particularly about properties versus individuals, notoriously raise sticky questions of metaphysics, which I hope to avoid. The point is to understand how two different kinds of descriptions relate, one of which seems peculiarly recondite and less directly accessible than the other. I believe that we can explore this relation independently of what view we maintain about properties. In particular, I do not want to imply in any way that forces are less real than distances or charges. Indeed, from one philosophical perspective—what is more real is what plays a role in the more fundamental, or more universal, theories in physics—the converse will be the case.

This is in part why I resist taking on Le Poidevin's fact reductionism and his analogy with token physicalism. Token physicalism, recall, has two aspects. First, although mental predicates cannot be defined in a physicalist vocabulary, still each instance or token of a true mental description is an instance of a true physical description. Second, in a special way the counterpart physical description fixes the mental one: the mental description supervenes on the physical; that is, whenever the same physical description obtains, so too will the same mental one. The abstract-concrete relation is similar in the first respect and in certain cases in the second as well. But we must beware. Often in discussions of supervenience there is the assumption that nature must "operate" at the more concrete level, hence Le Poidevin's formulation that statements about F's (e.g., mental states) are "made true by facts describable by statements about G's" (e.g., physical states). Obviously, that assumption is inappropriate in the case of the abstract-concrete relation.

As an empirical hypothesis, consider the claim that light rays traverse geodesics. *Geodesic* is a highly abstract concept; in different geometries, it takes on different more concrete forms. Imagine a spherical geometry. Is the statement that the light rays there travel on great circles made true by the fact that they traverse geodesics, or vice versa? I believe that, at least in a less careful version that we sometimes have in mind, the question does not even make sense: do the light rays travel on the paths they do *because* these paths are great circles or because they are geodesics? The point is that if one does think that some version of this question makes sense, it need in no way be the more concrete happenings that nature disposes of directly, with the supervening general descriptions following in train, rather than vice versa.[1]

I maintain that to say of mass m that it is distance r from another mass M is a concrete description to which corresponds the familiar more abstract description "Mass m is subject to a force of size GmM/r^2." When the little mass accelerates towards the bigger, is that because it is distance r away from it? Is it because it is subject to a force of size GmM/r^2? Or what? Compare: "being subject to force GmM/r^2" is an abstract description, like "geodesic"; more specifically, what being subject to that force consists in on a given occasion is typically being a mass m located a distance r from another mass M, just as, more concretely, to travel on a geodesic may be to travel on a great circle or on a Euclidean line. In each case, both descriptions are equally true and both are equally relevant to the motion dictated by the general law; one is just more concrete than the other.

3 Fables and Models

For Lessing, the fable is a genre intended neither for entertainment nor for the communication and direction of emotions but rather for elucidation. He subscribes to the distinction usual in the German Enlightenment between *intuitive cognition*, in which we tend directly to our ideas of things, and *symbolic cognition*, where we tend not to our ideas but rather to symbols we have substituted for them. The moral of the fable is a general claim, and hence is symbolic. For us to understand it clearly (and also for it to motivate us to act), it must be made visualizable; it must be given a concrete form. That is the purpose of the fable, and every element of the fable is there to serve that purpose. Hence the characteristic features of the fable in its traditional form from Aesop and Phaedras: it is short, precise, and unadorned. This sparseness and singularity of purpose makes the fable, as represented by Lessing, an ideal analogue for the scientific model. Consider $f = ma$ and Kuhn's models for it: the pendulum, the coupled harmonic oscillators, the gyroscope, plus the most familiar one, which I have added, the two-body system. Like the fable, these models are thin and diagrammatic: they contain exactly what they need to provide a concrete form for the force. The subsequent trajectories that the objects in the model are supposed to follow provide concrete cases of the general law of motion. *This* is what $f = ma$ amounts to in a particular situation.

The relation between the moral and the fable, according to Lessing, is that of the abstract to the more concrete. One of the primary alternative accounts that he wants to combat takes the fable to be an allegory. Lessing introduces the following fable to argue his point:

A marten eats the grouse;
A fox throttles the marten; the tooth of the wolf, the fox.
The weaker are always a prey to the stronger.

Allegories say not what their words seem to say, but rather something similar. where is the allegory in the fable of the grouse, the marten, the fox, and the wolf? "What similarity here does the grouse have with the weakest, the marten with the weak, and so forth? Similarity! Does the fox merely *resemble* the strong and the wolf the strongest, or *is* the former the strong and the latter the strongest. He *is* it" (Lessing 1967, 73). For Lessing, similarity is the wrong idea to focus on. The relationship between the moral and the fable is that of the general to the more specific. It is "a kind of misusage of the words to say that the special has a similarity with the general, the individual with its type, the type with its kind" (Lessing 1967, 73). Kuhn tells us that students come to see a resemblance among the various models they work with. That is appropriate to the abstract-concrete relation. The concrete cases that fall under a single abstract description resemble one another in some way, but none resembles the abstract itself.

The point comes up again when Lessing protests against those who maintain that the moral is hidden in the fable, or is at least disguised there. That is impossible on his view of the relationship between the two. Lessing argues, "How can one disguise (*verkleiden*) the general in the particular; that I do not see at all. If one insists on using a similar word here, it must be at least *einkleiden* rather than *verkleiden*" (1967, 86). *Einkleiden* is to fit out, as when you take the children to the department store in the fall and buy them new sets of school clothes. So the moral is to be "fitted out" by the fable. Although I've introduced this idea through a discussion of the fable and its moral, it marks an entirely commonplace feature of language. Most of what we say, and say truly, uses abstract concepts that want "fitting out" in more concrete ways. Of course, the more concrete descriptions may themselves be abstract when compared to yet another level of discourse in terms of which they can be more concretely fitted out in turn.

Precisely this idea of fitting out shows us why, as Kuhn argues, students need practice problems. There is no suggestion that the concepts of *force, mass,* and *acceleration* hve no meaning on their own, that they need to be defined by more concrete concepts like *separation* and *charge.* Rather, we need, case by case, to understand more concretely what being subject to a force of a given size consists in. We can adapt what Lessing says of the characters in his fables: "Does mass m's being a distance r from M merely *resemble* being subject to a force GmM/r^2, or *is* it being subject to such a force? It *is* it." I may give you the abstract advice "Be careful," but until you know more concretely what being careful consists in for different situations, this will be of little help to you. It is just this kind of insight that the students gain when they solve their practice problems and rehearse the specific force functions that describe specific models. They are not learning coordinating definitions; they are not reducing the theoretical to the observable; nor indeed are they doing anything that bears very directly on confirming the theory or testing it experimentally, since models are a long way from the world. Rather, they are fitting it out in more concrete forms to see what it really amounts to.

4 Models and Reality

Models make the abstract concepts of physics more concrete. They also help to connect theory with the real world. How does this work? Typically, we design our experiments to look as much as possible like the models we have available. Then we know what specific forms our general laws should take. Obviously, one single model will not serve; we expand it by piecing in others. That usually doesn't work with fables, and that is in part why Lessing keeps the characters in his fables so thin and featureless. If he were to fill in extra details, the characters would fall under new, different abstract concepts, which may suggest different behavior from the first and perhaps even contradictory behavior. Usually there is no general moral that pertains to both sets of behavior at once. Forces are nice because they add. We can piece together a complicated model from simpler parts, and the specific form for the force in the whole is just the sum of the specific forms for the pieces.

What happens when we run out of models? Clearly, we hunt for new ones. Are we sure to find them? Here attitudes will diverge, depending on how central and ubiquitous we take forces, for example, to be. To quote an example from Otto Neurath, one of the founders of the Vienna Circle, "In some cases a physicist is a worse prophet than a [behaviorist psycholgist], as when he is supposed to specify where on Saint Stephen's Square a thousand-dollar bill swept away by the wind will land, whereas a [behaviorist] can specify the result of a conditioning experiment rather accurately" (1987, 13). We may argue that here is a real situation, only part of which can be reasonably described by the models we have at hand. Other parts, wind turbulence, for instance, are obviously relevant to the motion, but we have no model that fits them very well. It may be, then, that there is no specific form for the abstract concept of force to take in the case of the thousand-dollar bill in Saint Stephen's Square. The general law of motion is true when its concepts apply, but we do not know that they apply here. Perhaps mechanics has met its limits.

Others, who expect universal theories with limitless domains, will not welcome this constriction of the dominion of force. It seems that they have two choices:

1. They may insist that for every situation there is some appropriate concrete description that can be linked in a systematic way with the abstract concept of force, even though we may have a very hard time discovering these descriptions.

2. They may decide to stop using force as an abstract concept and treat it as a concrete one, alongside the concepts of location, distance,

charge, and duration. In this case the concept of force becomes logically independent of the others and can in principle appear in any combination with them, just as the concept of charge is logically independent of that of separation.

If the first line is to be pursued, we must be careful about how laws are to be understood. We could fairly immediately find ourselves (as Richard Boyd and Richard Miller have made me see) in contradiction with empirical fact if we thought that some features relevant to motion did not fall in any way under the concept of force and yet simultaneously maintained the following:

a. Whenever the concrete characteristics of one of our models are fairly closely duplicated in an object, the object experiences a component force of the corresponding type.

b. The total force in a given situation is the sum of the component forces obtaining there.

c. $f_t = ma$, where f_t represents the total force.

My own views are that the concept of *natures* plays a special role in modern physics. Laws like $f = ma$ typically express how it is in the nature of the relevant properties to behave, and that in general is how they *will* behave *in ideal conditions*, i.e., in whatever conditions are appropriate for the expression of their nature (see Cartwright 1992). Hence, my own inclination, because it fits with this doctrine of natures, is to reject (c) and to say instead that $f_t = ma$ *whenever no other factors relevant to the motion are at work*. This allows that the law of motion applies in physics experiments where we are at pains to shield against all the relevant factors that do not fall under the concept of force, or, as we say from the other point of view, "all the factors we don't know how to model." Yet it does not commit us to the assumption that the laws of motion rule everywwhere nor the assumption that every feature that matters to a motion must satisfy the precise and limited conditions that theoretically makes something a force.

We are living in a period of pluralism. Universalisms—whether of physics, of culture, or of morality—are in disrepute all over. Perhaps in such a time alternative 1 will seem more attractive. It is important to be clear that this alternative does not make forces out to be unreal or ineffective, nor does it take Newton's law of motion to be false. If there are, as there appear to be, bits of nature that the known models of mechanics fit relatively well, like laboratory experiments or the planetary system or the pendulum in grandfather's clock, then the law of motion is both true (or as true as anything else) and nonempty. That says nothing about its scope. There is a tendency to feel that the scope must be universal. After all, we all expect more or less regularly that heavy bodies unsupported in the vicinity of the earth will fall, and when they do not, as in the case of airplanes or rockets or hail stones in the wind, there is a good reason why. But that is a large step from admitting that the highly exact and detailed concept of force applies, or that the very precise and theoretically embedded law of motion rules there. The fact that heavy bodies fall is what makes Newton's law promising for application outside the laboratory; it is no guarantee that the promise can be fulfilled.

Note

1. My thanks to Lorraine Daston for pressing the point that this assumption might be naturally associated with kinds of claims I want to defend. Thanks also to Conrad Wiedemann for help with Lessing's fable theory.

References

Cartwright, Nancy. "Fables and Models, I," *Aristotelian Society*, sup. vol. 45 (1991).

Cartwright, Nancy. 1992. "Natures and the Experimental Method." In J. Earman, ed., *Inference, Explanation, and Other Frustrations: Essays in the Philosophy of Science*. Los Angeles: University of California Press.

Duhem, Pierre. 1974. *The Aim and Structure of Physical Theory*. N.Y.: Atheneum.

Kuhn, Thomas. 1970. *The Structure of Scientific Revolutions*. 2nd ed. Chicago: University of Chicago Press.

Le Poidevin, Robin. "Fables and Models, II," *Aristotelian Society*, suppl. vol. 45 (1991).

Lessing, G. E. 1967. *Abhandlungen über die Fable*. Stuttgart. Reclam.

Neurath, Otto. 1987. "Unified Science and Psychology." In B. F. McGuiness, ed., *Unified Science*. Dordrecht: Reidel.

Working in a New World:
The Taxonomic Solution

Ian Hacking

Of all Thomas Kuhn's themes, the one that is most resolutely avoided in the polite company of American philosophers is precisely the one that most attracts the casual reader dipping into *The Structure of Scientific Revolutions*. I mean the idea that after a revolution we live in a different world. *Structure* incited a great brouhaha about incommensurability, theory change, paradigms, rationality, and the like. For all the noise and irritation, the effect was excellent, because there is now a surprising consensus on the power of these ideas and on what they amount to. It is true that they emerged from a stance that is "antirealist," to use the most vulgar of terms for dividing philosophical dogma. Idealism and nominalism were in the offing. Hence the ideas will long be deployed as counters in metaphysical debate and exposition. But they do not vex philosophers as they did twenty years ago. Undergraduates in a general class, not specialists, now tell me, "We can see how this was some sort of radical book once, but for us it is all kind of obvious, eh?" It has become a small educational mission to show that *The Structure of Scientific Revolutions* is not at all obvious.

There remains the question of "living in another world." Kuhn's own statements of it are very cautious and hedged (like so much of the book). Chapter 10 of *Structure* begins, "The historian of science may be tempted to exclaim that when paradigms change, the world itself changes with them" (p. 111).[1] Elsewhere the idea is equally guarded, and yet this is evidently a temptation to which the author dearly wanted to succumb. Relevant passages are the following:

The very ease and rapidity with which astronomers saw new things when looking at old objects with old instruments may make us wish to say that, after Copernicus, astronomers lived in a different world. In any case their research responded as though this were the case. (P. 117)

At the very least, as a result of discovering oxygen, Lavoisier saw nature differently. And in the absence of some recourse to that hypothetical fixed nature that he "saw differently," the principle of economy will urge us to say that after discovering oxygen Lavoisier worked in a different world. (P. 118)

When [the chemical revolution] was done, even the percentage composition of well-known compounds was different. The data themselves had changed. That is the last of the senses in which we may want to say that after a revolution scientists work in a different world. (P. 135)

The primary verb is either "live" or "work." The modalities are "may wish to say," "urge us to say," "may want to say." A summary statement poses a problem. *"Though the world does not change with a change of paradigm, the scientist afterwards works in a different world. . . . I am convinced that we must learn to make sense of statements that at least resemble these"* (p. 121). I call this the new-world problem. How shall we make sense of the idea of living or working in a different world after a change in paradigm?[2]

Despite *Structure*, revolutions and changes in paradigm are not necessarily abrupt discontinuities. Kuhn's recent work "points towards a significant reformulation" of some of those ideas.[3] He is now lukewarm about discontinuity, holding, plausibly enough, that even if some revolutions occur in a trice, many others do not. Likewise, he abandons the talk of "gestalt-switch" as descriptive of what happens to scientists who effect or follow a revolution. He apologizes for his earlier enthusiasm for gestalts and explains it. The historian reading old texts may do a sudden flip on understanding meanings and intentions different from anything we think now. But it is wrong to infer that dead scientists who effected the transition had a similar experience. Moreover, the gestalt switch is something that happens to the individual historian. The science itself is a property of the community, and it is at best a metaphor to speak of communal gestalt switches. (Less generously, it is a category mistake.) The new-world problem is not about working in a new world after a moment or a week of illumination and transformation. I will instead be discussing scientific enquiry in what we think of as the same field, but "at widely separated times," between which there occurred a paradigm shift. Scientists on either side of the time span are said to work in different worlds. What does that mean?

1   The Solution Sketched

The solution can be thought of as very "normal" philosophy in one of the oldest of philosophical paradigms, namely nominalism. Nominalist programs, however various, have a hard core. There are individuals in the world, but over and above the individuals, there are not any sets, kinds, universals, classes. Universals can exist in things, *in re*, but there are none prior to things, *ante rem*.[4] Thanks to nature's ways, the things in nature distinguish themselves into various kinds, but there are not kinds over and above the distinctions found in things.[5] In the set-theoretic form reinvigorated by W. V. Quine and Nelson Goodman, there do not exist sets over and above their members as further entities to be listed in the metaphysical census of the universe.

How shall we "make sense of statements that at least resemble" (a) "The world does not change with a change of paradigm" and (b) "The scientist after [a scientific revolution] works in a different world"? The nominalist replies, (a) the world is a world of individuals; the individuals do not change with a change of paradigm. But a nominalist may add, (b) the world in which we work is a world of kinds of things. This is because all action, all doing, all working is under a description. All choices of what to do, what to make, how to interact with the world, how to predict its motions or explain its vagaries is action under a description: all these are choices under descriptions current in the community in which we work and act and speak. Descriptions require classification, the grouping of individuals into kinds. And that is what changes with a change in paradigm: the world of kinds in which, with which, and on which the scientist works.

That's not enough, for if classifications and descriptions are intertranslatable, any choice of action before a scientific revolution can be described as a choice of action after one as well. Hence we must go back to the idea of incommensurability, but now in a very special and more coherent form. The natural-kind terms current in an old science cannot be translated into natural-kind terms in a new science.

I'll replace the phrase "natural kinds" by "scientific kinds," for we're as much concerned with kinds of instruments, apparatus, and artificial phenomena as we are with kinds found in nature. The argument for untranslatability has two parts. First, it is claimed that the kinds investigated in a branch of science can be arranged in taxonomic trees. Scientific kinds in a taxonomy never overlap; either one is properly contained in the other, or they are mutually exclusive. Using a logic of kinds, it is argued that a natural-kind term from one taxonomy can never be translated into one from another taxonomy.

Second, it is observed that scientific terms, names for scientific kinds in a taxonomy in use by a community, are projectible, in Nelson Goodman's sense (and spelling). It is argued that although members of a community might well "understand," i.e., have learned or not forgotten, the scientific terms of an earlier paradigm, they can no longer use these terms projectibly in their community, precisely because they are not translatable into present projectible scientific terminology. This combination of a logical theory of taxonomy and a linguistic theory of projectibility furnishes the basis of Kuhn's new limited and highly specific surrogate for the old idea of incommensurability.

This chapter is organized into the following sections. The core of (my version of) Kuhn's new theory is stated in sections 4–8.

2. *Diagnosing the New-World Problem.* Kuhn's diagnosis and mine.

3. *Nonstarters.* Why a number of approaches to the problem won't work.

4. *Some Logic of Kinds.* Definitions of the key ideas of "kind of," taxonomy, overlap, antitaxonomies, categories, and infima species, with examples.

5. *Scientific Kinds, Scientific Terms.* Why we should not here speak of natural kinds and natural-kind terms.

6. *Kuhn's Conditions.* The theses that scientific kinds are taxonomic and have infima species, and that scientific terms are projectible.

7. *Consequences.* The argument for untranslatability between paradigms. Further consequences: the refutation of self-refutation and a

response to Donald Davidson. Relations between paradigms: the nesting of taxonomies.

8. *Resort to J. S. Mill.* How Mill's theory of real Kinds avoids counterexamples to Kuhn's conditions.

9. *Kinds and Generality.* Kinds seem to be an old-fashioned, out-of-date way to characterize contemporary physical science or even that of the nineteenth century, where measurement and differential equations were the order of the day. They are surely not the whole story, but they remain, I argue, the key to any idea of universality and general knowledge, i.e., science.

10. *Working in a New World.* A recapitulation of the nominalist taxonomic solution.

2 Diagnosing the New-World Problem

Nonphilosophers have little trouble with living or working in a number of worlds. Philosophers do. Kuhn diagnosed our difficulty. We still work within "a philosophical paradigm initiated by Descartes." Philosophy, psychology, linguistics, art history "all converge to suggest that the traditional paradigm is askew." In 1962 history of science was joining "these crisis-promoting subjects." I don't quite agree with the diagnosis.

We must pause over the term "paradigm." In posing the new-world problem, I used the word in the strict sense of "Second Thoughts on Paradigms": a paradigm is constituted by an achievement and by a disciplinary matrix of one hundred or so workers. When Kuhn spoke of a paradigm initiated by Descartes, however, he was using the word in the popular, journalistic (but entirely viable) sense that caught on after 1962.[6] What part of our "Cartesian" philosophical paradigm did he mean? Perhaps the idea that our thoughts about the world are a passive reflection of something external. But that can't be the whole story. Many of us have abandoned that way of thinking yet still resist the idea that after a paradigm shift we live and work in a new world. Hence I would look for another diagnosis.

Elsewhere I made use of a very different author exclaiming, after reading Descartes, "He must live in a completely different world from ours." That was Bertolt Brecht, stunned by the fact that a man would prove his existence from his thinking.[7] It is what we do that proves our existence; it is our labor within which and by which we exist, thought Brecht. Brecht and Descartes do indeed represent two very different stances. The difference is not one pointed to by "psychology, linguistics, art history," and other reflective metaenterprises, nor are those subjects "crisis-producing." The relevant difference is between what Dewey called the spectator's view of the world, as against the vision of the agent doing something. Brecht was schooled in the philosophy of the worker, the actor, the speaker, the artist. Such agents do things. They are not and in principle cannot be solitaries like Descartes, proving his existence by meditation. Materialist Brecht knows, with Kuhn, that there is just one material world (proposition (a) of the new-world problem). But it remained natural for him, as for Kuhn, to speak of living and working in different worlds (proposition (b)).

To address the new-world problem, we don't have to change a paradigm so much as, in the words of C. S. Peirce, "stand upon a very different platform from that of Descartes," namely what Peirce called (in capital letters) that of COMMUNITY.[8] (With all this talent bad-mouthing Descartes, why does he stand up so well? He'll outlive us all.) Kuhn's current writing bearing on the new-world problem repeatedly mentions the community and the role of the classifications current in and authorized by the community. The new-world problem is to be addressed in terms of changes in the kinds of things and stuff with which a community works. A communal platform is not enough. What are important are the active working community and the tools.

3   Nonstarters

Mine is an unusual approach to Kuhn's past and even present writing. I should say why a number of more obvious attempts to solve the new-world problem go nowhere.

Interpreting the given

As Kuhn saw his problem in 1962, the easiest way out was a mistake. It was no good saying that although the world remains the same, we interpret it differently. To Kuhn that meant that there is something given, either materially or in experience, that is then interpreted by theories. Kuhn wanted no truck with "the myth of the given."

Changing the world

Of course, the world changed after, for example, the chemical revolution: we have changed it, and we have changed it because of that revolution. That rather despised subject, chemistry, has affected our daily lives and certainly our environment more than any other branch of natural science. Chemists used facts about the world to change the world: to create better bleaches, then better dyes, etc. The unsympathetic pragmatist/industrialist can say, there's no new-world problem; the world changes! I have much sympathy for this riposte, but it passes Kuhn by. In one of my quotations introducing the problem, Kuhn writes of the "ease and rapidity with which astronomers saw new things when looking at old objects with old instruments." The world did not change, not even those professional spaces in it occupied by instruments, yet the astronomers, Kuhn "wanted to say," lived in a new world. Kuhn also believes in changes internal to a science, changes that alter the world in which the scientist works before selling the patents and altering the environment.

Construction

Another important "different worlds" story, one that can be inferred from two celebrated propositions, also fails to help with Kuhn's problem. Let us follow fads and propose that "scientific facts are constructed." This first proposition is meant fairly literally. For each item now counted as a scientific fact, there was no fact until it was constructed. The tense structure of European languages may force us to say that it always "was" a fact, but this distorts the description of scientific practice. Combine this doctrine with the famous sentence "The world is made up of facts, not of things." It would follow that there is a new world every time a new fact comes into being. This won't do, because it makes no distinction between scientific revolutions and every other type of discovery. Kuhn thought that there was something peculiar about those revolutions he examined in *Structure* that justifies talk of working in a new world. In section 10, I will nevertheless show how to abuse this second proposition out of context and apply it to the taxonomic solution of the new-world problem.

The social

But what of a more radical story in which the traditional forms of metaphysics are ignored? Bruno Latour has been telling us that all the world is a social world, constantly being reformed by the interplay of what he calls "actants," be they bacteria or bacteriologists. Even if this story were to prove so compelling that it reorganized future philosophy of science, it would not exclude other stories. Kuhn has always made clear that despite the attraction of his work to those who pursue social studies of science, he has been chiefly concerned with what used to be called internal history, with the theoretical and experimental difficulties and projects that confront the scientific group at work. Network theory and an analysis of power plays might be needed to explain how the revolutionary group establishes allies essential to victory. But there is something else that happens, at first within the disciplinary matrix itself.

Language

What does one fall back upon? Perception, society, interpretation, the material world of objects and artifacts—these are left to languish. Language seems left. But the philosophy of language immediately prompted by *Structure* (and by the contemporaneous ideas of Hanson and Feyerabend) was quite unsavoury. There was the idea that the meanings of such words as "mass" altered with the introduction of a new theory. Whatever Kuhn's intentions, this idea became entangled with the older notion that the meaning of a theoretical term could be given only in the context of the theory in which it occurs. Whatever your views about incommensurability and meaning change, those ideas did not help with the new-world problem. If every theoretical change involves some change in the meaning of words, then revolutions are not singled out. To repeat, Kuhn was concerned only with revolutions and local changes that result from them. We owe a real debt to Putnam's doctrine of natural-kind terms, according to which reference is part of meaning but reference of a term is largely independent of any particular scientific theory.

Language learning

Barry Barnes and David Bloor have invoked language in a different way. Widely regarded as advocates of the "strong program in the sociology of science," they have more recently become philosophers of language who make much use of Wittgenstein's name. They hold that all words are radically "open-textured" (to use Friedrich Waismann's phrase, not theirs). That is, no matter how much language acquisition and usage there has been in the past, the future application of a word is never determined and is always open to be decided only in practice. This work emerged rather independently of Kripke's sceptical analysis of rule following but has a number of affinities with it. There are now debates about how to carry on with these ideas, whimsically described in terms of "left and right Wittgensteinians."[9] These discussions arise from perplexities about following a rule. They are not germane to the new-world problem, which arises from a revolutionary break in old rules. The problem is not about old vocabularies but about new ones, to which I now turn.

4 Some Logic of Kinds

Relevant kinds

Most kinds that interest us are not natural, nor are they kinds of artifacts. There are symphonies as well as sugars, experiments as well as explosives, carburetors as well as crows, not to mention hopes and hoplites and hobgoblins. Nelson Goodman wisely preferred to talk of relevant kinds. Some of his reasons should attract Kuhn.[10] But we need not go that route here. Kuhn is preoccupied by only one type of relevant kinds: scientific kinds. These are kinds that some group of scientific specialists finds relevant for its investigations during some period. They include far more than what we find in nature. We did not find plutonium, we made it, yet who would exclude plutonium from the natural kinds? Many of our celebrated effects—the Compton effect, the Zeeman effect—are created in the laboratory and are at best purifications of nature's phenomena. And among the most important scientific kinds are kinds of apparatus and instruments. Let us first look at the logic of kinds in general and then turn in section 5 to scientific kinds.

Kinds as sets

The natural kinds we distinguish form only a small proportion of kinds we find relevant, but even they are ontologically various. They include kinds of individuals, like the horse and the spectroscope; kinds of stuff, like phosphorous; and miscellaneous kinds, like colors. Quine treats kinds as sets. This has at least two advantages. First, we have a succinct way to state a pressing problem, namely "Which sets are natural kinds?" Second, we have the apparatus of elementary naive set theory, namely relations between kinds, such as proper containment (subset).

The language of sets directly applies only to kinds of individuals, however. Otherwise, we have to regiment and reconstruct. *Word and Object* shows one way to do that, turning stuff and substances into sets, but we have to rely on students of count and mass nouns. Nor are even colors happy. Are colors kinds of things: the spinning top and the Speedo swimsuit? Or are they kinds of surfaces: spots and stripes and patches and reflections?

Kind of

The logical relation between kinds that we most require has, in the case of sets of individuals, the structure of proper containment. The set of Clydesdales is contained in the set of draft horses, in turn contained in the set of horses, in turned contained in the set of mammals. The sets of refracting and reflecting telescopes are subsets of the set of telescopes. We have a locution in English that to some extent mimics proper containment but that applies equally to kinds that are sets of individuals and to kinds that are not. It is "kind of."

We may say that the Clydesdale is a kind of draft horse, that a draft horse is a kind of horse, that a horse is a kind of mammal. Reflecting and refracting telescopes are kinds of telescopes.[11] "Kind of" works quite well for other types of kinds. The radioactive isotope

^{32}P is a kind of phosphorous, which is in turn a kind of nonmetallic mineral. Scarlet is a kind of color. I don't say that "kind of" is entirely idiomatic in these constructions, but it is a good generic label for the relation that, in the case of kinds of individuals, has the structure of the proper-containment relation. "Kind of" and proper containment are both asymmetric. I will extend "kind of" so that it is transitive. This is a modest abuse. We do say that a Clydesdale is a kind of horse but not that it is a kind of mammal.

Proper containment by no means coincides with "kind of," even for kinds of individuals. Set theory is extensional, and subsets are determined by membership alone. If J and L are distinct kinds of individuals, and anything J is L, we have proper containment. Quine treats kinds as sets, so J is properly contained in L. But J need not be a kind of L. It may just happen—and there may even be a causal explanation of why it happened—that everything of kind J is of kind L. Take the class of national armies. These may be grouped in numerous kinds, for example, L = volunteer armies, J = anglophone armies, etc. J is at present properly contained in L, and we can give some not uninteresting historical explanation of why, but anglophone armies are not a kind of volunteer army.

Of course, "volunteer army" is not a "natural kind." Maybe set-theoretic proper containment and the "kind of" relation coincide for natural kinds of individuals. That is, if we have two natural kinds from what I later call the same category, then if one of them is set-theoretically contained in the other, it is also a kind of the other. Kuhn's account of kinds is partly motivated by that hunch.

Aside from making the "kind of" relation transitive, I will not force much onto our common way of speaking. I will in one respect be strict and not allow any instance of "a is a kind of Z" when a is an individual. Unlike Quine, I never identify individuals with sets that happen to be the unit sets of individuals. "Mother of God" may denote a kind of person that has at most one instance (at least until the second coming), but it remains the name of a kind of person. "The Virgin Mary" is believed to name an individual, and it, in contrast, does not denote a kind at all.

I am well aware of many colloquial uses of "kind of." We have not only "Hobbes was kind of boring," but also "Hobbes was kind of a bore." Hence apparently what is not merely colloquial but also inat-

Ian Hacking

tentive: "Hobbes was a kind of bore." I believe these useful ways of speaking will not prove confusing. Now for some definitions.

Taxonomy

A *taxonomy* is determined by a class of entities C and a transitive asymmetric relation K. $\{C, K\}$ is a taxonomy if and only if (1) it has a *head*, a member of C that does not stand in the relation K to any member of C but such that every other member of C stands in the relation K to it; (2) every member of C except the head stands in the relation K to some member of C.

Taxonomic Classes

$\{C, K\}$ is *taxonomic* if it breaks up into disjoint taxonomies. That is, there is a finite partition of $\{C, K\}$ into taxonomies $\{C_1, K\}, \ldots, \{C_n, K\}$ such that no member of C bears the relation K to two distinct heads in C.[12]

Categories

If K is a "kind of" relation, the head of each C_i is a *category*. When K is given or assumed, for short I will say that the class itself is taxonomic. I may also take each individual taxonomy in C as a category, named by its head—the category of colors or experiments or mammals, for example, in some suitably chosen C.

Antitaxonomies

Here is an example to which I will return. Consider the class of kinds {poison, arsenic, hemlock, vegetable, mineral} and "kind of" as used in common speech. Arsenic is a kind of mineral and a kind of poison. Hemlock is a kind of poison and a kind of vegetable. Within this class we have three heads, namely poison, vegetable, and mineral, but the trees beneath them are not mutually exclusive. Thus this class of kinds is not taxonomic. This simple antitaxonomy is shown in figure 1, with the arrow signifying the "kind of" relation. It is convenient to say that poisons *overlap* minerals. Set-theoretically, this means that some poisons are minerals, and some are not, but here I mean something about the "kind of" relation: that some kinds of

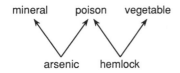

Figure 1
The antitaxonomy {poison, arsenic, hemlock, vegetable, mineral}

poisons are kinds of minerals and some kinds of poisons are not kinds of minerals.

Examples

Classes can be taxonomic with respect to an asymmetric and transitive relation for all sorts of reasons. A class of mutually exclusive sets is trivially taxonomic with respect to the subset relation; every set is a trivial head. Any arbitrary class consisting of disjoint sets and some subsets of those sets is taxonomic with respect to the relation of proper containment. Taxonomies derive their interest from the relation K. If K is "kind of" in English regimented only as above (it is transitive), then the class {color, scarlet, yellow, red, mammal, horse, human} is taxonomic, with categories color and mammal. The class itself is arbitrary, but the relations between elements in the taxonomies are logical, conceptual, or perhaps lexical. It is not a merely contingent fact that horses are mammals or that scarlet is a color.

There are other modalities. When C is any finite class of women and K is the relation of being descended from or a daughter of, then {C, K} is taxonomic. It is contingent that the members of C are related as they are, that the Virgin Mary is the daughter of Anne. But it is a logical or conceptual fact that the mother-daughter relation yields a taxonomy.[13] This is the inverse of the preceding example, where it is the logical relations between the elements in C, between scarlet and color, and not the relation K itself that determines the taxonomy.

Systematics

Natural history and systematics aim at a class of kinds and a "kind of" relation that is taxonomic. Here we have another and more interesting modality. It is not an accident that the taxa of natural history are taxonomic. We make them that way; we "force" our classifications to be taxonomic. I must emphasize that Kuhn's kinds have nothing to do with systematics, even if it is sometimes tempting to use a kind of animal or plant for an example, just as I have followed J. S. Mill in using the example the kind *horse* and have used the kind *Clydesdale*, a breed but not a species, a stock breeders' kind, not a zoological one.

Why avoid systematics?[14] The very idea of taxonomy comes from the study of living beings. Aristotle was the first and last great contributor to both logic and natural history. Porphyry's account of the five predicables, Mill's rejection of it, Mill's introduction of the philosophical term "kind," Russell and Quine on natural kinds—these are all descended from Aristotle's logical studies. Biological taxonomy and systematics descend from his natural history. Despite common ancestry, the logical study of natural kinds should stay away from systematics. Mill may have been the first to insist on this. He dissociated himself from the life sciences, noting that in his day two groups of sexually reproducing organisms were said (by one group of theorists, precursors of Ernst Mayr) to be of the same species if they could interbreed. Let the biologists define species to suit themselves, Mill wrote, we logicians don't have to follow in train.

Biological taxonomy is a structure of genus and species. This can be represented in a set-theoretical way, but the relationship of species to genus, and indeed the entire structure of biological taxa, is a matter of what we may most generally call the theory of life. Whether with Linnaeus we attend to form, whether with Adanson we attend to function, whether with the numerical taxonomists we arrange the taxa by counting similarities, or whether we construct a taxonomy that is homomorphic to an evolutionary tree, we are in every case invoking a conception of living beings. Different schools of systematics disagree about how to analyze the "kind of" relation. Taxa are defined and forced into a taxonomy in virtue of one or another theory about the nature of living beings. This has nothing to do with the logical theory of classification. It is nevertheless no easy matter to distinguish a logical taxonomy of kinds from a biological one, because there is a sort of logical skeleton in the biological tree. It is not a contingent fact that the horse and the whale are mammals, although most of the finer points of systematics do deal with contingencies.

Forcing

Fred Sommers's extremely interesting theory of types forces predicates of a natural language onto a taxonomic tree in a manner totally different from that of biology.[15] It does so by declaring, in a systematic way, that when predicates don't tidily fit, they are ambiguous. Sommers was not presenting a theory of natural kinds, but formally his theory of types is analogous to part of Kuhn's theory.

Sommers and I use the word "category" differently, so I will use the invented label "P set" for his "category." The P set of a predicate is the set of individuals of which it may be either true or false. To repeat Sommers's example, the P set of the predicate "expensive" is the set of all individuals that may truly or falsely be called expensive (including, I suppose, those middling individuals that aren't exactly expensive but aren't exactly not expensive either). A predicate is said to *span* its P set. Chairs, dinners at the Ritz, and breakfast at McDonald's are spanned by "expensive," but questions are not, says Sommers. So chairs and dinners are in a P set that excludes questions.

Sommers claims to prove a "law of categorial inclusion": if S_1 and S_2 are P sets, then either S_1 and S_2 have no members in common, or S_1 is included in S_2, or S_2 is included in S_1. This means that the class of P sets is taxonomic with respect to the relation of (proper) set inclusion. What then of obvious counterexamples? Chairs and questions are spanned by "hard," so the P sets of "hard" and "expensive" have members in common, but neither is properly included in the other. Sommers forces a taxonomy by declaring, plausibly in this case, that "hard" is ambiguous, two meanings for one word. In a similar way Kuhn's taxonomies, we will see, force scientific kinds from distinct paradigms to be untranslatable, which may also be a case of two meanings for one word.

Infima species

Following Mill, Kuhn has an interest in taxonomic trees that bottom out in a nontrivial way. What this means is as follows: Let K be a "kind of" relation. A member of C in a taxonomy $\{C, K\}$ is an infima species only if no member of C bears K to it and it is not of necessity a kind instantiated by at most one individual. It is possible to construe Leibniz on individual substance as holding that there are no infima species: all taxonomies bottom out with what he called individual concepts.

5   Scientific Kinds, Scientific Terms

Kuhn has been writing of natural kinds and natural-kind terms. I now urge him to break with that practice. We already in common speech talk about scientific terms. Let us also talk about scientific kinds, the kinds distinguished in a branch of science. Why? Because "natural kind" is, if not strictly ambiguous, at least a portmanteau word for two different ideas, one more philosophically dogmatic than the other. And "natural-kind term" is, to put it bluntly, devious, shunting back and forth across the ambiguity.

First the multiple senses of "natural kind."[16] In much natural-kind philosophizing, the natural kinds are supposed to be somewhat cosmic.[17] In the literature we find highbrow examples of kinds, such as phosphorous and electricity, or exotic ones, like tigers and lemons. We never find mud and dung cited as natural kinds. Mud is surely a kind that we find in nature, familiar to parents scrubbing children's clothing, to football players, and to ditchdiggers in temperate climes. What arrogance will insist that the plowman is not in touch with nature's kinds: mud and dung? Are these not natural kinds? Mundane kinds, let us say.

Most natural-kind philosophers have more cosmic kinds in mind. During the Enlightenment a cosmic kind was one that featured in the "inner constitution of nature." Philosophers nowadays seem to prefer kinds that figure in fundamental laws of nature. Some natural-kind philosophers are convinced that there must be cosmic kinds; these souls are heirs of Plato's theory of forms. Others, such as Quine, believe we are born with a potential for distinguishing certain mundane kinds, a potential selected by evolution. These kinds do, unsurprisingly, correspond to certain "functionally relevant groupings in nature," alerting us to food and predators. As we and our sciences

mature, we make more sophisticated sortings that help us understand, explain, predict, and intervene in the course of nature. These
scientific kinds become increasingly cosmic in scope, although we
have, in Quine's philosophy, no right to say that the world *must* come
equipped with cosmic kinds.

Many disputes about natural kinds seem to come to this: "The
world, any world remotely like ours, *must* have cosmic kinds. Nature,
to use Plato's unseemly metaphor, must have joints. Science aims at
carving her at the joints."—"There are no joints (thank goodness),
but we have certainly uncovered some kinds that are vastly more
fundamental for our scientific interests than others, and we should
continue in the project of looking for more."

No matter how one lines up with these two attitudes, the cosmic
and the mundane, we have all been infected with fallibilism. The
mundane and the cosmic philosopher agree that we do seem to have
revealed some pretty fundamental kinds, such as hydrogen. But even
the cosmic philosopher grants that we might be wrong. Hydrogen
could turn out to be far less fundamental then we now think—a
merely mundane kind, in the end. How to express this hesitation?
Most forthright, perhaps, would be to speak of "putative" or "supposed" or "current candidates for" natural, i.e., cosmic, kinds. There
is no doubt that hydrogen, like mud, is a mundane natural kind. No
future learning will convince people that there is no hydrogen. What
is merely putative is that this distinction will continue to loom large
on the cosmic scale.

Recent natural-kind philosophers prefer semantic ascent to putative kinds: ascent from kinds to natural-kind terms. "Natural-kind
term" is a devious phrase. It elides the distinction between the cosmic
and the mundane. We are commonly given as examples names for
mundane kinds ("water," "tiger," "lemon," "heat"), names that at the
same time are supposed to pick out more cosmic kinds, or kinds
structured by putative cosmic kinds (H_2O, the chromosomal structure
of tigers, the motion of molecules). Seldom are we given names that
originated in science, although there are a few—"electron" and "multiple sclerosis" come to mind.[18] We get all the virtues of the mundane—no doubt that those *are* kinds; no need to insert a "putative"
here—and the excitement of the cosmic. That is an equivocation.

Kuhn should stay away from "natural-kind term." All he needs is "scientific term"—no neologism that. A scientific term is a term used chiefly in some branch of science. Let me narrow the phrase in one way and liberalize it in another. Since we are discussing Kuhn, let us restrict ourselves to the terms used in postparadigmatic scientific disciplines or specialties. But since any specialty may use some pre-scientific terms, spruced up for immediate purposes, I allow "scientific term" to include terms from common language when they have a role in the specialty in question. And I will follow my admonition to stay away from systematics by fiat, not including the (doubtless "scientific") terms of a classificatory science among the scientific terms.

As for a scientific kind, let that be a kind denoted by a scientific term. Scientists doubtless hope that their kinds are somewhat cosmic, and they may or may not be right. "Scientific" bears on its face the connotation of "putatively cosmic." Something is scientific because of human aspirations and interests; it satisfies when we seem to be fairly right about the cosmos.

6 Kuhn's Conditions

Kuhn's theory uses three conditions, but there is a difficulty in stating them. It is not clear whether they are or should be conditions on scientific kinds or on scientific terms. Are we talking about language, the world, or how we conceptualize the world? This problem may be endemic to this and similar enterprises. Sommers was criticized on the ground that his "theorem" about categorial inclusion equivocated between words ("expensive") and things (the class of expensive things).[19] I will tread warily between kinds and terms.

Kuhn tends to write about words and the "lexicon" of a paradigmatic science. That will be a lexicon of scientific terms. He writes of a structured lexicon. The structure to which he refers has to do with kinds and the "kind of" relation. An ordinary dictionary informs us that the lotus is a kind of tropical water lily. So the "kind of" relationship is recorded in real-life lexicons to hand. But beware! Dictionaries are full of highly contingent information: that the lotus was supposed to induce a dreamy languor and forgetfulness, that the flower of the lotus that grew in Egypt is white, while that from India is blue. Being in a dictionary does not imply that a fact is "lexical," and, of course, Kuhn uses "lexicon" in a specialized sense current among some linguists.

Some facts, like the lotus being a kind of water lily, are "lexical," or at any rate conceptual. I will nevertheless state Kuhn's first condition as a claim about kinds that we distinguish, rather than about terms that we use. It can readily be made a lexical constraint, because it is about taxonomy, and any tree of kinds can be presented as a tree of names of kinds. It seems to me that his second condition is less readily interpreted as lexical. But his third condition is essentially about words. Here are the three conditions:

1. *Scientific kinds are taxonomic.* The relations between the kinds that create the taxonomy are logical, conceptual or lexical. This condition refers only to kinds in any branch of science with a paradigm. I conclude section 7 by discussing the meshing of contemporaneous paradigms.

2. *Scientific-kind taxonomies have infima species.* There are only finitely many words and descriptions in use in any branch of science. Hence a taxonomic array of names for scientific kinds will always bottom out, from exhaustion. To be interesting, this claim must be more subtle, that there are kinds such that, as a presupposition of the science, no subdivisions of those kinds can count as scientific kinds. It will be very hard to find explicit expression of any such condition. Maybe it is implicit. Pure water was once thought of as an infima species. Now we have ordinary water and heavy water, itself dividing into deuterium oxide and tritium oxide. I take it that this subdivision needed a new paradigm, physical rather than chemical. Water was thought of as ultimate stuff and is no longer so thanks to radically new science. At present deuterium oxide is an infima species.

3. *Scientific terms are projectible.* Names for scientific kinds are projectible in the language of the scientific community that employs them. The names are used in making generalizations, forming expectations about the future (or unexamined events in the past or distant present). They can be used in counterfactual conditionals. They occur in lawlike sentences.

7   Consequences

Meaning

Kuhn and Feyerabend put in circulation the idea that theory change implies meaning change. This was one of the sources of the idea of incommensurability. The noun "mass" has different meanings in relativistic mechanics and classical mechanics, or so it was said. The ensuing discussion was one of the less elevating incidents in the philosophy of science. It has been one of Kuhn's projects to provide an account of incommensurability that does not explicitly use even the idea of meaning. He now argues that a scientific term of a prerevolutionary paradigmatic branch of science cannot be translated into the lexicon of taxonomic scientific terms of its postrevolutionary successor.

We are not concerned primarily with meaning change of individual terms. According to condition 1, every scientific lexicon has a taxonomic structure. Part of knowing a special science is knowing what kinds of things it deals with and the basic logical relations between those kinds. The terms of an old science will not translate into the terms of the new one not because of something about the individual terms but because of something about their taxonomic relations. This doctrine is holistic, yes, but note that it is not waffling all-purpose holism. It is about one well-defined type of holistic structure. Kuhn's present work replaces generic meaning, incommensurability, and wholes with specifics.

Failure of translation

I now consider, in abstract terms, the problem of translating the language of a prerevolutionary paradigm into the language of a present branch of science. If prerevolutionary terms don't apply to anything to which postrevolutionary terms apply, there is little inclination to think that we have a successor science. Demonology has no evident successor; if you fancy its subject matter, you have to learn how to talk about demons the good old-fashioned way. When the lure of translation does exist, then for each kind term of the old paradigm, there are three possibilities: overlap with a kind of the new paradigm, subdivision, and coincidence.

a. *A kind overlaps a scientific kind in the new science.* Then by condition 1, the kind in the old science cannot be a kind in the new science. Hence the name of this kind cannot be translated into any expression in the new science that denotes a scientific kind. This argument is closely related to Sommers's forcing: condition 1 forces a term from an earlier science to have a meaning not expressed by the new scientific terms.

b. *A kind subdivides a kind in the new science that has no subkinds.* By condition 2, the kind in the new science is an infima species with no scientific subkinds, and so the old name cannot be translated into any expression in the new science that denotes a scientific kind.

c. *Although the lexicons of the old science and the new one differ taxonomically, a kind in the old science coincides with a kind in the new science.* This is the case with "water" before and after isotopes. We should not, in my opinion, argue for untranslatability. There are no adequate grounds for saying that "water" does not translate "aqua" or that it is now only a homonym for the "water" of Bacon or the King James Version.

Projectibility

The idea that scientific terms are projectible is the sleeping partner among conditions 1 to 3, and with good reason. We tend to forget it because it is a tautology, literally a tautology—the predicate, "projectible," just repeats part of the subject, "scientific term." To be a scientific term, a term must be used in a science, and to be a science is to be concerned with knowledge of general truths. Science is concerned not only with what is the case but also with what would be the case under different material conditions. A scientific revolution has the consequence that we no longer speculate, conjecture, predict, explain, and most importantly, work on the world using the old classifications. It is we in the community who project. "Projectible" doesn't mean rightly projected from some cosmic point of view: it is a humane concept implying only that a class of terms is used by a community for making lawlike statements, forming general conjectures, picking and making things with expectations about what they will do and how they will work, and saying what would have happened had we not done so and so. To call a statement lawlike is to call it not a law but only a candidate for a law. To call a term projectible is not to say that generalizations made with it are well founded or justified. It is to say only that such generalizations can be justified in the community. Projectibility defines the class of possibilities envisioned or capable of being taken seriously by a science at a time.

The refutation of self-refutation

Putnam is only the most prominent of many authors who have said that Kuhn's and Feyerabend's doctrines of incommensurability are self-refuting. The argument goes like this. Learned and literate philosophers tell us that old science is incommensurable with our science, that it cannot even be expressed in our language, and then they go on to explain the old science to us. Their practice refutes their dogma!

The standard reply, given by Kuhn but more forcefully by Feyerabend, is that the doctrine of incommensurability never said you can't learn another language. After learning the language, you can convey old science to others, not by fully teaching the language, but by conveying ideas by metaphors and encouraging a hermeneutic attitude.

Feyerabend is right, but he takes a wrong turn. Learning an old science, learning to mimic its sentences, doesn't mean you can use and project its terms. Here is a personal example. Like Feyerabend, I have an inveterate curiosity about other ways of thinking, especially those that lie in our own past. I once had the fantasy that I could not only understand Paracelsus but that if I exerted myself, I could forge a volume of pseudo-Paracelsus, work from the atelier of Paracelsus, such that, if I hired a good enough antiquarian to write out my words using Renaissance script and writing materials, I could fool half the learned. "A new volume of Paracelsus or one of his student-rivals," they'd say, admiringly. I could, I fantasized, string the words together so that they were internally coherent, but I couldn't live

Paracelsus. I could not seriously project the generalizations there inscribed. Of course, I could have bought into Paracelsus seriously, but then I would hardly survive with my present habits. I recall Herbert Butterfield, in *The Origins of Modern Science,* writing in the late 1940s of what is now a long-gone tradition of renaissance scholarship: the best of those scholars seemed to become "tinctured" with the madness of the science that they studied. Either I can't project Paracelsus's terms, or I do and drop out of my community.

Thus the original presentations of incommensurability really did leave something out. Dewey made wise fun of spectator-oriented philosophies. If translation is just replacing some words by others upon the page, it becomes a spectator sport. A translator as envisaged in much philosophy does not speak or use words but merely displays some words instead of others. Incommensurability theses should not insist that one cannot translate in the sense of "convey the intentions of the other." The point is that one cannot speak the old science while using the projectible scientific-kind terms of present science. ("Projectible" is inserted here as a reminder; scientific kinds are tautologously projectible in the science of which they are a part.) Conversely I can't use the old kind terms even when I understand them well. I can write a pastiche of Paracelsus. I can act out a Paracelsian role. But I can't do Paracelsian things. I act under my descriptions, which use words that my community projects.

The very thought of a conceptual scheme

The preceding remarks bear on Donald Davidson's famous paper. I do not think it wise to reintroduce "conceptual scheme" into philosophy. Quine used the expression to refer to a (global) set of sentences held for true. If one were to use the expression, it should surely be for a set of sentences up for grabs as true or false. A conceptual scheme would be a scheme of possibilities, not beliefs. My work on styles of reasoning introduces a theory about one way in which new sentences, new classifications, new objects, new laws, and so forth, can come into being. Kuhn alludes to this and rightly both compares and contrasts it with his own ideas.[20] One contrast is that I was speaking of a process that has in fact been largely cumulative in the development of Western science. Another contrast is that I was speaking of global developments in scientific reasoning, almost at the level of generality of "the scientific revolution," as opposed to the scientific revolutions that are the topic of *Structure*. Kuhn's taxonomies are thoroughly local.

Incommensurability as originally enunciated may have deserved Davidson's strictures. But Kuhn has now given us a local untranslatability to which Davidson's objections do not apply. We are provided with an analytic framework in which we see how we can understand, and only to that extent translate, earlier scientific conceptions. We see that understanding is not enough. We cannot use, project, work in a former body of organized thoughts while we preserve our own. There is nothing arcane about our inability to use another taxonomy alongside ours. Goodman's "grue" taught us that the problem is none other than that well-understood logical vice inconsistency. (But, of course, I can act out different lives, work in different worlds, splice genes in the morning and go to the homeopath in the afternoon. I have nothing but Feyerabendian praise for multiplicity of character.)

Meshing

One of the residual problems of *Structure* has been the relationship between different paradigms, different revolutions. Branches of science don't exist in isolation but feed into each other in innumerable ways. One theory serves as an analogy for another. An instrument devised for one body of knowledge is transplanted as what Bruno Latour calls a "black box" to another laboratory. Now, of course, there is much nonunderstanding by different specialties, even when they borrow from each other. There are also transfers of power: the humble grower of phages loves the authority inherited from the nuclear magnetic resonator. Nevertheless, quite aside from the powerful lust for unified science, there is sense in which scientists try to get their ideas to cohere with those of colleagues in numerous disciplinary matrices. A revolution in one area does affect others. How? *Structure* does not tell. I believe that the idea of taxonomic kinds can be used to develop an instructive answer. One kind of normal science has the effect of tinkering with taxonomies so that taxonomies of different fields mesh with each other. Once again I must allude to

Quine's great paper "Natural Kinds," which on its final pages introduces the question of meshing of kinds in the mature sciences.

8   Resort to J. S. Mill

Are the three conditions true? Condition 3 is an instructive tautology. I'm not sure about condition 2, partly because I can't make very clear nontrivial sense of it. Condition 1 does the work. Is it true? Are the scientific kinds of a branch of science taxonomic? The short answer is no. A longer answer is that a counterexample can, in the usual Popperian way, lead us to a further condition on "scientific kinds." So we might get a more restricted, less evidently false version of condition 1.

I've already stated a counterexample: the toxic antitaxonomy of figure 1. Arsenic is a kind of poison and a kind of mineral. Hemlock is a kind of poison and a kind of vegetable. So poison overlaps vegetable and mineral. Here we have a nontaxonomic class of scientific kinds, or else poison is not a scientific kind. Surely not the latter: we have a veritable science of poisons, toxicology!

There is no doubt that poison is a (mundane) "natural" kind. (Many poisons are manufactured and do not exist in nature on their own, but that shows nothing. Many thoroughly "natural" mundane kinds apply, inter alia, to artifacts: some swimsuits are red.) The fact that poison is a mundane kind does not stop it from being scientific. Gold is both mundane and scientific. The word "poison" is projectible. Forensic science deploys it in counterfactual conditionals powerful enough to free one man from jail and to electrocute another.[21] We need a further condition on scientific kinds to eliminate counterexamples of this sort. It is due to J. S. Mill.

Real Kinds

Mill thought that there was just one truth in the doctrine of the five predicables inherited from scholastic logic. There are two distinct sorts of kinds in nature. There is a distinction in nature between yellow things and those of other colors; yellow is a thoroughly natural, if mundane, kind. The color, wrote Mill, is distinguished and named by people, but what is distinguished is the work of nature, not art.

Ian Hacking

Yet, yellow things have very little in common except that they are yellow or that they possess features that follow from the fact that they are yellow. It is different with sulphur. There is, urged Mill, a virtually inexhaustible number of properties of sulphur that are not consequences of the marks by which we distinguish sulphur. "A hundred generations have not exhausted the common properties of animals or plants, of sulphur or phosphorous; nor do we suppose them to be exhaustible."[22] Sulphur, Mill said, is a "real Kind"; yellow is not: small r, capital K. Arsenic is and poison is not a real Kind. There is an inexhaustible number of things to find out about arsenic. There is nothing much common to poisons except what puts them in the class in the first place, namely the potential for killing people after being ingested. We could then offer a fourth condition as follows.

Scientific kinds are real Kinds

Mill's real Kinds have some good features. They are humane, balancing human interests and abilities against natural constraints and adversity. As well as being a fact about nature, it is a fact about us that phosphorous has an unlimited number of properties that do not follow from the marks by which we distinguish it but that we can endeavour to find out. As Laplace famously said of probability, this feature is determined in part by our knowledge and in part by our ignorance. And I would add, it is relative to our inquisitive habits and to our ingenuity in probing and intervening in the course of nature.

Yet real Kinds are not free of difficulty. Peirce objected, What do we mean by "follow from"? Among our scientific enterprises is the project of providing a general structural description of phosphorous from which all its properties follow. Were science successful, there would be no real Kinds![23] I do not think this is problematic for someone as humane as Mill, or someone who takes the attitude to scientific deductions evinced in Cartwright's paper in this volume. They would agree that we qualitatively understand why phosphorous behaves in many situations as it does. Yet even the quantitative facts recorded in standard tables of constants cannot be deduced from any known structure of phosphorous. There are and will remain innumerable determinations of the phenomena of phosphorous. I was once berated by the head of a chemistry department for encouraging his students to take a useless course in the philosophy of science. "Our students," he informed me, "are expected to publish ten papers before they graduate," and he showed me some of these contributions to knowledge, minute determinations of further properties of sulphuric and phosphoric compounds, as it happened. Had he the benefit of Mill, he could have continued that his students "proceed to new observations and experiments in the full confidence of discovering new properties which were by no means implied in those we previously knew."[24]

Very well, but the very indefiniteness that makes Mill's real Kinds attractive is also a disadvantage. How extensive a taxonomy is there for real Kinds? Mill cites *horse* and *phosphorous* as real Kinds. What about Clydesdales and ^{32}P? There are certain marks, criteria, that determine the kind of horse and the isotope of phosphorous. Are there indefinitely many things to find about Clydesdales? Perhaps we happily conclude that Clydesdales don't form a scientific kind. What about the radioactive isotope? Are there indefinitely many things to find out about it? I'm not sure it is a real Kind. Kuhn might be troubled here too.

There are other queries. It is one thing to exclude a counterexample, another to prove that counterexamples are impossible. Can we prove real Kinds are taxonomic, that no antitaxonomic examples are possible? Perhaps we can if we answer some preceding questions quite strictly and define the antitaxonomic structures of science out of existence. Likewise, we can answer those questions so that all real Kinds do have infima species, but again, I am unpersuaded that we should do so.

Analytic/synthetic

As observed earlier, "kind of" relations come in lots of modalities. We can have an arbitrary class of kinds of individuals (cosmic, mundane, or whatever) and regard the subset relation as a purely extensional and contingent "kind of" relation. In natural history and systematics the "kind of" relation may be empirical and contingent, as in numerical taxonomy, or highly theoretical. Taxonomies can be forced by systematics, Sommers or Kuhn. The relation that Kuhn himself intends is to be lexical or conceptual. But isn't "lexical" just a way of concealing a discredited adjective? If J and L are scientific kinds, isn't J a kind of L if and only if it is analytic that all J are L?

The demolition of analyticity can be used to teach two different lessons. The one most commonly taught is that there is no distinction. I prefer to learn that there are many distinctions. We certainly reject the bland positivist ditty "Analytic propositions are true in virtue of the meaning of the words used to express them." That is plausible for "No bachelor is unmarried." It hardly covers the necessity of "Scarlet is a kind of red," or "Red is a kind of color." It does not bear on "The horse is a kind of mammal" or "^{32}P is a kind of phosphorous." I know of no philosophically satisfying account of the "kind of" relation, but we are not entitled to dismiss by waving the denunciatory placard "Analytic/synthetic"!

9 Kinds and Generality

Quine urged that as sciences mature, they make decreasing use of the notion of kind. He was not saying that advanced knowledge no longer uses kinds. Of course they do; the muon is a kind of lepton. His point is that as science develops, we no longer need the notion of being of the same kind, or similarity, to characterize the sets that are (in his parlance) kinds.[25] Muons aren't defined as a set of particles similar to a given muon. Quine was pleased then to discover that the notion of kind was not needed for defining the kinds of mature science. He did not say that mature science has no kinds.

Nevertheless, doesn't mature science make little use of kinds? Measurement and differential equations were the hallmark of classical physics. Kinds seem like the leftover residue of the sciences of earlier eras. Logicians, ever Aristotelian, may assure us that predicate logic, even first-order logic, is adequate for expressing all of natural science, and predicates denote kinds. So what? Real-life physicists don't speak predicate logic. We dearly need to have up-to-date examples of taxonomies of kinds; old ones have come too readily to Kuhn's expositions. We also want practical examples in which the kinds are not "natural" kinds but are scientific kinds produced by experiment, and also examples of the kinds of instruments and apparatus that we construct for experiment. Jed Buchwald, in a paper that is a sequel to his contribution to this volume, provides one detailed example.[26] We need more such to see how Kuhn's project can be understood.

Whether or not we will extract spritely taxonomies of kinds from spectroscopy or whatever, there is no doubt that taxonomies still carry some cachet. Protosciences do try to use them to put their ramshackle conurbations of tarpaper shacks in order. The successive editions of the *Diagnostic and Statistical Manual* of the American Psychiatric Association are a case in point. The second edition was published in 1967, the third edition in 1980, and the third edition, revised, in 1987. These are authoritative and authoritarian attempts to produce a standardized psychiatric lexicon, in Kuhn's sense. All mental maladies are clamped to taxonomic trees. The trees were almost completely rebuilt between the second and third editions, and many of the disorders were placed in new relations. Thus "multiple personality disorder" had completely disappeared from the second edition but has a slot in the third, and was enlarged in the revised third edition, much to the delight of those psychiatrists sardonically called "multipliers" by their opponents, who in turn were dismayed. The multipliers have acknowledged many times in print that the entrance of multiple personality disorder into the official taxonomy enormously strengthened their case and dramatically increased their caseload. (As if its getting into the third edition had nothing to do with them. Ah what feigned innocence here!) A disorder with virtually no instances in the 1960s has passed from "rare" in 1980 to "surprisingly common" in 1990. For better or worse, such are the effects of forcing a taxonomy onto a body of practitioners.[28] (Patients with the disorder never appear in the clinics of European psychiatrists, who are not under the sway of *DSM* and whose patients see little of it in the popular press, on television, or in the movies.) The philosopher of the natural sciences may feel that the example shows that kinds and taxonomies are not enough. Physics and psychiatry may both have structures of taxonomic kinds, but do not the kinds of physics arise from the science and not from committee vote and lobbying?

On the contrary, I think it is well to compare the physical sciences and psychiatry from the point of view of kinds. They are not so different in point of lobbying. They are different with respect to effects. The introduction of the kind "pulsar" enabled astrophysicists to see a great many pulsars, even by reexamining their old data. The new world of kinds created a new world in which to work. But (to invoke Kuhn's conservatism) the world did not change. The number of pulsars out there was the same, even if the world in which the astrophysicist worked was new. In the case of multiple personality disorder in the United States, the world itself has changed. There are a lot more people out there evincing multiple personalities, just because this has become a possible and reinforced way to behave. The acceptance and cultivation of a new kind has led to a new kind of behavior. The natural and human sciences do not differ in point of making light or heavy use of kinds. They differ in the effect of kinds, in the feedback effect of new classifications, peculiar to and perhaps definitive of the strictly human sciences. Kuhn has recently remarked that the relative stability of the heavens "cannot be expected when the unit under study is the social or political system." Nor when it is the human psyche, not just because the human psyche is constantly changing, but because our very classifications change the people and behaviors classified. That, I think, is an interesting difference between the natural and human sciences, and it is a worthy reason to study kinds carefully.

There is a deeper reason too. Kinds are at the heart of all knowledge. Knowledge, the old scientia, begins with generality. The understanding of generality has long been befouled by logical artifice that began long before first-order logic. Whewell remarked that there never has been and never will be a definition of "dog."[27] Yet his peers and many of us believe there must be a definition. There must be a set of necessary and sufficient conditions for doghood. Whewell offered as diagnosis that without a definition, one could not expect syllogistic reasoning to succeed. Thus, according to Whewell, the demand for definitions specifying necessary and sufficient conditions was a result of the syllogistic paradigm. We are at an even greater disadvantage than our predecessors. First-order logic has made us think that generality itself is the product of the quantifiers. The Aristotelian syllogism had its all's and no's, but the subject and predicate, expressed with kind terms, were readily apparent. In first-order logic, the predicate variables are discreetly anonymous, and the quantifiers seem responsible for generality.

Enlightenment logicians (like those of many other periods, including, in our times, some students of Richard Montague) thought of important generalizations as being stated not in the form "All people are rational animals" but rather in the form "Man is a rational animal." That helps. Quantifiers are convenient for conveying the complexities of various sorts of generalizations and indispensable for delicate deductions. But what we need for the very existence of generalizations are kinds and a language in which to express them: general terms, scientific terms, natural-kind terms, common nouns, what you will.

This idea is especially reinforced by my recent thinking about American Sign Language in use by the deaf.[29] Hobbes remarks that a deaf and mute person of sufficient intelligence can know any particular instance of a Euclidean theorem, perhaps even by following its construction. But the person can't know the general proposition, the universal truth. That is because, said Hobbes, generality requires some mark outside the mind that can serve as a mark of a number of different things indifferently. Generality requires a kind and a mark, sign, or term for that kind. What Hobbes meant about the deaf mute is right. The person without the sign cannot have the idea of the general proposition. Hobbes was wrong about the potentialities of the deaf mute, but in a way that elegantly proves his point. The deaf mute who has American Sign Language or some other natural sign language has just as good a general mark as Hobbes, albeit one in a different, unspeakable, medium. That person can have the general idea. Thus the signing deaf serve not to refute Hobbes's point but to illustrate it.

This leads us back to projectibility. Goodman may have raised his wonderful problem backward. He asked which terms are projectible and answered by saying that entrenchment was the criterion that marked off projectible terms from others. He also briefly suggested that entrenchment could be used to distinguish natural kinds from artificial kinds.[30] If Kuhn is right, a scientific revolution can introduce a projectible term with no entrenchment. Revolutions override entrenchment. Projectibility does not need a record of past usage. But it needs something precious close to that. It needs communal usage, which is brought about by a revolution.

10 Working in a New World

As stated at the outset, a suspiciously easy nominalist resolution to the new-world problem has been to hand all along. The world does not change, but we work in a new world. The world that does not change is a world of individuals. The world in and with which we work is a world of kinds. The latter changes; the former does not. After a scientific revolution, the scientist works in a world of new kinds. In one sense, the world is exactly the same. A change in the class of sets of individuals that correspond to scientific kinds of things is not a change in the world at all. But in another sense the world in which the scientist works is entirely different, because what we work in is not a world of individuals but of kinds, a world that we must represent using projectible predicates.

When young, Wittgenstein wrote that the world is made up facts, not of things. Scientific facts are relations between kinds. The propositions, the sentences that state possible facts, are stated using scientific terms, names for scientific kinds. Thus my travesty of the *Tractatus* becomes, The world in which we work is made up of facts and possibilities, not of things. The way to continue abusing the book, with its *Sätze* and names, is pretty obvious.

Yet the emphasis on facts and possible facts would be wrong. What we make in the laboratory are (1) measurements and phenomena and (to switch categories) (2) instruments designed to fall under various descriptions. The prototype of an instrument may be one of a kind, as we say, but it is just that, "of a kind." The world is full of stuff and things indifferent, in their existence and activities, to what we say about them. But scientists never build anything that is not of a kind. I don't mean that they can succeed in building two of the kind—that is the myth of the repetition of experiments. I mean only that what they do makes sense only under the description of a kind. Artists have been trying to create what is not of a kind. They have been driven partly by the romantic impulse, itself so fed by the wild idiosyncrasies of nature. But the self-conscious desire to defy kinds arose from the mechanical success of the sciences these hundred years. That produced the art of antiscience, so different from the art of earlier days. The various manifestations of the new art have little in common except this: they wittingly strive for a particularity that is not a kind of anything.

Scientific apparatus is of a kind. Philosophers of theory emphasize that the apparatus is built only according to the intellectual blueprints that make sense of the object being built. Philosophers of experiment insist that the very experience of making the prototype influences the kind. It will prove to be a rich field of enquiry, for future philosophers of experimental science, to study how the introduction of a kind of instrument alters the world in which the experimenter works not by having a new pile of physical stuff held together with string and sealing wax but by having an instrument of a new kind, with which certain types of intentional behavior become possible.

Notes

1. All references are to the second edition of *Structure* (Chicago, 1970).

2. Unlike other contributions to this volume, this chapter is about Kuhn's present and future work. Even when I go off on tangents that he turns out to reject, it is important to get some reaction to these stimulating new ideas. I particularly refer to "The Presence of Past Science," the Shearman Memorial Lectures at University College, London, 23–25 November 1987. My copy is marked, "Draft: not for distribution, quotation or paraphrase." Occasionally I describe ideas found there almost to the point of paraphrase. It is to be understood that I am less describing Kuhn's final considered opinions than exploring some of his suggestions. The examples used in the Shearman lectures occur in an earlier paper, "What Are Scientific Revolutions?" in L. Krüger et al., *The Probabilistic Revolution* (Cambridge, Mass., 1976). See also his "Possible Worlds in History of Science," in Sture Allén, ed., *Possible Worlds in Humanities, Arts, and Sciences: Proceedings of Nobel Symposium 65* (Berlin, 1989), pp. 9–32, 49–51. The continued faith in "living in a new world" was stated in "The Natural and the Human Sciences", preprint of a talk given in 1989 under the auspices of the Greater Philadelphia Philosophy Consortium.

3. "Possible Worlds," p. 49.

4. The Latin tags are explained in H. H. Price, *Universals* (London, 1954).

5. Nominalism is not to be confused with "name-ism," the doctrine that things have nothing in common except their names. Did anyone ever hold such a view? Ordinary nominalists commonly say with Locke, "I would not here be thought to forget, much less to deny, that Nature, in the Production of Things, makes several of them alike: there is nothing more obvious ." The sorting under names is indeed the "Workmanship of the Understanding", but it is based upon "the similitude it observes" on (*An Essay Concerning Human Understanding* III.iii.13).

6. T. S. Kuhn, "Second Thoughts on Paradigms," *The Essential Tension* (Chicago, 1977), 293–319. My point is not that Kuhn is referring to a philosophical rather than a scientific paradigm. There are plenty of philosophical paradigms in the strict sense of

Ian Hacking

"Second Thoughts." Long ago G. E. Moore called Bertrand Russell's theory of definite descriptions a paradigm of philosophy. André Gombay tells me of one philosophical paradigm (in the narrow sense of the word) initiated by Descartes. In Dutch academic literature of the late seventeenth century, Descartes's *Principles of Philosophy* were distilled into a list of about twenty principles, each of which had a name and each of which was subjected to Cartesian scrutiny. Scholars would write whole books about (principle) "number 16" without even bothering to state what was in question. There were one hundred or so workers and a clear model of how to proceed. There was a shared understanding of what a compelling argument is, which we and equally, one suspects, the students of the day can figure out only from leading examples of the genre. There were institutional procedures of initiation and exclusion. When Kuhn says that we are still working within a "philosophical paradigm initiated by Descartes" he has nothing remotely like this in mind.

7. See the second parable of my "Five Parables" in R. Rorty et al., eds., *Philosophy in History* (Cambridge, 1984), 103–124.

8. C. S. Peirce, "Some Consequences of Four Incapacities," (1868) in C. Hartshorne and P. Weiss, eds., *Collected Papers of Charles Sanders Peirce* (Cambridge, Mass., 1934), 5, pp. 156–189.

9. David Bloor, "Left and Right Wittgensteinians," in Andrew Pickering, ed., *Science as Practice and Culture* (Chicago, 1992), pp. 266–282.

10. "I say 'relevant' rather than 'natural' for two reasons: first, 'natural' is an inapt term to cover not only biological species but such artificial kinds as musical works, psychological experiments and types of machinery; and second 'natural' suggests some absolute or psychological priority, while the kinds in question are rather habitual or traditional or devised for a purpose" (Nelson Goodman, *Ways of Worldmaking* [Indianapolis and Cambridge, Mass., 1978], p. 10).

11. On another occasion one might worry more about names for kinds in different languages. In French, only reflecting telescopes are *télescopes,* refracting ones being, like spectacles, *lunettes.*

12. In consequence, if *I, J,* and *L* are in *C,* then if *I* stands in the relation *K* to both *J* and *L,* then either *J* stands in the relation *K* to *L* or vice versa. Why not say what Kuhn has written so much more simply: if two natural kinds have members in common, they are related as genus to species? Because that is too powerful. Natural kinds that fall into different categories may contingently apply to some of the same things. Otherwise there would be no true interesting universal statements of the form "Everything that is both of kind *I* and of kind *J* is also of kind *L.*" It may be true that if two "natural kinds" of the same category (i.e., from the same taxonomy, with the same head) have members in common, then they are related as genus to species.

13. In his theory of rigid designation, Kripke is inclined to make the Virgin Mary a daughter of Anne in all possible worlds. Even so, the modality is different (arguably, S_4 as opposed to S_5) than the necessity with which "descended from" is transitive. I chose the example to be able to emphasize, as an aside, the enormous number of different permutations of modalities that can arise in taxonomy. The interesting ones arise from "forcing," described below, be it by Sommers, the numerous sects of systematicists, or Kuhn.

14. Good reasons for distinguishing systematics from theories of "natural kinds" were given by John Dupré, "Natural Kinds and Biological Taxa," *Philosophical Review* 90 (1981): 66–90.

15. Fred Sommers, "Types and Ontology," *Philosophical Review* 72 (1963): 327–363.

16. Here I speak more of implicit senses than explicit theories. A résumé of some of the more nominalist theories from the time of William Whewell and J. S. Mill is given in my "Tradition of Natural Kinds," *Philosophical Studies* 61 (1991): 109–126.

17. I take the word from Quine's remarkable paper "Natural Kinds," in *Ontological Relativity and Other Essays* (New York, 1969), p. 138. He calls colors intuitive kinds. They, like all the secondary qualities of Englightenment philosophy, are important for people in childhood and in the early days of the species. But they would not count as kinds on a cosmic scale. One suspects that Nelson Goodman, cagey dealer in paintings by colorists and patron of the Cleveland Art Gallery, might have a more generous view of what's cosmic than Quine.

18. Keith Donellan has remarked that in Kripke's and Putnam's theory, nearly all natural-kind terms are of just this sort and are not drawn from the sciences at all, despite the fact that we would expect the theory to apply primarily to the host of scientific terms ("Kripke and Natural Kind Terms," in Carl Ginet, ed., *Knowledge and Mind: Philosophical Essays* [New York, 1981], pp. 84–104. "Multiple Sclerosis" is an example of Putnam's ("Reference and Truth," in his *Philosophical Papers*, vol. 3 [Cambridge, 1986], p. 71). Avishai Margalit has noticed that kinds of disease don't at all behave as Putnam or Kripke would imply ("Sense and Science," in E. Saarinen et al., eds., *Essays in Honour of Jaakko Hintikka* [Dordrecht, 1986], pp. 17–48).

19. R. B. De Sousa, "The Tree of English Bears Bitter Fruit," *Journal of Philosophy* 63 (1966): 37–46; David Massie, "Sommers' Tree Theory: A Reply to De Sousa," *Journal of Philosophy* 64 (1967): 185–193.

20. Kuhn's allusion appears in a footnote in his Shearman Memorial Lectures. I introduced "styles of scientific reasoning" in "Language, Truth, and Reason," in M. Hollis and S. Lukes, ed., *Rationality and Relativism* (Oxford, 1983), pp. 48–66. See my "'Style' for Historians and Philosophers," *Studies in the History and Philosophy of Science* 23 (1992): 1–20.

21. Is poison a natural kind for the cosmic philosopher of natural kinds? If natural kinds are, as some contend, those that occur in laws of nature, we have to ask whether "poison" occurs in statements of laws of nature. Perhaps not, but to prove that, we would have to be vastly more precise about "laws of nature" than the most deft of the logical empiricists or their heirs. We could take another tack. Maybe no substances are rigidly designated by "poison," there always being, to use the metaphor, some possible world in which deadly substances are benign.

22. J. S. Mill, *A System of Logic* (1843), in J. M. Robson, ed., *The Collected Works of John Stuart Mill*, vols. 7, 8 (Toronto, 1973), p. 122.

23. "Mill says that if the common properties of a class thus follow from a small number of primary characters, 'which, as the phrase is, *account for* all the rest,' it is not a real kind. He does not remark, that the man of science is bent upon ultimately thus accounting for each and every property that he studies" (C. S. Peirce, "Kind," *Baldwin's Dictionary of Philosophy and Psychology* [New York, 1903], vol. 1, p. 60).

24. Thus ends Mill's sentence cited in note 22 above.

25. I argue against Quine that we never need the notion of kind in the way he thinks we do, so that this contrast between mature science and infant learning vanishes.

Quine believed that the notion of kind was indispensable because of his theory of quality spaces and his wish not to multiply anything beyond necessity. See I. Hacking, "Natural Kinds," in R. Gibson, ed., *Perspectives on Quine* (Oxford, 1989), pp. 129–142.

26. Jed Buchwald, "Natural Kinds and the Wave Theory of Light," paper presented at the conference New Trends in the Historiography of Science, Corfu, 27 May 1991.

27. William Whewell, *The Philosophy of the Inductive Sciences Founded upon Their History* (London, 1840), 475 f. In addition, Whewell is excellent on the connection between universality and kinds.

28. See my "Multiple Personality Disorder and Its Hosts," *History of the Human Sciences* 5 no. 2 (1992): 3–31.

29. "Signing," *London Review of Books*, 5 April 1990.

30. Nelson Goodman, *Fact, Fiction, and Forecast*, 4th ed. (Cambridge, Mass.), p. 118.

Afterwords

Thomas S. Kuhn

Rereading the papers that make up this volume has recalled the feelings with which, almost two years ago, I rose to present my original response to them. C. G. Hempel, who for more than two decades has been a beloved mentor, had just delivered the remarks with which this volume now opens. They were the penultimate event of an intense day-and-a-half conference characterized by splendid papers and warm constructive discussion. Only a few personal occasions deaths, births, and other salient comings together or apart— had moved me as deeply. When I reached the podium I was not sure I would be able to speak, and it took a few moments to find out. After the conference my wife said to me that I'd never be the same again, and time is proving her right. On this occasion, as on that, I begin with heartfelt thanks to those who made the occasion possible: its inventors, organizers, contributors, and participants.[1] They have presented me with a gift I did not know existed.

In accepting that gift, I begin by returning to the remarks made by Professor Hempel. I too remember our first meeting: I was at Berkeley but considering an attractive invitation from Princeton; he was in residence across the Bay at the Center for Advanced Study in the Behavioral Sciences. I called on him there to ask for clues to what life and work at Princeton might be like. If that visit had gone badly, I might not have accepted Princeton's offer. But it did not, and I did. Our meeting in Palo Alto was only the first of a still-continuing series of warm and fruitful interactions. As Professor Hempel (for me he long ago became Peter) has said, our views at the start were very different, far more so than they have become through our interactions. But they were perhaps not quite so different as we both then thought, for I had begun to learn from him almost fifteen years before.

By the late 1940s I was deeply convinced that the received view of meaning, including its various positivist formulations, would not do: scientists did not, it seemed to me, understand the terms they used in the way described by the various versions of the tradition, and there was no evidence that they needed to do so. That was my state of mind when I first encountered Peter's old monograph on concept formation. Though it was many years before I saw its full relevance to my emerging position, it fascinated me from the start, and its role in my intellectual development must have been considerable. Four essential elements of my developed position are, in any case, to be found there: scientific terms are regularly learned in use; that use involves the description of one or another paradigmatic example of nature's behavior; a number of such examples are required for the process to work; and, finally, when the process is complete, the language or concept learner has acquired not only meanings but also, inseparably, generalizations about nature.[2]

A more general, broader, and deeper version of these views appeared a few years later in the classic paper that Peter significantly entitled "The Theoretician's Dilemma."[3] The dilemma was how to preserve a principled distinction between what he then still called "observational" and "theoretical terms." When, a few years later still, he began instead to describe the distinction as one between "antecedently available terms" and those learned together with a new theory, I could see him as having implicitly adopted a developmental or historical stance. I am uncertain whether that change of vocabulary occurred before or after we first met, but the basis for our convergence was by then clearly in place and our rapprochement perhaps already underway.

After I got to Princeton, Peter and I talked regularly, and occasionally we also taught together. When I later briefly took over the course in which I had assisted him, I opened by telling the class that my object was to show them the extra benefits of putting to work, within the historical or developmental approach to philosophy of science, some of the splendid analytic tools developed within the more static logical-empiricist tradition. I continue to think of my philosophical work as pursuing that goal. There are still other products of my interactions with Peter, and I shall turn to a significant one below. But what I primarily owe him is not from the realm of ideas. Rather it is the experience of working with a philosopher who cares more about arriving at truth than about winning arguments. I love him most, that is, for the noble uses to which he puts a distinguished mind. How could I not have been deeply moved when I followed him once more to the podium?

These remarks should suggest that I have known from the start of my meddling with philosophy that the historical approach I joined in developing owed as much to difficulties encountered by the logical-empiricist tradition as it did to history of science. Quine's "Two Dogmas" provides a second, for me formative, example of what I took those difficulties to be. About all of this Michael Friedman's elegant sketch is quite right, and I look forward to the fuller version he promises. In his original conference paper he added another telling remark, one that is here elaborated by John Earman in appropriate, but for me excruciating, detail. Whatever role the problems encountered by positivism may have played in the background for *The Structure of Scientific Revolutions,* my knowledge of the literature that attempted to deal with those problems was decidedly sketchy when the book was written. In particular, I was almost totally innocent of the post-*Aufbau* Carnap, and discovering him has distressed me acutely. Part of my embarrassment results from my sense that responsibility required that I know my target better, but there is more. When I received the kind letter in which Carnap told me of his pleasure in my manuscript, I interpreted it as mere politeness, not as an indication that he and I might usefully talk. That reaction I repeated to my loss on a later occasion.

Nevertheless, the passages which John quotes to show the deep parallels between Carnap's position and mine also show, when read in the context of his paper, a correspondingly deep difference. Carnap emphasized untranslatability as I do. But, if I understand Carnap's position correctly, the cognitive importance of language change was for him merely pragmatic. One language might permit statements that could not be translated into another, but anything properly classified as scientific knowledge could be both stated and scrutinized in either language, using the same method and gaining the same result. The factors responsible for the use of one language rather than another were irrelevant both to the results achieved and, more especially, to their cognitive status.

This aspect of Carnap's position has never been available to me. Concerned from the start with the *development* of knowledge, I have seen each stage in the evolution of a given field as built—not quite squarely—upon its predecessor, the earlier stage providing the problems, the data, and most of the concepts prerequisite to the emergence of the stage that followed. In addition, I have insisted that some changes in conceptual vocabulary are required for the assimilation and development of the observations, laws, and theories deployed in the later stage (whence the phrase "not quite squarely" above). Given those beliefs, the process of transition from old state to new becomes an integral part of science, a process that must be understood by the methodologist concerned to analyze the cognitive basis for scientific beliefs. Language change is *cognitively* significant for me as it was not for Carnap.

To my dismay, what John not unfairly labels my "purple passages" led many readers of *Structure* to suppose that I was attempting to undermine the cognitive authority of science rather than to suggest a different view of its nature. And even for those who understood my intent, the book had little constructive to say about how the transition between stages comes about or what its cognitive significance can be. I can do better on these and related subjects now, and the book on which I am presently at work will have much to say about them. Obviously, I cannot even sketch the book's content here, but I shall use my license as commentator to suggest as best I can what my position has become in the years since *Structure*. I will, that is, use the papers in this volume as grist for my current mill. To my great pleasure, all of them contribute to my purpose, though the treatment that results is inevitably incompletely balanced.

I begin with some anticipatory remarks on the topic that dominates my project: incommensurability and the nature of the conceptual divide between the developmental stages separated by what I once called "scientific revolutions." My own encounter with incommensurability was the first step on the road to *Structure,* and the notion still seems to me the central innovation introduced by the book. Even before *Structure* appeared, however, I knew that my attempts to describe its central conception were extremely crude. Efforts to understand and refine it have been my primary and increasingly obsessive concern for thirty years, during the last five of which I've made what I take to be a rapid series of significant breakthroughs.[4] The earliest of these first surfaced in a series of three unpublished Shearman lectures delivered in 1987 at University College, London. A manuscript of those lectures is, as Ian Hacking says, the primary source for the taxonomic solution to what he calls the new-world problem. Though the solution he describes was never quite my own and though my own has developed substantially since the manuscript he cites was written, I take immense pleasure in his paper. I will presuppose acquaintance with it in this attempt to suggest what my position has become.

First, though natural kinds provided me with a point of entry, they will not—for reasons Ian cites—resolve the full range of problems that incommensurability poses. The kind-concepts I require range far beyond anything to which the phrase 'natural kinds' has ordinarily referred. But, for the same reason, Ian's "scientific kinds" will not do either: what is required is a characteristic of kinds and kind-terms in general. In the book I will suggest that this characteristic can be traced to, and on from, the evolution of neural mechanisms for *re*identifying what Aristotle called "substances": things that, between their origin and demise, trace a lifeline through space over time.[5] What emerges is a mental module that permits us to learn to recognize not only kinds of physical object (e.g., elements, fields, and forces), but also kinds of furniture, of government, of personality, and so on. In what follows I shall refer to it frequently as the lexicon, the module in which members of a speech community store the community's kind-terms.

That required generality reinforces, though it does not cause, a second difference between my position and the one Ian presents. His nominalist version of my position—there are real individuals out there, and we divide them into kinds at will—does not quite face my problems. The reasons are numerous, and I mention only one here: how can the referents of terms like 'force' and 'wave front' (much less 'personality') be construed as individuals? I need a notion of 'kinds', including social kinds, that will populate the world as well as divide up a preexisting population. That need in turn introduces a last significant difference between me and Ian. He hopes to eliminate all residues of a theory of meaning from my position; I do not believe that that can be done. Though I no longer speak of anything so vague and general as "language change," I do talk of change in concepts and their names, in conceptual vocabulary, and in the structured conceptual lexicon that contains both kind-concepts and their names. A schematic theory intended to provide a basis for talk of this kind is central to my projected book. With respect to kind terms, aspects of a theory of meaning remain at the heart of my position.

Here I can hope only to sketch what my position has become since the Shearman lectures, and the sketch must be both dogmatic and incomplete. Kind-concepts need not have names, but in linguistically endowed populations they mostly do, and I will restrict my attention to them. Among English words, they can be identified by grammatical criteria: for example, most of them are nouns that take an indefinite article either by themselves or, in the case of mass nouns, when conjoined with a count noun, as in 'gold ring'. Such terms share a number of important properties, the first set of which was enumerated in my earlier acknowledgement of my debt to Peter Hempel's work on concept formation. Kind-terms are learned in use: someone already adept in their use provides the learner with examples of their proper application. Several such exposures are always required, and their outcome is the acquisition of more than one concept. By the time the learning process has been completed, the learner has acquired knowledge not only of the concepts but also of the properties of the world to which they apply.

Those characteristics introduce a second shared property of kind-terms. They are projectible: to know any kind-term at all is to know some generalizations satisfied by its referents and to be equipped to look for others. Some of these generalizations are normic, admit exceptions.[6] "Liquids expand when heated" is an example even though it sometimes fails, e.g., for water between 0 and 4 degrees centigrade. Other generalizations, though often only approximate, are are nomic, exceptionless. In the sciences, where they mainly function, these generalizations are usually laws of nature: Boyle's law for gases or Kepler's laws for planetary motions are examples.

These differences in the nature of the generalizations acquired in learning kind-terms correspond to a necessary difference in the way the terms are learned. Most kind-terms must be learned as members of one or another contrast set. To learn the term 'liquid', for example, as it is used in contemporary non-technical English, one must also master the terms 'solid' and 'gas'. The ability to pick out referents for any of these terms depends critically upon the characteristics that differentiate its referents from those of the other terms in the set, which is why the terms involved must be learned together and why they collectively constitute a contrast set. When terms are learned together in this way, each comes with attached normic generalizations about the properties likely to be shared by its referents. The other sort of kind-term—'force' for example—stands alone. The terms with which it needs to be learned are closely related but not by contrast. Like 'force' itself, they are not normally in any contrast set at all. Instead, 'force' must be learned with terms like 'mass' and 'weight'. And they are learned from situations in which they occur together, situations exemplifying laws of nature. I have elsewhere argued that one cannot learn 'force' (and thus acquire the corresponding concept) without recourse to Hooke's law and either Newton's three laws of motion or else his first and third laws together with the law of gravity.[7]

These two characteristics of kind-terms necessitate a third, the one at which this exercise has been aimed. In a sense that I will not further explicate here, the expectations acquired in learning a kind-term, though they may differ from individual to individual, supply the individuals who have acquired them with the meaning of the term.[8] Changes in expectations about a kind-term's referents are therefore changes in its meaning, so that only a limited variety of expectations may be accommodated within a single speech community. So long as two community members have compatible expectations about the referents of a term they share, there will be no difficulty. One or both of them may know things about those referents that the other does not, but they will both pick out the same things, and they can learn more about those things from each other. But if the two have incompatible expectations, one will occasionally apply the term to a referent to which the other categorically denies that it applies. Communication is then jeopardized, and the jeopardy is especially severe because, like meaning differences in general, the difference between the two cannot be rationally adjudicated. One or both of the individuals involved may be failing to conform to standard social usage, but it is only with respect to social usage that either of them can be said to be right or wrong. What they differ about is, in that sense, convention rather than fact.

One way to describe this difficulty is as a case of polysemy: the two individuals are applying the same name to different concepts. But that description, though correct as far as it goes, fails to catch the depth of the difficulty. Polysemy has a standard remedy, widely deployed in analytic philosophy: two names are introduced where there had been only one before. If the polysemous term is 'water', the difficulties are to be lifted by replacing it with a pair of terms, say 'water$_1$' and 'water$_2$', one for each of the concepts that previously shared the name 'water'. Though the two new terms differ in meaning, most referents of 'water$_1$' are referents of 'water$_2$' and vice versa. But each term also refers to a few items to which the other does not, and it was about the applicability of 'water' in such cases that the two community members disagreed. Introducing two terms where there was one before appears to resolve the difficulty by enabling the disputants to see that their difference was simply semantic. They were disagreeing about words, not about things.

That way of resolving the disagreement is, however, linguistically unsupportable. Both 'water$_1$' and 'water$_2$' are kind-terms: the expectations they embody are therefore projectible. Some of those expectations are different, however, which results in difficulties in the region where both apply. Calling an item in the overlap region 'water$_1$' induces one set of expectations about it; calling the same item 'water$_2$' induces another, partly incompatible, set. Both names cannot apply, and which to choose is no longer about linguistic conventions but rather about matters of evidence and fact. And if the matters of fact are taken seriously, then in the long run only one of the two terms can survive within any single language community. The difficulty is most obvious with terms like 'force' that bring with them nomic expectations. If a referent lay in the overlap region (say between Aristotelian and Newtonian usage), it would be subject to two incompatible natural laws. For nomic expectations the prohibition must be slightly weakened: only terms which belong to the same contrast set are prohibited from overlapping in membership. 'Male' and 'horse' may overlap but not 'horse' and 'cow'.[9] Periods in which a speech community does deploy overlapping kind-terms end in one of two ways: either one entirely displaces the other, or the community divides into two, a process not unlike speciation and one that I will later suggest is the reason for the ever-increasing specialization of the sciences.

What I have just been saying is, of course, my version of the solution to what Ian has dubbed the new-world problem. Kind-terms supply the categories prerequisite to description of and generalization about the world. If two communities differ in their conceptual vocabularies, their members will describe the world differently and make different generalizations about it. Sometimes such differences can be resolved by importing the concepts of one into the conceptual vocabulary of the other. But if the terms to be imported are kind-terms that overlap kind-terms already in place, no importation is possible, at least no importation which allows both terms to retain their meaning, their projectibility, their status as kind-terms. Some of the kinds that populate the worlds of the two communities are then irreconcilably different, and the difference is no longer between descriptions but between the populations described. Is it, in these circumstances, inappropriate to say that the members of the two communities live in different worlds?

I have so far been discussing what Ian calls scientific kinds or at least the kinds that nature exhibits to the members of a culture, and I will be returning to them when discussing Jed Buchwald's paper in the next section. But it will help to consider first an example of the significance of the no-overlap principle for social kinds. John Heilbron's and Noel Swerdlow's papers provide a central illustration.

John's "Mathematicians' Mutiny" is a splendid example of the historian's craft. It is also thoroughly relevant to the old paper of mine to which he applies it. Though he fits to that paper an even narrower straitjacket than the one I made myself, I've learned from and accept in full the more complex and nuanced studies of the development and interrelations of scientific fields that he here and elsewhere provides. Taken together, his studies constitute a major and still advancing achievement. But John's methodological remarks about the vocabulary the historian requires in order to describe the phenomena he or she studies seem to me mistaken in ways that have often damaged historical understanding.

The fundamental product of historical research is narratives of development over time. Whatever its subject, the narrative must always begin by setting the stage. If its subject is beliefs about nature, it must open with a description of what beliefs were accepted at the time and place from which the narrative begins. That description must include as well a specification of the vocabulary in which natural phenomena were described and beliefs about them stated. If, instead, the narrative deals with group activities or practices, it must open with a description of the various practices recognized at the time the narrative begins, and it must indicate what was expected of those practices both by practitioners and those around them. In addition, the stage setting must introduce names for those practices (preferably the names used by practitioners) and display contemporary expectations about them: how were they justified and how criticized in their own time?

To learn the nature and objects of these beliefs and expectations, the historian deploys the techniques that I once sketched under the rubric of translation but would now insist are directed to language learning, a distinction to which I will return below. To communicate the results to readers, the historian becomes a language teacher and shows readers how to use the terms, most or all of them kind-terms, current when the narrative began but no longer accessible in the language shared by the historian and his or her readers. Some of those terms—'science' or 'physics', for example—still exist in the readers' language but with altered meanings, and these must be unlearned and replaced by their predecessors. When the process is complete, or sufficiently complete for the historian's purposes, the required stage setting has been provided, and the narrative can begin. It can furthermore be related entirely in the terms taught at the start or in their successors, the latter being introduced within the narrative. It is only in the initial teaching operation, in setting the stage, that the historian must, like other sorts of language teachers, make use of the language that readers bring with them. (See John's remark about my use of anachronistic terminology in the title of my paper: "Mathematical and Experimental Traditions in the Development of Physical Science.") It is, of course, always tempting and sometimes irresistibly convenient to use later, already familiar, terms or other terms that, like John's synchronic usages, depart from those deployed at the time. Circumlocution is avoided and the result is not invariably damaging. But the price of convenience is always great risk: exquisite sensitivity and great restraint are required to avoid damage. Experience suggests that few historians develop these qualities in sufficient measure; certainly, I have repeatedly fallen short myself.

The danger in using the names of contemporary scientific fields when discussing past scientific development is the same as that of applying modern scientific terminology when describing past belief. Like 'force' and 'element', 'physics' and 'astronomy' are kind terms, and they carry behavioral expectations with them. These and other names of individual sciences are acquired together in a contrast set, and the expectations that enable one to pick out examples of the practice of each are rich in characteristics that differentiate the examples of one practice from the examples of another, poor in characteristics that examples of a single practice share. That is why one has to learn the names of a number of sciences together before one can pick out examples of the practice of any one of them. Interjection of a name not current at the time thus usually results in a violation of the no-overlap principle and generates conflicting expectations about behavior. That I take to be the lesson to be learned from John's Lavoisier and Poisson examples. His three orthographically distinguished usages have the merit of showing why the debates arose, but they neither point the way to nor play a role in their resolution. To me, those examples suggest not the need for the three uses in historical descriptions but the trouble caused by failure to avoid two of them.

The most obvious danger results from what John calls diachronic usage—the resort to modern terminology—which he suggests might be distinguished by italics. Noel Swerdlow's paper is a strikingly successful and still badly needed attempt to undo one prominent example. When I entered history of science, it was customary, largely due to the influence of Pierre Duhem, to speak of "medieval science," and I often used that highly questionable phrase myself. Many people, likely still including me, regularly spoke also of "medieval physics" and sometimes of "medieval chemistry" as well. Some experts spoke also of "medieval dynamics" and "kinematics," drawing a distinction for which I could find neither need nor basis in the texts. At its narrowest, this introduction of modern conceptual distinctions led to misreading, and some of these directly influenced the understanding of figures as recent as Galileo. At its broadest, represented by phrases like "medieval science" and "medieval physics," the use of a modern vocabulary led to debates about whether the Renaissance had played any role in the origin of modern science, a debate that, though never conclusive, regularly minimized the role of the Renaissance in scientific development. Though the situation has considerably improved in the forty years since I entered the field, important residues of that debate still linger, residues that Noel is insisting must be set aside. For all my pretense of a position beyond anachronism, I've learned important lessons from his paper and am sure others will do the same.

So far I have been discussing the use of the field names that John labels "diachronic," not the use he calls "synchronic" and employs to refer to "a science or sciences within a narrowly restricted period of time." That use presents subtler problems than the diachronic, but they are problems of the same general sort. In a letter to me John motivates the introduction of synchronic names for fields by pointing out that "contemporary usage is seldom uniform, even at a single historical moment and certainly not over a period long enough to interest the historian," and I know what he has in mind. But the historian has no need to introduce special terms that average out variations in usage with, say, time, place, and affiliation. As with the very similar differences between the idiolects of different individuals, the averaging process takes care of itself. If variations in usage, whether from individual to individual or group to group, did not, at the time under study, interfere with successful communication concerning problems relevant to the narrative, the historian may simply use the terms deployed by his or her subjects. If the variations did make a historical difference, the historian is required to discuss them. In neither case is averaging appropriate. The same is true of variation over time. If the variation is systematic and large enough to make it hard for members of a later generation to understand predecessors who matter to them, then the historian must show how and why those changes came about. If understanding is not affected by the passage of time, then there is no more reason to introduce a new term than to choose whether to use the older or the newer version. Indeed, in the latter case, it is hard to see in what sense there are then two versions to choose between.

I am not, let me be clear, suggesting that the historian is required to report every change of usage, whether from place to place, from group to group, or from time to time. Historical narratives are by their nature intensely selective. Historians are required to include in them only those aspects of the historical record that affect the accuracy and plausibility of their narrative. If they ignore such items— changes in usage included—they risk both criticism and correction. But omission of changes and acceptance of the consequent risk is one thing; introducing new terms is another. As with John's diachronic use, so with his synchronic, new terms can disguise problems that historians are required to face. License to alter the descriptive language of the times they describe should, I think, be denied.

Jed Buchwald's rich and evocative paper returns the topic from social to scientific kinds, and Norton Wise's raises the issue of the relationship between them. The most obvious and direct ties between Jed's paper and the problematic I've been developing are his brief discussions of the difference between the concepts of light rays and of polarization as they are found in the wave and in the emission theory of light.[10] (For rays, geometric optics is also relevant.) Jed's discussions make no reference to kinds or to the no-overlap principle, and none is required. But these examples can easily be recast. 'Ray', for example, is used as a kind-term by both the wave and emission theories: the overlap between its referents in the two cases (together with the overlap between the kinds of polarization appropriate to the two theories) cause the difficulties that Jed's paper discusses. In a brilliant paper that takes off from the one he presented to the conference, Jed has now systematically analyzed numerous aspects of the transition from the emission to the wave theory of light as the result of changes in kinds. His paper is likely, I think, to introduce a new stage in the historical analysis of episodes involving conceptual change.[11]

The second point of contact between Jed's paper and my remarks on kinds concerns translation. In *Structure* I spoke of meaning change as a characteristic feature of scientific revolutions; later, as I increasingly identified incommensurability with difference of meaning, I repeatedly referred to the difficulties of translation. But I was then torn, usually without quite realizing it, between my sense that translation between an old theory and a new one was possible and my competing sense that it was not. Jed quotes a long passage (from the Postscript added to the second edition of *Structure*) in which I took the first of these alternatives and described, under the rubric of translation, a process through which "participants in a communication breakdown" could reestablish communication by studying each other's use of language and learning, finally, to understand each other's behavior. With what he says in discussing that passage, I fully agree: in particular, though the process described is vital to historians, scientists themselves seldom or never use it. But it is important also to recognize that I was wrong to speak of translation.[12] What I described, I now realize, was language learning, a process that need not, and ordinarily does not, make full translation possible.

Language learning and translation are, I have in recent years been emphasizing, very different processes: the outcome of the former is bilingualism, and bilinguals repeatedly report that there are things they can express in one language that they cannot express in the other. Such barriers to translation are taken for granted if the matter to be translated is literature, especially poetry. My remarks on kinds and kind-terms were intended to suggest that the same difficulties in communication arise between members of different scientific communities, whether what separates them is the passage of time or the different training required for the practice of different specialties. For both literature and science, furthermore, the difficulties in translation arise from the same cause, the frequent failure of different languages to preserve the structural relations among words, or in the case of science among kind-terms. The associations and overtones so basic to literary expression obviously depend upon these relations. But so, I have been suggesting, do the criteria for determining the reference of scientific terms, criteria vital to the precision of scientific generalizations.

The third way in which Jed's paper intersects my remarks on kinds relates to Norton Wise's paper as well, and the relationship is in both cases more problematic and more speculative than those I've so far been discussing. Jed's paper speaks of an unarticulated core or sub-structure which he contrasts with an explicitly articulated superstructure. People who share substructure, he suggests, may disagree about appropriate articulations, but people who differ with respect to substructure will simply misunderstand each other's points, usually without realizing that anything different from disagreement is involved. These properties echo those of the mental module I called "the lexicon" when updating my solution of the new-world problem, the module in which each member of a speech community stores the kind-terms and kind-concepts used by community members to describe and analyze the natural and social worlds. It would be too much to suggest that Jed and I are talking about the same thing, but we are certainly exploring the same terrain, and it is worth specifying that shared terrain more closely.

For brevity I restrict attention to the most populous part of the lexicon, the one that contains concepts learned in contrast sets and carrying normic expectations. What this part of their lexicons supplies to community members is a set of learned expectations about the similarities and differences between the objects and situations that populate their world. Presented with examples drawn from various kinds, any member of the community can tell which presentations belong to which kind, but the techniques by which they do so depend less on the characteristics shared by members of a given kind than on those which distinguish members of one kind from those of another. All competent community members will produce the same results, but they need not, as I've previously indicated, make use of the same set of expectations in doing so. Full communication between community members requires only that they refer to the same objects and situations, not that they have the same expectations about them. The ongoing process of communication which unanimity in identification permits allows individual community members to learn each others' expectations, making it likely that the congruence of their bodies of expectations will increase with time. But, though the expectations of individual community members need not be the same, success in communication requires that the differences between them be heavily constrained. Lacking time to develop the nature of the constraint, I will simply label it with a term of art. The lexicons of the various members of a speech community may vary in the expectations they induce, but they must all have the same *structure*. If they do not, then mutual incomprehension and an ultimate breakdown of communication will result.

To see how closely this position maps onto Jed's, read my lexically induced "expectations" as Jed's "articulations" of a core. People who share a core, like those who share a lexical structure, can understand each other, communicate about their differences, and so on. If, on the other hand, cores or lexical structures differ, then what appears to be disagreement about fact (which kind does a particular item belong to?) proves to be incomprehension (the two are using the same name for different kinds). The would-be communicants have encountered incommensurability, and communication breaks down in an especially frustrating way. But because what's involved is incommensurability, the missing prerequisite to communication—a "core" for Jed, a "lexical structure" for me—can only be exhibited, not articulated. What the participants in communication fail to share is not so much belief as a common culture.

Norton's paper is also concerned with the commonalities that make a shared scientific culture, and his mediating balances behave somewhat like Jed's core in that they isolate characteristics shared by items located at the various different nodes of his network. In this case, however, the likenesses are between items in the social world rather than the natural. They hold, that is, between practices in the various scientific fields as well as between them and the larger culture (note Norton's introduction of the figure of Republican France). Having been on record for many years as deeply skeptical about the extent to which grounded ties of this wide-ranging sort could be found, I feel bound to announce that I've been largely converted, mostly as a consequence of Norton's work, especially that on nineteenth-century Britain.[13] These ties between the practices Norton discusses cannot, I think, be coincidental or mere imaginative fabrications. They signify, I am convinced, something of great importance to an understanding of science. But at this early stage of the development of these ties, I'm deeply uncertain what they can signify: essential parts of the story his points require seem to me to be missing.

In the first place, I do not know what Norton's "rationalist scientific culture" is, how the practices between which his balances mediate are recognized or selected. My early readings of his paper suggested that they were simply the sciences as practiced in the national culture of late nineteenth-century France, but Norton has assured me that that is not at all his intent. Not all French scientific practices belong to his network, he insists, and some of the practices that do are located in other national cultures. But neither can his network be identified simply by the bridge with respect to which its nodes are similar: an arbitrarily selected set of practices will ordinarily be similar in some respect or other. I am not suggesting that a *definition* of rationalist scientific culture is required, but I do feel the need for some description of its salient characteristics, of characteristics that would collectively permit me to pick out some practices as exemplifying the cultures and others as not. The point is not that I want to check Norton's story. He recognizes the practices involved, and I've great confidence in his judgment. I'm sure he will, in the long run, provide an answer to my question. But until I know something about how Norton recognizes the practices which are the nodes of his network, I am quite literally not going to understand what he's trying to tell me.

That difficulty is aggravated by another, one that I'm far less sure can be resolved and to which I'm especially sensitive because it exemplifies a trap into which I fell again and again in *Structure*. Norton illustrates the bridges supplied by his balances by showing a small number of individuals interacting across them: Lavoisier with Laplace, Condillac with Lavoisier, Lavoisier with Condorcet. But he uses those illustrations to suggest that the bridges link not only people but also practices—chemistry, physical astronomy, electricity, political economy, and others—and that suggestion leads to three difficulties to which I'll point in order of increasing importance.

The first difficulty I mention only for one of its consequences. To generalize these bridges from individuals to the various scientific practices, one would need to show that they operated for a considerable number of practitioners and to illustrate the differences their existence made to the practices that they linked. I am not, however, the one to throw stones at people who overgeneralize, and the point I'm after is different. The speed of Norton's transition from individual to group obscures another possible explanation of the individual behavior he reports. Perhaps the mediating balances are not characteristics of the various scientific cultures but rather of the larger culture within which those practices take place. That could make bridges available to individuals without affecting the mode of group practice at all. Such an explanation may not be right, but room needs to be allowed for its consideration. And considering it would, among other desiderata, provide the room needed to ask what's right about the position of the diehards who insist, for example, that chemistry is chemistry, physics physics, and math math in whatever culture they occur.

The third difficulty is of another and more important sort. In my view Norton's passage from individual to group involves a damaging category mistake, the one of which I was repeatedly guilty in *Structure* and which is endemic also in the writings of historians, sociologists, social psychologists and others. The mistake is to treat groups as individuals writ large or else individuals as groups writ small. It results, at its crudest, in talk of the group mind (or group interest) and, in its subtler forms, in attributing to the group a characteristic shared by all or most of its members. The most egregious example of this mistake in *Structure* is my repeated talk of gestalt switches as characteristic of the experience undergone by the group. In all these cases the error is grammatical. A group would not experience a gestalt switch even in the unlikely event that every one of its members did so. A group does not have a mind (or interests), though each of its members presumably does. By the same token, it does not make choices or decisions even if each of its members does so. The outcome of a vote, for example, may result from the thoughts, interests, and decisions of group members, but neither the vote nor its outcome is a decision. If, as has traditionally been taken for granted, a group were nothing but the aggregate of its individual atomic members, this grammatical error would be inconsequential. But it is increasingly recognized that a group is not just the sum of its parts and that an individual's identity in part consists in (not simply, is determined by) the groups of which he or she is a member. We badly need to learn ways of understanding and describing groups that do not rely upon the concepts and terms we apply unproblematically to individuals.

I do not command the required ways of understanding, but in recent years I have made two steps towards doing so. The first, which I've already mentioned but still lack space to explain, is the distinction between a lexicon and a lexical structure. Each member of a community possesses a lexicon, the module that contains the community's kind-concepts, and in each lexicon the kind-concepts are clothed with expectations about the properties of their various referents. But though the kinds must be the same in the lexicons of all community members, the expectations need not be. Indeed, in principle, the expectations need not even overlap. What is required is only that they give the lexicons of all community members the same structure, and it is this structure, not the varied expectations through which different members express it, that characterizes the community as a whole.

My other step forward is the discovery of a tool that I've as yet scarcely learned to put to use. But I am currently being much instructed by the discovery that the puzzles about the relation of group members to group have a quite precise parallel in the field of evolutionary biology: the vexing relation between individual organisms and the species to which they belong. What characterizes the individual organism is a particular set of genes; what characterizes the species is the gene pool of the entire interbreeding population which, geographical isolation aside, constitutes the species. Understanding the process of evolution has in recent years seemed increasingly to require conceiving the gene pool, not as the mere aggregate of the genes of individual organisms, but as itself a sort of individual of which the members of the species are parts.[14] I am persuaded that this example contains important clues to the sense in which science is intrinsically a community activity. Methodological solipsism, the traditional view of science as, at least in principle, a one-person game, will prove, I am quite sure, to have been an especially harmful mistake.

That Norton's paper calls forth thoughts of this sort is an index of the seriousness with which I take it. He is, I'm quite certain, on his way to significant discoveries, and I look forward to them with considerable excitement. But those discoveries are still emerging. So far, I am finding them extraordinarily difficult to grasp.

I come at last to relativism and realism, issues central to the papers of Ernan McMullin and Nancy Cartwright but implicit in several of the other papers as well. As in the past Ernan proves to be among my most discerning and sympathetic critics, and I will presuppose much that he has said in order to concentrate on points at which our views depart. Of these the most important to both of us involves what Ernan takes to be my antirealist stance and my corresponding lack of concern with epistemic (as against puzzle-solving) values. But that characterization does not quite catch the nature of my enterprise. My goal is double. On the one hand, I aim to justify claims that science is cognitive, that its product is knowledge of nature, and that the criteria it uses in evaluating beliefs are in that sense epistemic. But on the other, I aim to deny all meaning to claims that successive scientific beliefs become more and more probable or better and better approximations to the truth and simultaneously to suggest that the subject of truth claims cannot be a relation between beliefs and a putatively mind-independent or "external" world.

Postponing remarks on the nature of truth claims, I begin with the question of science's zeroing in on, getting closer and closer to, the truth. That claims to that effect are meaningless is a consequence of incommensurability. This is not the place to elaborate the needed arguments, but their nature is suggested by my earlier remarks on kinds, the no-overlap principle, and the distinction between translation and language-learning. There is, for example, no way, even in an enriched Newtonian vocabulary, to convey the Aristotelian propositions regularly misconstrued as asserting the proportionality of force and motion or the impossibility of a void. Using our conceptual lexicon, these Aristotelian propositions cannot be expressed—they are simply ineffable—and we are barred by the no-overlap principle from access to the concepts required to express them. It follows that no shared metric is available to compare our assertions about force and motion with Aristotle's and thus to provide a basis for a claim that ours (or, for that matter, his) are closer to the truth.[15] We may, of course, conclude that our lexicon permits a more powerful and precise way than his of dealing with what are *for us* the problems of dynamics, but these were not his problems, and lexicons are not, in any case, the sorts of things that can be true or false.

A lexicon or lexical structure is the long-term product of tribal experience in the natural and social worlds, but its logical status, like that of word-meanings in general, is that of convention. Each lexicon makes possible a corresponding form of life within which the truth or falsity of propositions may be both claimed and rationally justified, but the justification of lexicons or of lexical change can only be pragmatic. With the Aristotelian lexicon in place it does make sense to speak of the truth or falsity of Aristotelian assertions in which terms like 'force' or 'void' play an essential role, but the truth values arrived at need have no bearing on the truth or falsity of apparently similar assertions made with the Newtonian lexicon. Whatever I may have believed when I wrote the *Copernican Revolution*, I would not now assume (*pace* Ernan) "that the simpler, the more beautiful [astronomical] models are more likely to be true." Though simplicity and beauty provide important criteria of choice in the sciences (as does making causal sense of phenomena, which Ernan also cites), they are instrumental rather than epistemic where lexical change is involved. What they are instrumental for will be my closing topic below.

At least all this is the case if the sense of 'epistemic' is the one I take Ernan to have in mind, the sense in which the truth or falsity of a statement or theory is a function of its relation to a real world, independent of mind and culture. There is, however, another sense in which criteria like simplicity may be called epistemic, and it has already figured, implicitly or explicitly, in several of the papers in this volume. Its most suggestive appearance is also the briefest: Michael Friedman's description of Reichenbach's distinction between two meanings of the Kantian a priori, one which "involves unrevisability and . . . absolute fixity for all times" while the other means "'constitutive of the concept of the object of knowledge'." Both meanings make the world in some sense mind-dependent, but the first disarms the apparent threat to objectivity by insisting on the absolute fixity of the categories, while the second relativizes the categories (and the experienced world with them) to time, place, and culture.

Though it is a more articulated source of constitutive categories, my structured lexicon resembles Kant's a priori when the latter is taken in its second, relativized sense. Both are constitutive of *possible experience* of the world, but neither dictates what that experience must be. Rather they are constitutive of the infinite range of possible experiences that might conceivably occur in the actual world to which they give access. Which of these conceivable experiences occurs in that actual world is something that must be learned, both from everyday experience and from the more systematic and refined experience that characterizes scientific practice. They are both stern teachers, firmly resisting the promulgation of beliefs unsuited to the form of life the lexicon permits. What results from respectful attention to them is knowledge of nature, and the criteria that serve to evaluate contributions to that knowledge are, correspondingly, epistemic. The fact that experience within another form of life—another time, place, or culture—might have constituted knowledge differently is irrelevant to its status as knowledge.

Norton Wise's closing pages I take to be making a very similar point. Through much of his paper, technology (for his culture, the various balances) is seen as providing a culturally based mediator between instruments and reality at one end of his cylinder (his figure 18) and between instruments and theories at the other.[16] Except that the technologies are conceived as situated in local cultures, no traditional philosopher of science would find fault with that model. Of course instruments, including the sense organs, are required to mediate between reality and theory. To this point in the argument, no references to anything like a constructed or mind-dependent reality are called for. But Norton then folds his cylinder back on itself to form a doughnut, and the picture changes decisively (his figure 19). A geometry that requires a figure with two ends is replaced by one that calls for three symmetrically placed slices. Technology continues to provide a two-way route between theories and reality, but reality provides the same sort of route between theory and technology, and theories provide a third route between reality and technology. Scientific practice requires all three of these sorts of mediation, and none of them has priority. Each of his three slices—technology, theory, and reality—is constitutive for the other two. And all three are required for the practice whose product is knowledge. When Norton closes by describing what he has been doing as "depicting cultural epistemology," I think he gets it just right. But I would add "cultural ontology" as well.

Nancy Cartwright's exciting paper indicates ways to move further in the same direction, but for my purposes, its opening remarks on the theory/observation distinction need first to be somewhat recast. I agree that the distinction is needed, but it cannot be just that between the "peculiarly recondite terms [of modern science and] those we are more used to in our day-to-day life." Rather, the concepts of theoretical terms must be relativized to one or another particular theory. Terms are theoretical with respect to a particular theory if they can only be acquired with that theory's aid. They are observation terms if they must first have been acquired elsewhere before the theory can to learned.[17] 'Force' is thus theoretical with respect to Newtonian dynamics but observational with respect to electromagnetic theory. This view is very close to the third of the interpretations that Nancy suggests for Peter Hempel's phrase 'antecedently available', and that interpretation has been instrumental, as my opening remarks will have suggested, in Peter's and my very considerable rapprochement.

Replacing the concept of theoretical terms with that of antecedently available terms has three special advantages. First, it ends the apparent equivalence of "observational" and "non-theoretical": many of the recondite terms of modern science are both theoretical and observational, though the observation of their referents requires recondite instruments. Second, unlike its predecessor the distinction between theoretical and antecedently available terms is freed to become developmental, as I think it must: terms antecedently available, whether to an individual or a culture, are the base for the further extension of both vocabulary and knowledge. And third, viewing the distinction as developmental focuses attention on the process by which a conceptual vocabulary is transmitted from one generation to the next—first to young children being prepared (socialized) for the adult society of their culture, and second to young adults being prepared (again, socialized) to take their place among the practitioners of their discipline.

For present purposes the last point is the crucial one, for it will bring me rapidly back to realism. In the theory of the lexicon, to which I've already repeatedly referred, a key role is played by the process by which lexicons are transmitted from one generation to the next, whether from parents to children or from practitioners to apprentices. In that process the exhibition of concrete examples plays the central role, where the "exhibition" may be accomplished either by pointing to real-life examples in the everyday world or the laboratory or else by describing these potential examples in the vocabulary antecedently available to the student or inductee. What is acquired in this process is, of course, the kind-concepts of a culture or subculture. But what comes with them, inseparably, is the world in which members of the culture live.

Nancy omits the developmental context, which I take to be central. But the process is one she twice illustrates: for scientific kinds, by the passage about Newton's second law that she resurrects from the second edition of *Structure,* and for social kinds, by her discussion of fables. The pendulum, the inclined plane, and the rest are examples of $f = ma$, and it is being examples of $f = ma$ that makes them similar, like each other. Without having been exposed to them or some equivalents as examples of $f = ma$, students could not learn to see either the similarities between them or what it was to be a force or a mass; they could not, that is, acquire the concepts of force and mass or the meaning of the terms that name them.[18] Similarly, the three examples of the fable—marten/grouse, fox/marten, and wolf/fox—are concrete illustrations of what, lacking a better term, I shall call the "power situation," the situation in which terms like 'strong', 'weak', 'predator', and 'prey' function. It is their illustrating the same aspect of the situation that simultaneously makes them similar to each other and makes the situation into the one the fables convey. Without exposure to these or similar situations, a candidate for socialization into the culture that exhibits them could not acquire the social kinds called 'the strong', 'the weak', 'predators', or 'prey'.

Though other resources are available for acquiring social concepts like these, fables and the maxims that accompany them have the particular merit of simplicity, which is presumably why they have played so large a role in the socialization of children. Nancy speaks of them as "thin," a term she also applies to, say, models like the frictionless plane and the point pendulum. The latter are, if you will, physicists' fables (Newton's second law being the maxim juxtaposed to them), and it is their characteristic thinness which makes them so especially useful for socializing potential members of the profession. That is why they figure so prominently in science textbooks.

Except for my insistence on positioning them within the learning or socialization process, these points are all explicit in Nancy's paper, and so is another which she has given me new words to describe. Once the new terms (or the revised versions of old ones) have been acquired, there is no ontological priority between their referents and the referents of the antecedently available terms deployed in the acquisition process. The concrete (pendulum or marten) is neither more nor less real than the abstract (force or prey). There are, of course, both logical and psychological priorities between the members of these pairs. One cannot acquire the Newtonian concepts of force and mass without prior access to such concepts as space, time, motion, and material body. Nor can one acquire the concepts of predator and prey without prior access to such concepts as kinds of creature, death, and killing. But there are not, as Nancy puts it, relations of either fact or meaning reduction between the members of these pairs (between force and mass, on the one hand, and space, time, etc., on the other; or between predator and prey, on the one hand, and death, killing, etc., on the other). In the absence of such relations, there is no basis for singling out one or the other juxtaposed set as the more real. To insist on this point is not to limit the concept of reality but rather to say what reality is.

It is in our response to that shared analysis that Nancy's route and mine diverge, but in a way I am finding especially instructive. Both Nancy and I are pushed to a reluctant pluralism. But she would achieve hers by permitting restrictions on the universality of true scientific generalizations, suggesting, for example, that the truth of Newton's second law does not depend upon its applying to all of its potentially concrete models. The law's scope is, for her, uncertain: in one part of its domain it may be true, while in another some other law may obtain. For me, however, that form of pluralism is barred. 'Force', 'mass', and their like are kind-terms, the names of kind-concepts. Their scope is limited only by the no-overlap principle and is thus part of their meaning, part of what enables their referents to be picked out and their models to be recognized. To discover that the scope of a kind-concept is limited by something extrinsic, something other than its meaning, is to discover that it never had any proper applications at all.

Nancy introduces scope restrictions to account for the occasional failure of the search for workable models of laws she considers true. I would instead resolve such failures by introducing a few new kinds that displace some of those in use before. That change constitutes a change in lexical structure, one that brings with it a correspondingly changed form of professional practice and a different professional world within which to conduct it. Her pluralism of domains is for me a pluralism of professional worlds, a pluralism of practices. Within the world of each practice, true laws must be universal, but some of the laws governing one of these worlds cannot even be stated in the conceptual vocabulary deployed in, and partially constitutive of, another. The same no-overlap principle that necessitates the universality of true laws bars the practitioners resident in one world from importing certain of the laws that govern another. The point is not that laws true in one world may be false in another but that they may be ineffable, unavailable for conceptual or observational scrutiny. It is effability, not truth, that my view relativizes to worlds and practices. That formulation is compatible with transworld travel: a twentieth-century physicist can enter the world of, say, eighteenth-century physics or twentieth-century chemistry. But that physicist could not practice in either of these other worlds without abandoning the one from which he or she came. That makes transworld travel difficult to the point of subversion and explains why, as Jed Buchwald emphasizes, practitioners of a science almost never undertake it.

One more step will return me to Ernan's paper and take me to the last of the problems to be considered in this one. The developmental episodes that introduce new kinds and displace old are, of course, the ones that in *Structure* I called "revolutions." At the time I thought of them as episodes in the development of a single science or scientific specialty, episodes that I somewhat misleadingly likened to gestalt switches and described as involving meaning change. Clearly, I still think of them as transforming episodes in the development of individual sciences, but I now see them as playing also a second, closely related, and equally fundamental role: they are often, perhaps always, associated with an increase in the number of scientific specialties required for the continued acquisition of scientific knowledge. The point is empirical and the evidence, once faced, is overwhelming: the development of human culture, including that of the sciences, has been characterized since the beginning of history by a vast and still accelerating proliferation of specialties. That pattern is apparently prerequisite to the continuing development of scientific knowledge. The transition to a new lexical structure, to a revised set of kinds, permits the resolution of problems with which the previous structure was unable to deal. But the domain of the new structure is regularly narrower than that of the old, sometimes a great deal narrower. What lies outside of it becomes the domain of another scientific specialty, a specialty in which an evolving form of the old kinds remains in use. Proliferation of structures, practices, and worlds is what preserves the breadth of scientific knowledge; intense practice at the horizons of individual worlds is what increases its depth.

This is the pattern that led me, at the end of my remarks on Ian Hacking's paper, to speak of specialization as speciation, and the parallel to biological evolution goes further. What permits the closer and closer match between a specialized practice and its world is much the same as what permits the closer and closer adaptation of a species to its biological niche. Like a practice and its world, a species and its niche are interdefined; neither component of either pair can be known without the other. And in both cases, also, that interdefinition appears to require isolation: the increasing inability of the residents of different niches to crossbreed, on the one hand, and the increasing difficulty of communication between the practitioners of different specialties, on the other.

That pattern of development by proliferation raises the problem to which, in a more standard formulation, most of Ernan's paper is devoted: what is the process by which proliferation and lexical change take place, and to what extent can it be said to be governed by rational considerations? On those questions, more than on any of those I've discussed above, my views remain very close to those developed in *Structure*, though I can now articulate them more fully. Indeed, Ernan has already articulated most of them for me. There are only two points in his presentation of my position from which I've any inclination to dissent. The first is the use he makes of the distinction between shallow and deep revolutions: though revolutions do differ in size and difficulty, the epistemic problems they present are for me identical. The second is Ernan's understanding of the intent with which I refer to Hume's problem of induction: I share his intuition that the developmental approach to science will dissolve (not solve) Hume's problem; the object of my occasional references to it was simply to disclaim responsibility for a solution.

In other areas what I need to do is explain what Ernan sees as equivocations and inconsistencies in my position. That will require my presupposing, at least for the sake of the argument, that you have already set aside the notion of a fully external world toward which science moves closer and closer, a world independent, that is, of the practices of the scientific specialties that explore it. Once you have come that far, if only in imagination, an obvious question arises: what, if not a match with external reality, is the objective of scientific research? Though I think it requires additional thought and development, the answer supplied in *Structure* still seems to me the right one: whether or not individual practitioners are aware of it, they are trained to and rewarded for solving intricate puzzles—be they instrumental, theoretical, logical, or mathematical—at the interface between their phenomenal world and their community's beliefs about it. That is what they are trained to do and what, to the extent they retain control of their time, they spend most of their professional lives doing. Its great fascination—which to outsiders often seems an obsession—is more than sufficient to make it an end in itself. For those engaged in it, no other goal is needed, though individuals often have a number of them.

If that is the case, however, the rationality of the standard list of criteria for evaluating scientific belief is obvious. Accuracy, precision, scope, simplicity, fruitfulness, consistency, and so on, simply *are* the criteria which, puzzle solvers must weigh in deciding whether or not a given puzzle about the match between phenomena and belief has been solved. Except that they need not all be satisfied at once, they are the "defining" characteristics of the solved puzzle. It is for maximizing the precision with which, and the range within which, they apply that scientists are rewarded. To select a law or theory which exemplified them less fully than an existing competitor would be self-defeating, and self-defeating action is the surest index of irrationality.[19] Deployed by trained practitioners, these criteria, whose rejection would be irrational, are the basis for the evaluation of work done during periods of lexical stability, and they are basic also to the response mechanisms that, at times of stress, produce speciation and lexical change. As the developmental process continues, the examples from which practitioners learn to recognize accuracy, scope, simplicity, and so on, change both within and between fields. But the criteria that these examples illustrate are themselves necessarily permanent, for abandoning them would be abandoning science together with the knowledge which scientific development brings.

The pursuit of puzzle solving constantly involves practitioners with questions of politics and power, both within and between the puzzle-solving practices, as well as between them and the surrounding non-scientific culture. But in the evolution of human practices, such interests have governed from the start. What further development has brought with it is not their subordination but the specialization of the functions to which they are put. Puzzle-solving is one of the families of practices that has arisen during that evolution, and what it produces is knowledge of nature. Those who proclaim that no interest-driven practice can properly be identified as the rational pursuit of knowledge make a profound and consequential mistake.

Notes

1. My special thanks go to Judy Thomson, who conceived the voyage; to Paul Horwich, who captained the ship; and to my secretary, Carolyn Farrow, who has been his able first mate.

2. C. G. Hempel, *Fundamentals of Concept Formation in Empirical Science*, Encyclopedia of Unified Science, vol. 2, no. 7 (Chicago: University of Chicago Press, 1957). I could have found similar elements in the discussion of Ramsey sentences in R. B. Braithwaite's *Scientific Explanation* (Cambridge: Cambridge University Press, 1953), but my encounter with it came later.

3. Originally published in 1958, the paper is most conveniently available as chapter 8 in C. G. Hempel, *Aspects of Scientific Explanation and Other Essays in the Philosophy of Science* (New York: Free Press, 1965). That formulation I still regularly use in my teaching. A more fully articulated version of a similar position is explicitly applied to my views in Wolfgang Stegmüller's *The Structure and Dynamics of Theories*, trans. W. Wohlhueter (New York: Springer Verlag, 1976), a book whose influence is also reflected in some of my more recent work. In particular, see the paper cited in n. 7, below.

4. An excellent account of the earlier stages of those attempts is included in Paul Hoyningen-Huene, *Die Wissenschaftsphilosophie Thomas S. Kuhns* (Braunschweig/Wiesbaden: Vieweg, 1989), soon to appear in English translation from the University of Chicago Press.

5. As this sentence may suggest, a significant role in the recent development of my ideas has been played by David Wiggins, *Sameness and Substance* (Cambridge: Harvard University Press, 1980).

6. On "normic generalizations," see Michael Scriven's too much neglected paper "Truisms as the Ground for Historical Explanations," in Patrick Gardiner, ed., *Theories of History* (New York: Free Press, 1959).

7. On the problem of learning 'force' see my paper "Possible Worlds in History of Science," in Sture Allén, ed., *Possible Worlds in Humanities, Arts and Sciences* (New York:

de Gruyter, 1989), pp. 9–32. The paper also discusses, though in the context of concept development rather than of concept acquisition, the significance of the contrast set which contains 'liquid' to the determination of the referents of 'water'. A slightly abridged version of this paper is available in C. Wade Savage, ed., *Scientific Theories*, Minnesota Studies in the Philosophy of Science, vol. 14 (Minneapolis: University of Minnesota Press, 1990), pp. 298–318.

8. Explicating this sense of 'meaning' would require giving body to the claim that kind-terms do not have meanings by themselves but only in their relations to other terms in an isolatable region of a structured lexicon. It is congruence of structure that makes meanings the same for those who have acquired different expectations from their learning experience.

9. Reference to the contrast set containing 'male' and 'female' indicates both the difficulties and the importance of developing more refined versions of this "no-over-lap" principle. I think no individual creature is both *a* male and *a* female, though it may exhibit both male and female characteristics. Perhaps good usage also permits describing an individual as both male and female, the terms being used adjectivally, but the locution seems to me strained.

10. A more unified and correspondingly clearer presentation of these concepts is to be found in the introduction of Jed's book *The Rise of the Wave Theory of Light* (Chicago: University of Chicago Press, 1989), pp. xiii–xx.

11. See his "Kinds and the Wave Theory of Light," *Studies in the History and Philosophy of Science* 23 (1992): 39–74. The main diagrams in that paper were originally intended for an appendix to the one in this volume.

12. The same use of 'translation' is quoted by Ernan McMullin from another place. With what he has to say about the phenomena I refer to, I again fully agree, but again my reference should not have been to translation.

13. See especially Crosbie Smith and M. Norton Wise, *Energy & Empire: A Biographical Study of Lord Kelvin* (Cambridge: Cambridge University Press, 1989), and earlier articles of Norton's there cited.

14. See David Hull, "Are Species Really Individuals?" *Systematic Zoology* 25 (1976): 174–191.

15. Aristotle's discussions of force and motion did, of course, include statements that we can make in a Newtonian vocabulary and can then criticize. His explanations of the continued motion of a projectile after it leaves the mover's hand are especially well-known examples. But the basis of our criticisms is observations that could in many cases have been made by Aristotle, that were made explicit by his successors, and that led to the development of the so-called impetus theory, a theory that avoided the difficulties that Aristotle's had faced but that did not directly affect the Aristotelian conceptions of force and motion. This example is developed in my "What Are Scientific Revolutions?" in L. Krüger, L. J. Daston, and Michael Heidelberger, eds., *The Proba-bilistic Revolution*, vol. 1, *Ideas in History* (Cambridge: MIT Press, 1987), pp. 7–22.

16. Norton would probably use the term 'ideology' rather than 'theory', but he regularly conflates the two, as in figure 18, which illustrates the present point. Reasons for the difference in our choice of terms will be apparent.

17. For this view, see particularly Stegmüller, *Structure and Dynamics of Theories,* pp. 40–57. Its origin is in J. D. Sneed, *The Logical Structure of Mathematical Physics* (Dordrecht: Reidel, 1971), but its presentation there is widely scattered.

18. The point is more precisely made in the paper cited in n. 7 above. 'Force' can be acquired without 'mass' by exposure to examples of Hooke's law. 'Mass' can then be added to the conceptual vocabulary by presenting illustrations either of Newton's second law or of his law of gravity.

19. These points are elaborated in two papers of mine that Ernan cites: "Objectivity, Value Judgment, and Theory Choice," in my *Essential Tension* (Chicago: University of Chicago Press, 1977), pp. 320–329, and "Rationality and Theory Choice," *Journal of Philosophy* 80 (1983): 563–570. The themes developed in the second of these papers are still another product of my interactions with C. G. Hempel, the one to which, in the first section of this paper, I promised to return.

Contributors

Jed Z. Buchwald is Bern Dibner Professor of History of Science at the Massachusetts Institute of Technology and Director of the Dibner Institute for the History of Science and Technology. He is the author of *From Maxwell to Microphysics* (University of Chicago Press, 1985), *The Rise of the Wave Theory of Light* (University of Chicago Press, 1989), and the just completed book entitled *The Creation of Scientific Effects: Heinrich Hertz and Electric Waves.*

Nancy Cartwright is Professor in the Department of Philosophy, Logic, and Scientific Method at the London School of Economics. Her research interests center in the philosophy of science, with special attention to quantum mechanics and recently to the philosophy of economics. She has written two books, both with Oxford University Press: *How the Laws of Physics Lie* (1983) and *Nature's Capacities and Their Measurement* (1989).

John Earman is Professor of History and Philosophy of Science at the University of Pittsburgh. His primary research interests are in the foundations of physics and confirmation theory. He is the author of *A Primer on Determinism* (D. Reidel, 1986), *World Enough and Space-Time: Absolute versus Relational Theories of Space and Time* (MIT Press, 1989), and *Bayes or Bust? A Critical Examination of Bayesian Confirmation Theory* (MIT Press, 1992).

Michael Friedman is Professor of Philosophy and Research Professor of the Humanities at the University of Illinois at Chicago. He is the author of *Foundations of Space-Time Theories: Relativistic Physics and Philosophy of Science* (Princeton University Press, 1983) and *Kant and the Exact Sciences* (Harvard University Press, 1992). His current research concerns the relationship between the history of science and the history of philosophy from Kant through Carnap.

Ian Hacking teaches philosophy at the University of Toronto. His most recent book is *The Taming of Chance* (Cambridge University Press, 1990). He has written other books on probability and the philosophy of language and, in the philosophy of science, *Representing and Intervening* (Cambridge University Press, 1983). He is completing a book to be called *Kinds of Things and Kinds of People*.

J. L. Heilbron, who claims the honor of being T. S. Kuhn's first doctoral student, obtained his Ph.D. in history at the University of California at Berkeley in 1964. Apart from three years at the University of Pennsylvania, he has been at Berkeley ever since, in various capacities, currently as Vice Chancellor. He writes about the intellectual and institutional history of the physical sciences from the sixteenth into the twentieth centuries.

Paul Horwich is Professor in the Department of Linguistics and Philosophy at the Massachusetts Institute of Technology. His interests are in the philosophy of science and the philosophy of language. In these areas he has written *Probability and Evidence* (Cambridge University Press, 1982), *Asymmetries in Time* (MIT Press, 1987), and *Truth* (Basil Blackwell, 1990).

Thomas S. Kuhn has recently retired from the Laurance S. Rockefeller Professorship of Philosophy at the Massachusetts Institute of Technology. In earlier years he was also a member of the faculties at Harvard, Berkeley, and Princeton. He has served both as president of the History of Science Society and, more recently, as president of the Philosophy of Science Association. As a historian, Kuhn has published books on various topics, ranging from *The Copernican Revolution* (Harvard University Press, 1957) to *Black-Body Theory and the Quantum Discontinuity* (Oxford University Press, 1978). He is best known, however, for a more theoretical, philosophical volume entitled *The Struc-*

ture of Scientific Revolutions (University of Chicago Press, 1962) and for the accompanying volume of essays *The Essential Tension* (University of Chicago Press, 1977).

Ernan McMullin is Director of the Program in History and Philosophy of Science at the University of Notre Dame. He writes on philosophy of science and on the history of the philosophy of science. Among his recent works are *The Inference That Makes Science* (Marquette University Press, 1992), *Philosophical Consequences of Quantum Theory*, ed. with James Cushing (University of Notre Dame Press, 1989), and *The Social Dimensions of Science*, ed. (University of Notre Dame Press, 1992).

N. M. Swerdlow is Professor of Astronomy and Astrophysics and of History at the University of Chicago. His work is in the history of the exact sciences, principally astronomy, from antiquity through the seventeenth century. He is the author, with O. Neugebauer, of *Mathematical Astronomy in Copernicus's De Revolutionibus* (Springer-Verlag, 1984) and is currently preparing, with A. T. Grafton, a volume of translations of writings by Regiomontanus.

M. Norton Wise is Professor of History at Princeton Univesity. He is coauthor, with Crosbie Smith, of *Energy and Empire: A Biographical Study of Lord Kelvin* (Cambridge University Press, 1989). His research is in the history of physics in the modern period, especially electricity and magnetism, mechanics, and quantum mechanics. Methodologically, he treats physics as cultural history, an activity situated in space and time that draws on the larger culture for essential resources and meanings.

Index